Embedded and Fan-Out Wafer and Panel Level Packaging Technologies for Advanced Application Spaces

Embedded and Fan-Out Wafer and Panel Level Packaging Technologies for Advanced Application Spaces

High Performance Compute and System-in-Package

Edited by

Dr. Beth Keser
Intel Corporation
San Diego, California

Steffen Kröhnert
ESPAT-Consulting
Dresden, Saxony, Germany

IEEE PRESS
WILEY

Published by John Wiley & Sons, Inc., Hoboken, New Jersey.

For general information on our other products and services or for technical support, please contact our Customer Care Department within the United States at (800) 762-2974, outside the United States at (317) 572-3993 or fax (317) 572-4002.

Wiley also publishes its books in a variety of electronic formats. Some content that appears in print may not be available in electronic formats. For more information about Wiley products, visit our web site at www.wiley.com.

Library of Congress Cataloging-in-Publication Data is applied for

Cover Design: Wiley
Cover Image: © Volker Mai/Fraunhofer IZM

FSC
www.fsc.org
MIX
Paper from responsible sources
FSC® C013604

This book is dedicated to my parents, Robert Schneggenburger, Sr., and Carole Schneggenburger Bell. Without their love and encouragement, I would not have been able to personally or professionally accomplish half of what I have in my 50 years. Also, I would like to include my husband, Milan Keser, who gracefully "retired" from engineering in 2004 to raise our daughters, while I pursued my passion for advancing electronic package engineering.

Beth Keser

I would like to dedicate this book to my father, Bernd Kröhnert, who passed away in March 2021, at the age of 80 after having patiently endured his slowly progressing dementia disease during the last years. He always set a good example for me in persistently and purposefully pursuing his goals, with the close involvement of the people he had the privilege to work or live with. He was not deterred by setbacks but always looked ahead and drew strength from it, enabling him to take a different path, correcting the course and trying new things instead of conforming. He supported close cooperation between everyone, young and old, women and men in a culture of respect, acceptance, and recognition. Dare to do new things and keep what has proven itself. If we only do what we have always done, we will not develop any further.
These are character traits that we urgently need in our rapidly changing world today. Therefore, I wish the readers of this book these qualities, which will help to promote innovation and excellence for the benefit of humankind and despite all the competition that is needed to advance our society, not to forget about humanity.

Steffen Kröhnert

We are especially grateful to our friends and family for their patience and support while we took precious time away from them on weekends and after work hours in order to complete this work.

Beth and Steffen

Contents

1

Fan-Out Wafer and Panel Level Packaging Market and Technology Trends

Santosh Kumar[1], Favier Shoo[1], and Stephane Elisabeth[2]

[1] *Yole Developpement, Lyon, France*
[2] *System Plus Consulting, Nantes, France*

1.1 Introduction to Fan-Out Packaging

1.1.1 Historical Perspective

All great technological revolutions have been driven by relentless innovation and commercialized out of need. Each breakthrough compounds existing and past technologies to create a product that is much better than the previous one. Semiconductor packaging has been unfolding on many fronts for many decades. When surface mount technology (SMT) was established in the 1980s and ball grid array (BGA) was introduced in the 1990s, they enabled a major reduction in size of all electronic systems. In the twenty-first century, newly emerged end-user systems in consumer electronics and mobile devices have become the core pillars of modernized progress and economic growth. One major evolution of mobile devices is the smartphone. Diversified application needs coupled with the slowing down of Moore's law has turned the attention of the semiconductor industry to advanced packaging technology for enhanced system-level performance and functionality as well as a smaller form factor, reduced power consumption, and cost-effectiveness. In fact, key platforms of advanced packaging technologies, such as flip-chip, fan-out (FO), and fan-in, as illustrated in Figure 1.1, have enabled better performance with new functionality, all achieved in the smaller form factor that makes today's sleek yet powerful smartphone a reality.

Since 2000, there has been a widespread adoption of wafer-level packaging (WLP), in which most of the packaging and testing is accomplished in full wafer form. WLP does not require an intermediate integrated circuit (IC) substrate, thus allowing a thinner package profile, and it can be directly mounted onto the motherboard. Specific to fan-out packaging (FO) are interconnects that go

Embedded and Fan-Out Wafer and Panel Level Packaging Technologies for Advanced Application Spaces: High Performance Compute and System-in-Package, First Edition. Edited by Beth Keser and Steffen Kröhnert.

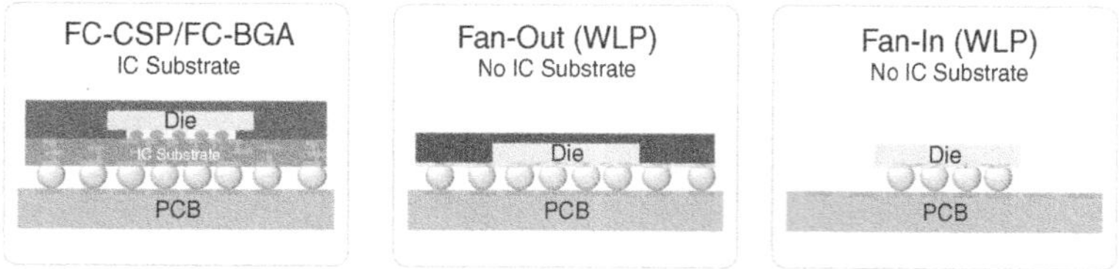

Figure 1.1 Schematic of flip-chip chip-scale package (FC-CSP) and flip chip BGA (FC-BGA), fan-out, and fan-in.

beyond the chip edge and allow multi-chip and 2.5D and 3D packaging solutions. FO technology can be used to fabricate a redistribution layer (RDL) interposer, which is a low-cost alternative to 2.5D packaging. Also, FO technology facilitates multi-die stacking in the vertical direction to enable 3D packaging solutions. These benefits of input/output (IO) density scalability and the integration of passive and active chips in the same package with drastic miniaturization potential by means of 2D, 2.5D, and 3D structures have made FO one of the top options in semiconductor packaging. Since then, there have been unprecedented levels of integration breakthroughs, with an overlap in design, manufacturing, and with companies between front-end (FE) and back-end (BE). Figure 1.2 represents a "middle-end" zone, where bumping and packaging can be executed at the wafer level, with a FO package teardown as an example.

The commercialization history of FO dates back more than a decade, as shown in Figure 1.3. In 2006, Freescale established its first 200 mm pilot line for the redistributed chip package (RCP) in Tempe, Arizona, USA. [1] Back then, RCP was viewed as a disruptive technology that could not only eliminate interconnects such as flip-chip bumps and wire bonds and IC substrates but also did not require

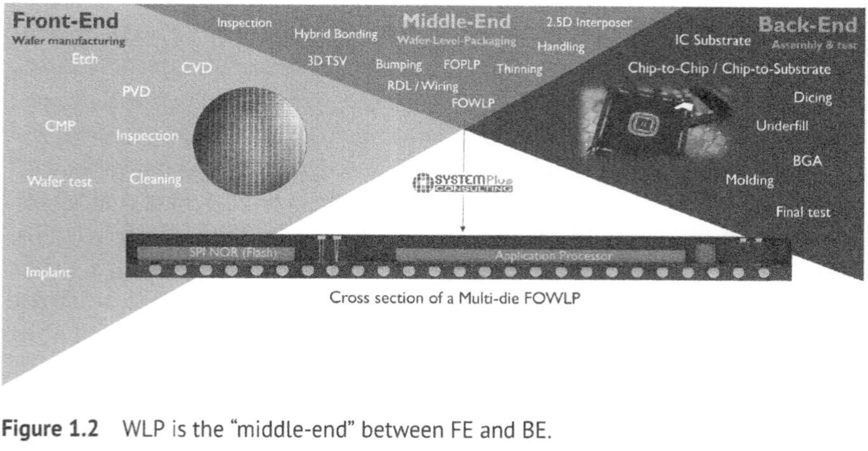

Figure 1.2 WLP is the "middle-end" between FE and BE.

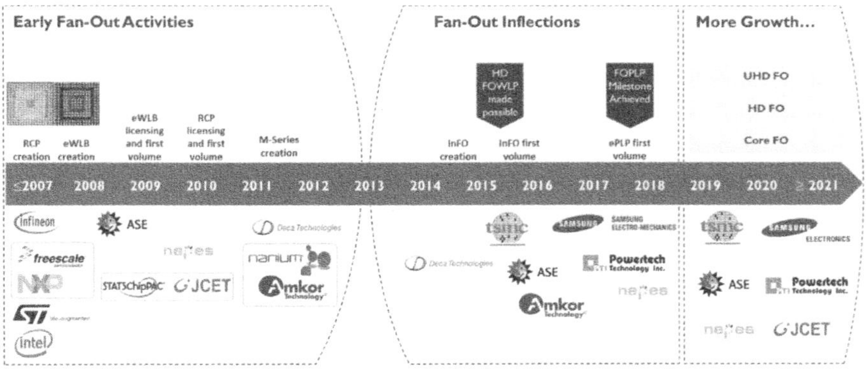

Figure 1.3 FO-WLP evolution.

thinned die to achieve thin package profiles. Then in 2010, Freescale and Nepes collaborated in a joint development effort to enhance further the capabilities of RCP technology, setting up a 300 mm RCP line in Nepes' Singapore facility. Nepes was Freescale's first large-scale partner for RCP licensing. In 2015, NXP acquired Freescale, so NXP now owns and manufactures the RCP technology.

In addition, Infineon is the originator of embedded wafer-level BGA (eWLB), a chip-first face-down solution that was developed from 2001 to 2008. [2] In 2007, Infineon and ASE declared an eWLB manufacturing partnership under a license model. [3] Knowing that a legal licensing partnership can combine the developed eWLB technology with the packaging know-how of an outsourced assembly and test (OSAT) supplier, Infineon started to license more liberally. Since 2007, Infineon has licensed eWLB solutions to ASE, Amkor Portugal (formerly known as Nanium), JCET Group (formerly known as STATS ChipPAC), and STMicroelectronics. [4, 5] The licensee has the right to manufacture eWLB. Intel, by virtue of its acquisition of Infineon's wireless business, also has access to eWLB technology. [6]

From 2008 to 2009, ASE installed a 200 mm production line for fan-out wafer-level packaging (FO-WLP) based on eWLB but subsequently stopped production in 2012 because the market at that time was not big enough to maintain manufacturing activities. In 2014, ASE reentered the market and invested in a 300 mm FO production line, which proved to be successful in gaining Qualcomm's business in 2016. ASE also developed and invested in other FO-WLP technologies and licensed Deca Technologies' solution in 2016. So ASE has a highly diversified FO-WLP portfolio with eWLB license, M-Series license, as well as their own in-house fan-out chip on substrate (FOCoS) and fan-out panel-level packaging (FO-PLP).

In 2009, STATS ChipPAC first qualified eWLB in its 200 mm production line to be a second source of Infineon's eWLB for its baseband products. JCET acquired STATS ChipPAC for $780 M in November 2014 and is driving the company to a major position in the FO-WLP market. STATS ChipPAC at the time led the market, and by May 2016 they had shipped more than 1 billion units of eWLB packages since 2009, reaching more than 2 billion units in 2018. [7] STATS ChipPAC is essentially focusing production on core FO-WLP volume applications, which means lower-end packages for baseband modems (BB), power management ICs (PMIC), and radio frequency (RF) devices, but is also positioned in the system-in-package (SiP) market. Due to cost reduction pressure, there are both 300 mm and 330 mm production lines. The 330 mm panel size allows for economy of scale.

In 2011, Nanium was the second company to qualify 300 mm eWLB in high-volume production and shipped 200 million eWLB components in less than two years. [8] The company quickly developed eWLB based on its 300 mm RDL capability, initially developed for the stacked dynamic random-access memory (DRAM) market. Nanium's eWLB packaging production lines (two 300 mm lines) have been fully operational since 2012. Nanium's business model was to develop more high-integration applications such as multi-chip package (MCP) and SiP in the mobile, industrial, medical, and automotive markets to create more added value and not rely only on single die packages in mobile, which was the main market and where competition only got tougher. In February 2017, Nanium was acquired by Amkor [9].

In 2011, Deca Technologies released their M-Series using their adaptive patterning (AP) technology. Deca's M-Series offers a different type of solution—it is a chip-first face-up solution using an AP software technology to solve pattern misalignment issues. [10] Deca Technologies reported in 2015 having sold more than 100 million units since 2011 and have achieved much larger volumes based on the investment of $60 M from ASE to license the technology and $50 M from Qualcomm. In 2019, Nepes acquired Deca's Philippines operation, liberating Deca from the manufacturing business. [12] Beyond 2019, Deca repositioned its business model from manufacturing to licensing.

In 2015, Amkor presented two chip-last solutions: SWIFT (silicon wafer integrated fan-out technology) and SLIM (silicon-less integrated module). [13] While SWIFT's target markets are sweet spots for FO-WLP (BB for mobile, RF, etc.) and high-density FO-WLP such as application process engines (APE), SLIM is more focused on applications requiring very high I/O counts or high-resolution connections, such as memory with central process units (CPU) or graphics processing units (GPU) or system-on-chip (SoC) partitioning, and complex SiP. Through the purchase of Nanium in 2017, Amkor redefined its strategy regarding FO-WLP. Nanium's know-how and eWLB technology will focus on core FO-WLP while SWIFT solutions will target more advanced applications, high-density (HD) FO-WLP or ultra-high-density (UHD) FO-WLP. Due to some overlap with SWIFT and with a higher cost, SLIM technology is no longer actively offered.

It was not until 2016 that FO-WLP experienced a significant inflection point when TSMC's integrated FO-WLP (InFO) technology first broke into the market, resulting in a whole new higher-end market segment, defined as high-density fan-out (HD FO) due to its embedding of large die (> 10 mm × 10 mm), with finer line and space (L/S) RDL and higher I/O (> 1000), commercialized for Apple's iPhone APE. [14] It was during this phase that the industry started to realize how important the use of FO-WLP is for higher-end applications. This inflection point was a massive boost for the FO-WLP platform as it was then proven for one of the most important applications in smartphones for one of the biggest original equipment manufacturer (OEM) brands, Apple.

Thus, a new wave of FO-WLP innovations followed, with newcomers entering the supply chain. Since 2016, InFO remains a key FO-WLP product offered by TSMC to support mobile smartphones, wearable smartwatches, and even high-performance computing with new InFO variants such as InFO on substrate (InFO_oS), which is defined as ultra-high-density fan-out (UHD FO).

The invention of HD FO was highly beneficial for the FO-WLP market space as another major player, Samsung Electronics, started to invest in Samsung Electro-Mechanics' (SEMCO) HD FO solution with panel-level packaging in 2017. [15] In 2018, SEMCO successfully developed the FO-PLP-based APE-PMIC solution and commercialized it in the Samsung Galaxy Watch. SEMCO's FO-PLP technology set an unprecedented milestone with the penetration of FO-PLP into the consumer market. This is viewed by many as direct competition against TSMC's InFO in the mobile market although FO-PLP does not have the L/S density of InFO. In 2019, Samsung Electronics acquired SEMCO's FO-PLP for $750 M. [16] The synergy between FE and packaging technologies may bring more efficiency under one umbrella, as with TSMC's model (FE + packaging).

In 2017, Powertech announced the production of different FO-WLP technologies: CHIEFS (chip integration embedded fan-out solution) and CLIP (chip-last integration package). This offer was rounded off by PiFO (pillars in fan-out) and BF^2O (bump free fan-out). Backed-up by their strong position in memory and field programmable gate array (FPGA) packaging, Power Technology Incorporated's (PTI) FO-PLP will be a promising alternative to replace part of the high-cost 2.5D and 3DIC market. They also targeted high I/O counts and multiple-chip integrations, such as SoC and memory, baseband, wireless modules, and wide I/O memory. Being one of the first companies to go to panel meant many challenges and risks in process making and reliability but if achieved, it could bring down the cost by 30 to 40%. In 2018, PTI announced a $1.6B investment plan to construct a new advanced packaging plant at the Hsinchu Science Park. [17]

1.1.2 Key Drivers: Why Fan-Out Packaging?

FO-WLP is a versatile semiconductor packaging technology that can be used for various key applications, such as departitioned large processor die, mobile APE, automotive radar and RF, audio codec, PMIC and potentially 5G antennae in package (AiP). It has the advantage of a thinner package size, improved RF performance, higher I/O density, and lower thermal resistance as compared with the conventional flip-chip. Besides fanning out of the electrical I/O, it can also be used for various 2.5D and 3D multi-chip integrations. Today, FO-WLP has evolved from a low-end packaging technology into high-performance and cost-effective integration platforms, as shown in Figure 1.4.

A strong demand for high performance compute (HPC) systems and data centers has emerged in recent years. Fabless and OEMs are delivering technology solutions to drive the hyper-scale data centers and accelerate the growth of artificial intelligence (AI) and machine learning (ML). 2.5D/3D heterogeneous packaging is known to greatly increase computing power. Due to process node scaling limitations in the Si wafer fab, performance scaling requires a bigger die, and yield-cost is a big concern simply because the die is bigger. Hence, the industry is now cutting the big die into smaller pieces and relying on advanced packaging solutions to integrate die together with the shortest and densest interconnections. Flip-chip BGA (FC-BGA) substrate manufacturing is a mature process. However, if we go below a L/S of 8 μm/8 μm for the multilayer FC-BGA substrate, the yield decreases. Moreover, reliability concerns have emerged when testing under high and low temperature and humidity conditions. Therefore, many fabless companies, such as MediaTek and HiSilicon, have been actively qualifying HD FO-WLP. Preliminary

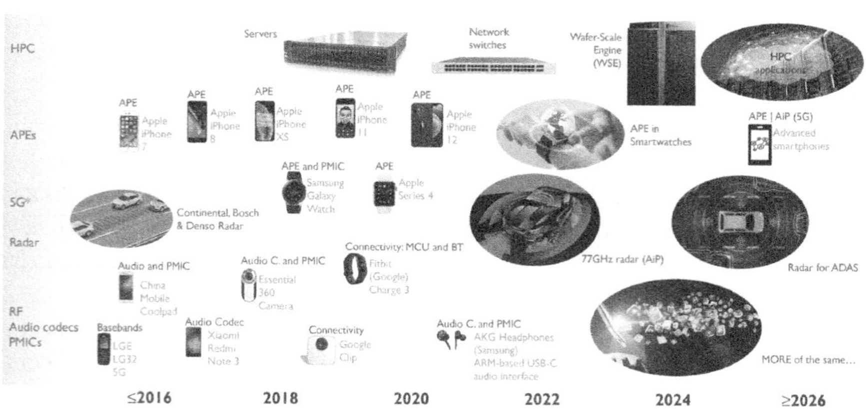

Figure 1.4 Roadmap of fan-out packaging market drivers.

data has shown that FO-WLP can handle the bandwidth and high speed SERDES signal. So FO-WLP can indeed increase the gross die size by combining smaller die in the package as an alternative to improving the wafer yield when the silicon die size is usually enormous and suffers from low yield issues, especially in advanced wafer processes.

In 2016, TSMC with its FE 10 nm logic and BE InFO package-on-package (InFO_PoP) technology (HD FO) successfully obtained orders for Apple's A11 APE, powering the iPhone 7. Apple APE continues to be packaged in InFO_PoP today. In 2018, Samsung used HD FO-PLP for its APE and PMIC for smartwatches as has Apple for its smartwatches. This trend will probably continue because HD FO will continue to be used for devices that need higher performance yet a reduction in total package dimension.

One of the early adopters of FO-WLP is automotive radar, especially at the higher frequencies of > 76–81 GHz. In 2012, the first 77 GHz radar monolithic microwave integrated circuit (MMIC) was commercialized by Infineon with their eWLB technology. Since then, the use of radar sensors has become widespread as companies have been trying to implement radar and other components in a package, thus integrating more die at a lower system cost. Advancements in packaging (especially FO-WLP) enables the integration of antenna and RF chips to perform at a higher efficiency level. This is opening the door to meet the demands of automotive radar for next-generation automotive capabilities such as advanced driver assistance systems (ADAS).

77 GHz radar sensors with FO-WLP are already well adopted in the automotive market for cabin or infotainment and ADAS safety. 5G mmWave, which operates at similar high frequency range > 60 GHz, requires higher data rates along with wider bandwidth. As we enter this new decade, one thing is certain: 5G mmWave adoption will continue to grow since companies are accelerating toward this common goal. To achieve wide bandwidth, a higher frequency (thus mmWave solution) is needed. However, this wavelength is similar to the interconnect length in an IC package. For this reason, signal loss from the RF chip to the antenna is expected. In order to reduce the loss with shorter interconnect lines, the antenna is designed into the package.

TSMC has published multiple technical papers to demonstrate how InFO_AiP can enable low transmission loss and high antenna performance for mmWave system integration, as compared to flip-chip IC substrates. [18, 19] However, the implementation of InFO_AiP will depend on Apple's decision. Apple is now internally developing the various 5G RF chipsets including modem and RFIC. Currently mmWave AiPs in iPhones use Qualcomm chipsets. Apple will likely implement FO-WLP AiP when they develop their own chipsets.

1.1.3 FO-WLP vs. FO-PLP

FO-WLP is still considered a costly platform as compared to a conventional packaging platform such as flip-chip or wire-bond. Hence, FO-WLP is adopted in limited applications, such as audio codecs, connectivity, microcontroller units (MCU), power management ICs (PMICs), and RF. It is still too expensive for many low-end applications.

Purchasers of FO-WLP have been trying to drive down the manufacturing cost. A significant cost reduction is highly attractive for the companies to achieve strong profit margins. Within the FO-WLP market, FO-PLP is a possibility for cost-reduction due to the economies of scale of its manufacturing process. Qualcomm and MediaTek are two prominent companies demanding it.

FO-PLP does not make sense if volume is not high, and capital expenditure (CapEx) and return on investment (ROI) are not justifiable. Some view FO-PLP as one of the ways to by-pass the intellectual property (IP) of FO-WLP, which is saturated. Some believe FO-PLP is the only way to reduce cost. Opinions are highly divided. Pro-FO-WLP companies are no longer underestimating FO-PLP penetration. In fact, more companies are currently being swayed toward FO-PLP. Pro-FO-WLP companies are concerned that if FO-PLP really takes off, then the price point of FO-WLP is no longer competitive. Coupled with a strong end-customer push for cost reduction, we are seeing more companies such as PTI, SEMCO, ESWIN, ASE, Nepes, and Deca venturing into FO-PLP. On the other hand, some companies prefer to wait and observe the progress first, unwilling to invest in FO-PLP without volume. To them, panel-level needs to see very-high volume to really take off.

To understand the theoretical benefit of FO-PLP, an equipment throughput model is established to compare the impact on the number of tools needed for production between FO-WLP vs. FO-PLP, as shown in Figure 1.5.

In this model, the production volume needed for the FO-WLP market is 1 941 605 wafers (300 mm equivalent) which Yole Développement reports was the total FO production volume for 2018. With similar throughput, FO-PLP needs fewer tools than FO-WLP when trying to produce the 1.941M wafers mentioned above. The difference in the number of tools required between FO-WLP and FO-PLP are more significant at a lower throughput of 10 carriers per hour (difference of 17 tools) as compared to 40 carriers per hour (difference of 4 tools). The difference in number of tool units is bigger when the process throughput is lower because more tools will be needed to meet the 1.941M wafer requirement at this low throughput. From a ratio point of view, FO-PLP needs about 3 times fewer tool units as compared to FO-WLP, assuming same yield and tool throughput.

1.1.4 Future of Fan-Out Packaging for Heterogeneous Integration

The trend is to go for more integration, and FO-WLP is a potential candidate for future scaling into heterogeneous integration. This opens the way for new

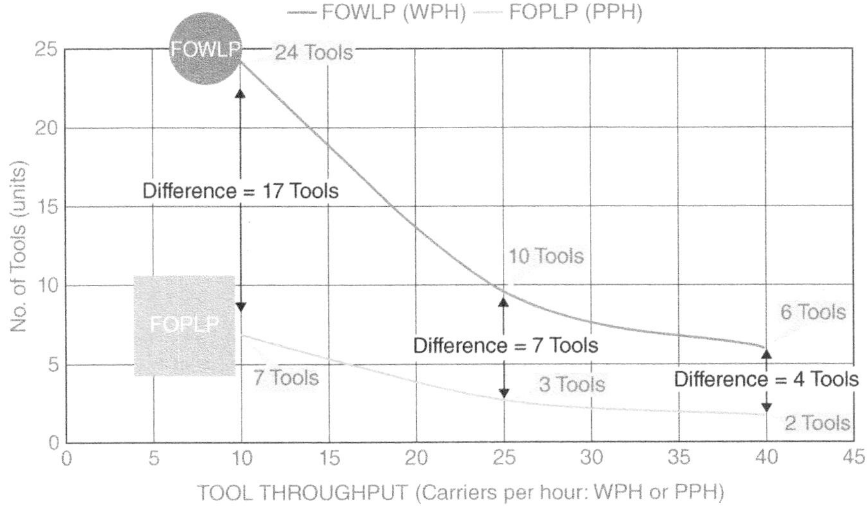

Figure 1.5 FO-WLP vs. FO-PLP: carrier impact on No. of tools.

MCMs (multi-chip module), PoPs (package-on-package), and SiPs. FO-WLP's differentiating characteristic is the ability to embed a wide range of die (independent of die type, size, or side) allowing many integration possibilities: small die, large die, stacked, or side-by-side multi-die; 2D solutions in single and multi-chip configurations; 2.5D interposer solutions; 3D SiP and PoP solutions that could include face-to-back (active die face to die back) or face-to face (active die face to active die face or F2F) options; or heterogeneous integration with both passive and active components.

FO-WLP is compatible with MCM architecture and enables the inclusion of different die of different types and sizes in the same package. It easily connects them on the same plane, allowing better coupling. FO-WLP solutions can also work in PoP architecture, which enables a 3D approach to reduce footprint and improve electrical and thermal performance even further. FO-WLP allows flexibility in the die to embed: standard IP in IC in the bottom package and multiple die with specific IP in the top package, for instance, to benefit from any appropriate node and die size and all with a small form factor. FO-WLP enables SiP architecture too, which means the capability of embedding and/or connecting numerous die and components in a flexible way to get a single system with only one package.

Moore's law is slowing down, and despite chips becoming smaller and smaller, there are accompanying cost, power management, and heat dissipation challenges. In such a context, more-than-Moore solutions help improve form factor, performance, and cost via innovative ideas. A promising way is to package different die together to gain space and performance at a lower cost. This trend is consistent with the FO solution and is even more favorable because of FO-WLP integration capabilities.

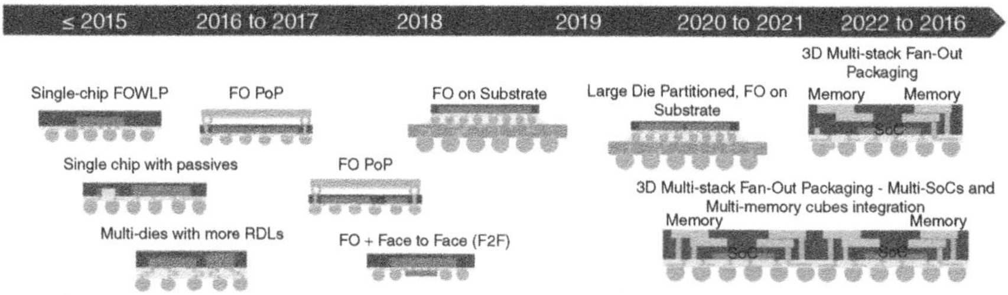

Figure 1.6 Timeline of FO-WLP integration capability.

From Figure 1.6, it is evident the path that FO-WLP will open new ways for MCMs, SiPs, and PoPs manufacturing. Mold compound embedding permitted flexibility in type and size of components to be embedded together in one package. Between 2016 and 2018, the industry further improved the technical capabilities of FO-WLP in the direction of thinner interconnection solutions (RDL, fab back-end of line or BEOL) and more RDL layers, up to 4 layers, moving into HD FO. A combination with RDL on top-face and multiple FO packages brings even thinner PoP, SiP, and face-to-face/face-to-back packaging capability. From 2019 onward, FO packages placed on a substrate that enables larger packages results in a 2.5D-like approach but without the expensive Si interposer with thru silicon vias (TSV) for the high-end application domain. Also, reducing the large single unit die size through the integration of several smaller die using FO-WLP helps to improve the gross die yield. Moving forward, more variants of 3D FO-WLP (such as PoP) are expected for yet thinner profiles, with high memory capacity and memory bandwidth.

1.2 Market Overview and Applications

1.2.1 Fan-Out Packaging Definition

A recognized characteristic of FO packages is that, as the name suggests, interconnections are fanned out on the chip, and because of that, bumping is not dependent on die surface. This means that FO has the potential to achieve any number of interconnects with standard pitches at any shrink stage of the wafer node technology. If the only definition of FO is a package from which connections and bumping are out of the chip scale, then almost all packages can be defined as FO, including FC-BGA, FC-CSP, and embedded die. To make a fair comparison the situation is clarified as shown in Figure 1.7. FO-WLP technologies should have at least one of these two key characteristics: (i) FO-WLP solutions that use mold compound to embed the die—not laminated materials, or (ii) FO-WLP solutions that do not use IC substrates to fan out I/O from chip area.

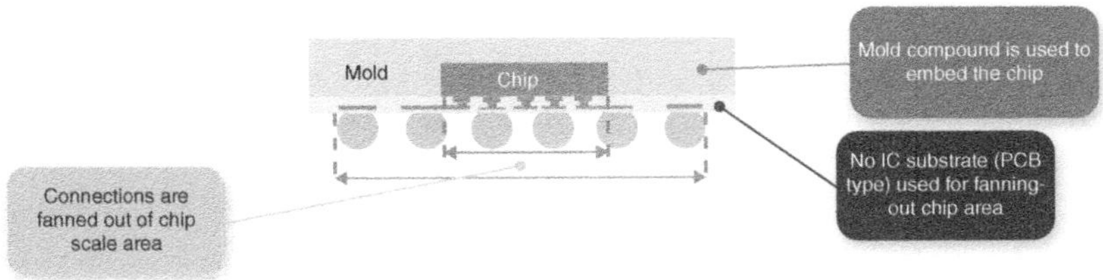

Figure 1.7 Definition of FO-WLP.

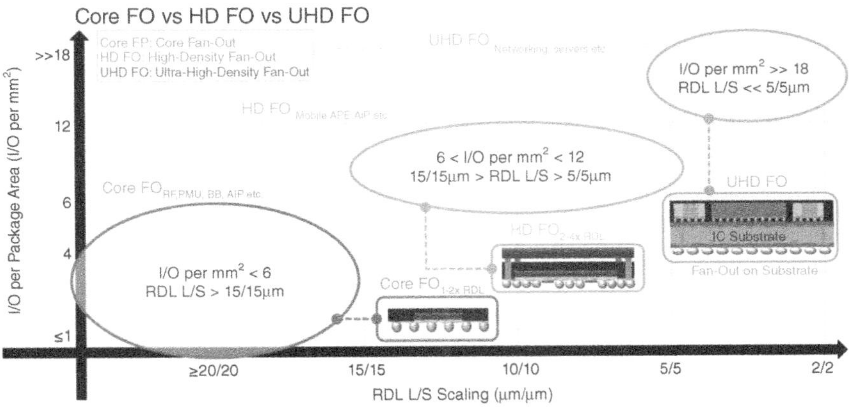

Figure 1.8 Market segmentation of FO.

1.2.2 Market Segmentation: Core FO vs. HD FO vs. UHD FO

FO packaging is penetrating higher-end IO densities, much greater than 18 per mm^2, and finer RDL, with L/S measurements far below 5 μm/5 μm. Along with the proliferation of new technologies and new commercialized products, industry companies are having difficulty segmenting the market.

Figure 1.8 illustrates the market class segmentation of FO packages by UHD FO, HD FO, and core FO. UHD FO is classified as $\gg$ 18 I/O per mm^2 and L/S $\ll$ 5 μm/5 μm, HD FO has between 6 and 12 I/O per mm^2 with L/S between 5 and 15 μm/15 μm, and core FO has < 6 I/O per mm^2 with L/S > 15 μm/15 μm. This technical definition allows the industry to understand and make informed decisions on developments and strategies to gain a competitive advantage within an individual market class and/or different individual market classes. This helps in understanding what capabilities each segment is looking for or offering.

1.2.3 Market Valuation: Forecast of Revenue and Volume

The FO platform is increasingly viewed as one of the top packaging options. FO remains a fast-growing market with revenue growth of 15.9% from 2019 ($1256 M) to 2025 ($3046 M). The revenue varies between different market classes; see Figure 1.9. UHD FO is valued at $504 M in 2019 and is set to reach $1523 M by 2025 at a 20.2% compound annual growth rate (CAGR). Projected at 15.8% CAGR, HD FO is estimated to grow from $534 M in 2019 to $1291 M by 2025. On the other hand, Core FO is expected to reach $231 M by 2025 with a moderately stagnant CAGR of 1% from $218 M in 2019. UHD FO is finding its own market space and is experiencing the fastest growth of 20.2% CAGR. HD FO is already dominant in the market, and the continued expansion indicates that HD FO is driving the overall FO market forward, in a league of its own. For core FO, growth is stagnant with limited expansion.

The current FO-WLP installed based is enough to sustain existing core FO demand. There is no significant expansion for existing end products. Yet, fabless companies continues to push packaging houses for wider adoption at a lower cost. FO production is expected to grow from 1703 kwspy (kilo wafer starts per year) in 2019 to 3419 kwspy by 2025, at a CAGR of 12.3%, as shown in Figure 1.10.

1.2.4 Current and Future Target Markets

Within FO, mobile, consumer, telecom, and infrastructure end markets are influencing the overall trend of FO. These trends are represented in Figure 1.11. Within FO, the mobile and consumer end market leads the market with a revenue of $741 M in 2019, projected to reach $1498 M by 2025 at a 12.5% CAGR. Likewise, the telecom and infrastructure end market is growing strongly with a 20.2% CAGR from $504 M in 2019 to $1523 M in 2025. Automotive and mobility is also

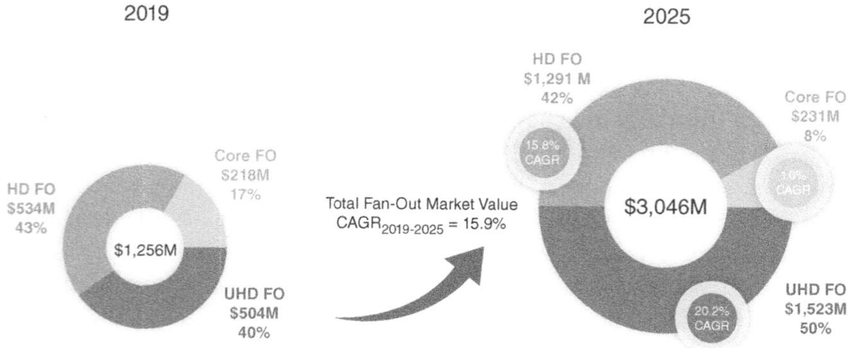

Figure 1.9 Revenue market forecast of fan-out packaging, split by market class.

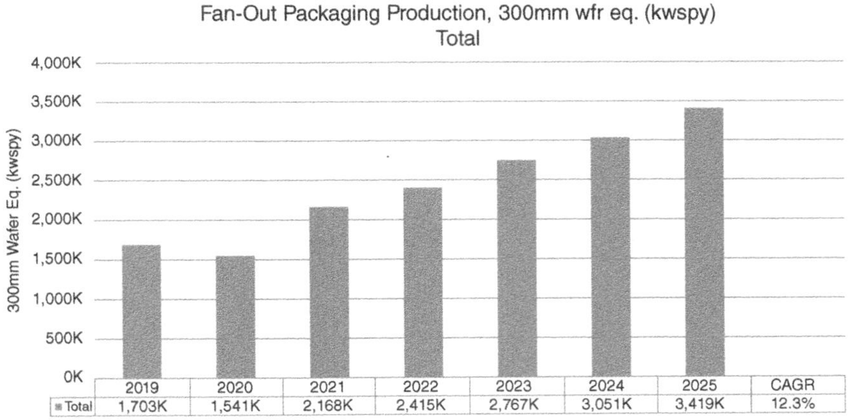

Figure 1.10 Production volume forecast of fan-out packaging.

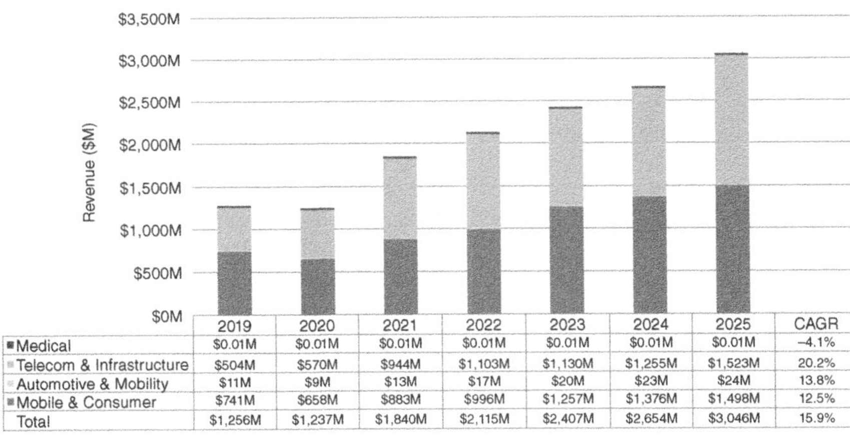

Figure 1.11 Revenue forecast of fan-out packaging, split by end market.

increasing significantly with 13.8% CAGR from $11 M in 2019 to $24 M in 2025. The medical end market, however, being a relatively smaller-sized market, is expected to see a decline in adoption.

There is an impact in 2020 due to COVID-19. In 2020, the world experienced an unprecedented pandemic, as this novel coronavirus, COVID-19, spread rapidly on a global scale. The number of infected cases continued to increase in the first half of 2020, which triggered lockdowns in various countries, one after another. This has caused a pandemic-driven contraction of economic activity, negatively affecting semiconductor development and production.

FO revenue is heavily dependent on the mobile and consumer end market, with 59% exposure as of 2019. Thus, it was inevitable for FO revenue to experience a decline in 2020, reflecting the COVID-19 effect on mobile and consumer applications. By 2021, on the other hand, data-driven end systems will grow in popularity as COVID-19 is being quelled. Industry companies are prepared to deliver more digital functionalities. Consequently, 5G-led adoption is expected to grow rapidly thereafter.

FO revenue for telecom and infrastructure improved moderately in 2020 and has bounced back strongly in 2021. In fact, FO is experiencing a massive uptick in the demand by telecom and infrastructure end market. End users are not only coming to recognize the crucial role of technology (from supporting remote working to scaling digital operations due to lockdown) but will also invest in and utilize it even more. Hence, more end-system demand will accelerate in HPC applications, shaped by new end-user behavior.

In summary, pent-up demand has returned in 2021. There is a light at the end of the tunnel as more technology-related business leaders are increasingly optimistic that businesses and consumers will stabilize at a new normal. FO will unquestionably ride this wave in 2022 and 2023.

1.2.5 Applications of Fan-Out Packaging

Figure 1.12 shows a comprehensive overview of the most widely used FO applications. There are different levels of classification highlighting key granularities, including end market, application, device, and product model/name. Through this classification, readers can have an overall understanding of FO adoption. With the evolution of FO technologies, the application range is broadening. Because of this, the technology can be targeted toward a wider range of applications.

Since the turn of the twenty-first century, FO has been expanding to encompass a wide range of devices in different applications. Within the mobile and consumer end market, the adoption of FO-WLP is often driven by the reduction in package thickness with higher level of performance integrity. For example, the APE device in Apple's iPhone 7 and beyond is packaged by TSMC InFO_PoP, and Samsung Exynos 9110 APE and PMIC (multi-die) is packaged using FO-PLP and is found in the Samsung Galaxy Watch. Apart from the APE, devices such as PMIC, audio codec, RF transceiver, BB, RF Chip and AiP, MCU and bluetooth (BT), APE and NOR memory, and PMIC are all packaged using FO for mobile smartphones, smartwatches, wearables such as headphones, and internet of things (IoT) applications.

For smartphones, APE packaging is the main contributor to package thickness. It is not strictly linear due to different options in the memory packaging thickness, because there is not much progress seen at memory packaging thickness level.

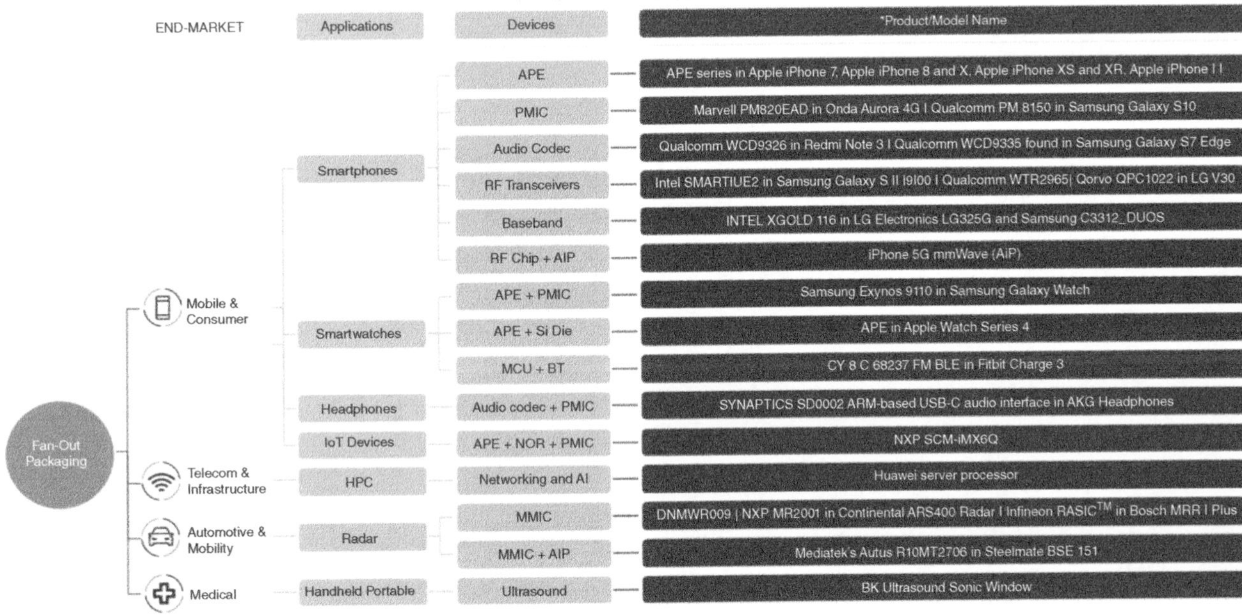

Figure 1.12 Application overview of fan-out packaging. *Non-exhaustive list of companies and products.

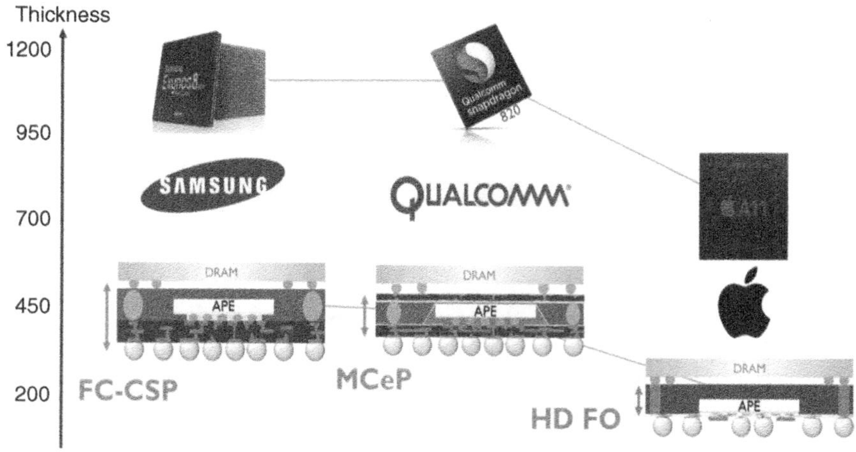

Figure 1.13 Application processing engine (APE): package thickness is significant.

Instead, it is the package choice between FC-CSP versus molded core embedded package (MCeP) versus HD FO that makes a difference. From Figure 1.13, it is clear that FO packaging enables a drastic reduction in thickness of full APE and memory PoP. The package thickness is reduced from 550–425 μm to 231 μm because of HD FO. Thickness is not the only parameter of interest for APE packaging, but it is of high importance for slimmer smartphone design because of a well-managed form factor. Trends in recent years show Apple has leapfrogged other leaders with TSMC's InFO_PoP (HD FO), packaging the APE at lower thickness. Other leaders are keen to decrease thickness to compete. It is known that Qualcomm prefers MCeP and drives innovation via the IC substrate. In 2018, however, Samsung utilized SEMCO's newly developed FO-PLP solution (ePLP) in Samsung Galaxy Watch's APE-PMIC. Although this is a smartwatch (not mobile) and still considered core FO in terms of IO, the APE package thickness is now comparable to TSMC's InFO. This is a milestone for Samsung Electronics on the path to a possible migration of FO technology into Mobile APE–HD FO.

In recent years, FO has been expanding beyond the mobile and consumer markets and penetrated the telecom and infrastructure markets. Currently, the 2.5D Si interposer is one of the best solutions to handle high-end device integration needs, but the 2.5D Si interposer is very expensive. FO offers an alternative solution to 2.5D Si interposer for mid-to-high-end HPC applications that seek cost/performance optimization. Significant cost reduction is achieved by splitting a large-advanced-node networking chip, such as network processor and network switch, into several small networking chips and reconstituting them with UHD FO packaging thus enabling high-density, high-speed interconnects on an IC

Preface

Embedded and fan-out packaging technologies have come a long way since our publication of *Advances in Embedded and Fan-Out Wafer Level Packaging Technologies* (Wiley) in February of 2019. These package technologies have become more ubiquitous. Where they used to be used only in mobile wireless devices such as smart phones, cell phones, and tablets, they are now found as packaging technologies for disaggregated solutions in network switches, network SERDES IP chips, artificial intelligence, 5G, and automotive solutions. Due to this revolutionary advancement in only three years, we felt the need to publish an update on this technology space in *Embedded and Fan-Out Wafer and Panel Level Packaging Technologies for Advanced Application Spaces: High Performance Compute and System-in-Package* (Wiley) in 2022.

This 11-chapter publication begins with a market analysis from Yole Développement entitled "Fan-Out Wafer and Panel Level Packaging Market and Technology Trends," which covers the past and future of this exciting new technology. Unique to this book is a cost analysis provided in Chapter 2 by SavanSys Solutions LLC. Also unique to this publication are two chapters by Taiwan Semiconductor Manufacturing Company, Ltd., detailing their integrated fan-out technology called InFO in the mobile phone and high-performance compute space. Deca Technologies Inc. is included again in this book where they detail their progress on M-Series design rules and scale-up from 300 mm round to a large panel manufacturing technology. Fraunhofer IZM then details their panel technology in Chapter 6. The book then focuses on embedded technologies by ASE Group, AT&S AG, and Phoenix Pioneer Technology. Finally, the last two chapters are written by premier thought leaders in future packaging technologies at the University of California, Los Angeles, and the Georgia Institute of Technology.

Beth Keser
Steffen Kröhnert

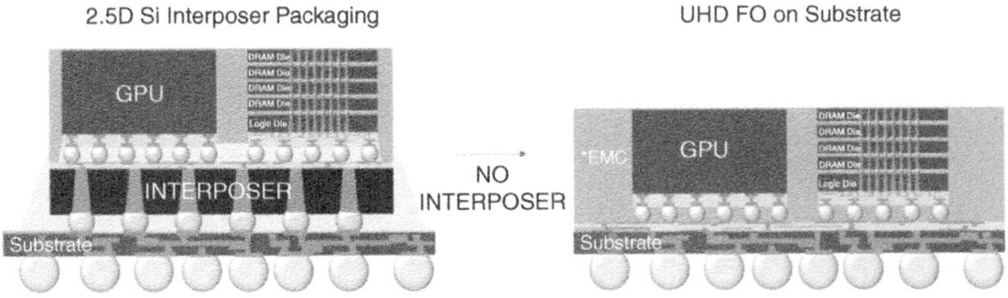

Figure 1.14 Architectural comparison between 2.5D Si interposer vs. UHD FO on substrate.

substrate. Similarly, UHD FO can integrate advanced node GPUs or ASICs with high bandwidth memory (HBM) to support HPC applications, such as fine pitch SERDES interconnects featuring high data rate and high memory data bandwidth communication between compute die and HBM memory. These products make up the heart of UHD FO. A comparison between 2.5D Si interposer vs. UHD FO on substrate can be observed in Figure 1.14.

Within the automotive end market, radar is the device that mainly utilizes FO, for RF performance to transmit and receive signals. Radar for automotive applications now has more functionalities and is becoming widespread. It can precisely locate vehicles in blind spots or approaching from behind. More and more radar sensors are required in cars to improve safety and driving assistance. So far, it is essentially used for adaptive cruise control (long range) and collision / blind spot detection (medium/short range). This trend will become even more pronounced as cars are becoming more and more autonomous. More recently, in 2020, new innovations in AiP FO can be found in MediaTek's Autus R10 MT2706 in the Steelmate BSE151, as shown in Figure 1.15. Key highlights of MediaTek's AiP radar with eWLB are:

- Printed circuit board (PCB) substrate has a ground plane underneath the antennas to bounce back the radiation from the antenna. Easy shielding solution to minimize signal loss.
- Package Area: 39.0 mm^2 | 25% is taken by die | 28% is taken by antennas | Min L/S: 13.6 μm/11.8 μm.
- With AiP, antenna being integrated into package, package area is significantly reduced: −83% as compared with TI antennae on package (AoP) radar and −27% vs. Acconeer AoP radar.
- Note that AoP is mainly done by FC with IC substrate currently.

Similarly, another potential application awaits, FO AiP in 5G applications. In 5G, the challenge is higher-frequency operation. All the packages must be redesigned to optimize signal gains for higher frequencies. It is understood that a

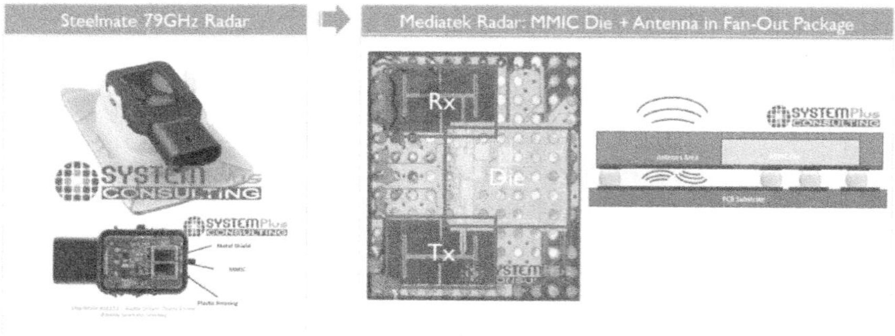

Figure 1.15 Steelmate 77 GHz radar and MediaTek chipset with AiP.

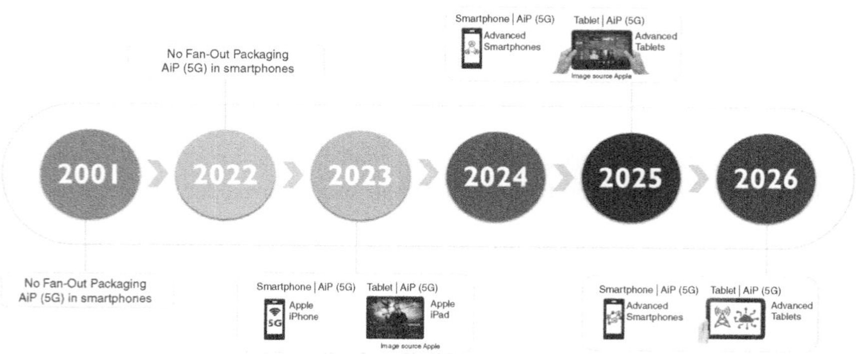

Figure 1.16 Roadmap of FO AiP.

one-millimeter wavelength (mmWave) 5G iPhone will require three to four AiP modules. FO packaging minimizes antenna signal loss in a thinner form factor offering desirable AiP in higher-end phones pursuing sleeker designs. Presently, it is understood that Apple may consider using a phased array antenna in its iPhone, with two parts that work together to form a radio signal beam. This beam can be electronically steered in different directions without moving the antenna. The modem chip and the antenna module work closely together to provide this functionality. We do note that mmWave itself is not new. For example, cars make use of mmWave radar chips operating at 77 GHz. These radar chips, essentially packaged by FO, are used for lane detection and other safety features typically in luxury cars. It is known that TSMC has been highly active in this regard. A potential roadmap is FO AiP being adopted in Apple iPhone, as shown in Figure 1.16.

1.3 Technology Trends and Supply Chain

1.3.1 Fan-Out Packaging Technology Roadmaps

FO is penetrating even higher-end I/O densities, much greater than 18 per mm^2, and finer RDL, with L/S measurements far below 5 μm/5 μm. TSMC and ASE with their FO on substrate technology InFO_oS and FoCoS, respectively, are already in production for networking applications. In the next five years, the package size will exceed 25 mm × 25 mm, and the L/S will go to 2 μm/2 μm and below. The number of RDLs will increase to 4 and above, and the package thickness (without BGA) will go to 150 μm. The FO technology roadmap is shown in the figure below. The roadmap described here in Figure 1.17 is for volume production and an expected average of the different technologies on the market.

The FO architectures with more integration, such as PoP and SiP, will be widespread as the technologies improve (higher I/O count, RDL number, TMV, chip-last, etc.). The FO technology roadmap for different end applications are shown in Figure 1.18.

1.3.2 Fan-Out Packaging Technology by Manufacturer

1.3.2.1 Amkor

Amkor Portugal, formerly Nanium, is well established in the FO landscape and one of the early adopters of eWLB technology. Nanium was purchased by Amkor in 2017 and is now officially referred to as Amkor Technology Portugal (ATEP). Thanks to its long history in the FO market (since 2009), Nanium has developed a wide portfolio of products based on eWLB technology. Willing to diversify from

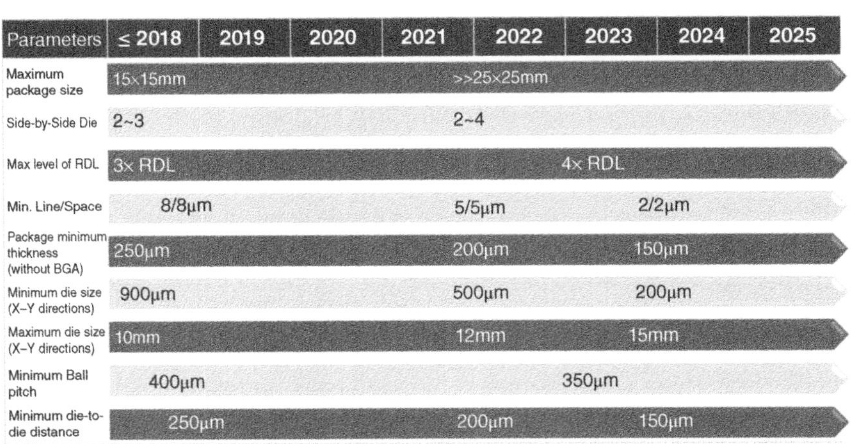

Figure 1.17 FO technology volume production roadmap.

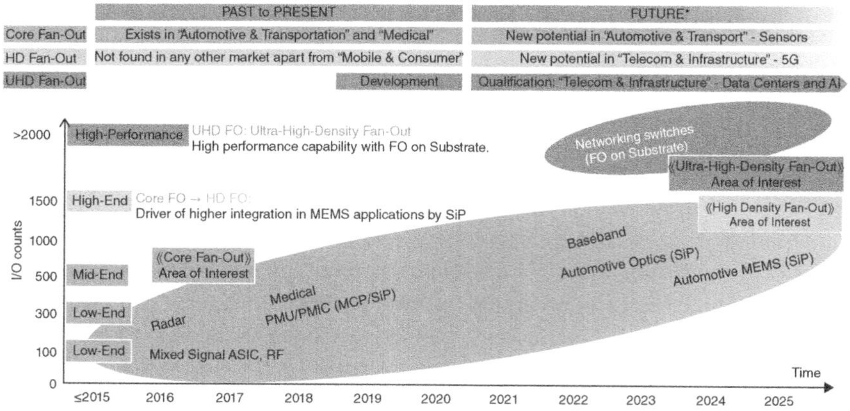

Figure 1.18 FO technology roadmap for different end applications.

single die packages, ATEP offers another level of FO integration, enabling them to embed more chips and passives and build more complex architectures (MCM, SiP, PoP, F2F). These complex architectures allow eWLB to compete with conventional substrate-based PoP and SiP and reach markets where high I/O count and complex products are necessary. These technologies are enabled by the development of processes to get more RDL (and so more connections), larger-sized die embeddable, and so forth. Amkor Portugal's brand name for FO-WLP based on the eWLB technology from Infineon is now WLFO (wafer-level fan-out), and all the products of their portfolio start with "WL": WLBGA, WL-land grid array (WLLGA), WLSiP, WL3D, WLPoP, WLMCM.

1.3.2.2 JCET

STATS ChipPAC, acquired by JCET in 2015, is the leading provider of eWLB and announced that it shipped more than 1.5B FO-WLP devices in March. Willing to diversify its markets and gain market share with the BB, PMIC, RF single die packages (sweet spot of FO-WLP), STATS ChipPAC has also established a roadmap and is offering MCM, PoP, and SiP. This includes hybrid offers in which FO packages are attached to substrates that have the potential for high-density applications. JCET/STATS ChipPAC is pushing development of higher integration with eWLB technology: MCM with several RDLs of different L/S as well as thin SiP and PoP with large die and high I/O count. JCET's advanced 3D eWLB technology provides a smaller form factor and performance value-add and is proving to be a new 3D SiP packaging platform that can expand its application range to various types of emerging mobile and IoT applications, including sensors, MEMS, and automotive applications.

1.3.2.3 NXP

NXP developed the RCP technology which, was licensed by Nepes. The main differences between eWLB and RCP are that RCP uses a dispensed encapsulant rather than a compression molded one and has an additional copper structure that helps limit the die shift during wafer molding and provides electromagnetic shielding and more rigidity to the package, but adds costs. eWLB has the advantage of an easier process due to fewer steps (no copper structure) and a long history with several companies in carrier and molding technology selection. RCP seems to have better performance in die shift but despite that had more limited success in the market, being provided only by Freescale (now NXP) and Nepes. With the maturity of the process and an approach similar to eWLB, RCP providers are promoting a second generation of products with more complex applications and higher integration (MCM, SiP, PoP). Freescale (NXP) is the main provider of RCP, and its packages are used for its own products. It has a large portfolio because of the high positioning of NXP in automotive, IoT, and wearables. Products benefiting from RCP's high electrical capability are seen in automotive, such as NXP MR2001 radar (77 GHz). Due to the higher integration, targeted applications for RCP are wider and include MEMS, RF, wearable, and IoT. RCP will probably remain a niche product though, since even NXP themselves use with other FO solutions, such as eWLB, in some applications.

1.3.2.4 DECA Technologies

Deca developed the M-Series FO technology. It is a chip-first face-up solution using an AP software technology. Deca has developed this breakthrough adaptive patterning™ approach to solve the die shift problem in FO-WLP technology by creating a unit-specific pattern for every device on every wafer or panel to perfectly match actual device locations. Deca has 40 issued and pending patents on AP. Deca has a complete production solution, in partnership with industry-leading electronic device automation (EDA) vendors and equipment suppliers. With this technology, the company reports that it has a fast product qualification capability and has solved issues of pattern misalignment that are critical in other chip-first FO technologies.

The AP software technology includes optical inspection and design capabilities to measure and correct for die shift and to recalculate device I/O to ball interconnects directly during processing. Connection from the die pads to the RDL are made by copper studs that can be placed either directly on the bond pad or routed along the RDL. This technology is licensable by Deca and is already licensed to a major manufacturer, ASE.

In February 2020, Deca officially became an independent technology development and licensing company. At the same time, Nepes acquired Deca's manufacturing operations in the Philippines, with licensed M-Series. With such a move, the

M-Series has the potential to take a large share of the market and become a new reference. Deca's M-Series is established in Taiwan, Korea, and the Philippines thanks to licensing from ASE and now Nepes. A notable development within the original application is the use of Deca's M-Series FO technology to create a protected WLP.

1.3.2.5 ASE

ASE has a multitude of packages using the FO nomenclature: chip-first, chip-last, face-up, face-down, PoP, SiP, and FO on substrate. They have also licensed various technologies: eWLB from Infineon since 2009 and M-Series from Deca Technologies since 2016, after investing $60 M for the license, and started to build a panel line in 2018.

Some technologies have already been accepted by the industry for a long time: chip-first face-down eWLB is the benchmark for all FO-WLP and demonstrated its potential for single die embedding and for further integration in high-volume manufacturing (HVM) first. Chip-first face-up M-Series demonstrated its capability in dealing with the die shift issue. Other platforms are addressing different markets and are using different technologies: FOCLP is a flip-chip of the die on a coreless substrate that are embedded in a molding compound in a subsequent step. FOCoS is an FO package on an advanced substrate. FO SiP is a chip-last solution. FOCLP addresses a market in which I/O count is limited. ASE FOCoS technology is starting to be positioned as a competitor for 2.5D Si interposer for large SoC applications at a lower price. Consequently, the ASE Group with its FOCoS is in a good position to follow that trend. FOCoS has already been in volume production since 2018. The RDL can be as narrow as 2 μm/2 μm L/S. It can be performed face-up or face-down before being placed on a substrate.

ASE is also working on panel FO with different approaches: They are building a 615 mm × 625 mm panel line to produce Deca Technologies' M-Series. ASE is also building a 300 mm × 300 mm panel line but with more aggressive L/S, targeting more advanced applications than those that M-Series will address.

ASE identified FO opportunities for high frequency, such as automotive radar (76–81 GHz), 5G backhaul (> 20 GHz), 5G fronthaul (> 20 GHz), WiGig (60 GHz), and AiP.

1.3.2.6 TSMC

TSMC was the first company to develop FO InFO technology to target application processor packaging for high-end smartphones. In 2015, they purchased a plant for $85 M in Taoyuan. TSMC then invested more than $500 M in equipment and refurbished the plant for InFO production. The investment reached $1 B in 2019. The first main product target was Apple's APE A10 for its iPhone 7, and InFO captured that market. The PoP structure was accepted by Apple because of its thin size versus standard flip-chip. Apple confirmed that choice with APE A11 in 2017, APE A12 in 2018, and again with APE12 in 2019. InFO has made TSMC's 7 nm process technology more competitive than its counterparts, and TSMC won all orders for A14 chips that Apple released in 2020 and A15 orders in 2021.

InFO_PoP challenges the standard FC-PoP market with thinner packages and better performance. InFO's market of interest is PoP and SiP for complex mobile applications such as APE and DRAM PoP, for which customers are always pushing for ever thinner packages. Customers are now asking for sub-0.8 mm package thickness, which is not possible with PoP based on current laminate substrate technology. A11 reverse engineering has revealed that InFO technology could cut the thickness of an APE package down to 0.23 mm, with 3 RDLs and a minimal L/S of 8 μm/11 μm. This allowed Apple to get a full PoP thickness (without BGA balls) below 800 μm. TSMC will continue reducing the thickness of these PoPs with higher L/S resolution, RDL on top of the bottom package, and other improvements. Another added value is electrical performance thanks to shorter interconnections compared to a FC-CSP PoP. The yield achieved by TSMC, and consequently the impact on the cost, are questionable. However, most likely, cost was not as important for TSMC as it could be for an OSAT since they can compensate for it with Si wafer foundry earnings.

TSMC has developed various variants of InFO such as InFO_oS (on substrate similar to FOCoS by ASE) and InFO_MS, InFO_AiP, and InFO MUST-in MUST, where MUST stands for multi-stack (InFO_MiM) for various applications. For 5G mmWave wireless communication, InFO_AiP integrates the dipole and patch antenna with a mmWave front-end module (FEM) chip leveraging HD RDL and fine pitch. [18] TSMC's low dielectric constant polymer passivation material and uniform RDL enables high gain and low loss. It has been revealed that TSMC has qualified InFO_AiP with a package dimension of 12 mm × 12 mm × 0.9 mm with 2 RDL layers [19].

TSMC is also positioning UHD InFO derivatives, such as InFO_oS and InFO_MS (memory on substrate) for networking and AI inferencing applications, respectively, in the mid- to high-end HPC market. InFO_oS leverages InFO technology and features higher density 2 μm/2 μm RDL line width and space to integrate multiple advanced logic chiplets for 5G networking applications. It enables hybrid pad pitches on system-on-a-chip (SoC) with a minimum 40 μm I/O pitch, minimum 130 μm C4 Cu bump pitch and > 2X reticle size InFO on > 65 mm × 65 mm substrates. TSMC is expected to integrate more chips as customers continue to accelerate the adoption of a chiplet packaging scheme for their next generation products.

UHD FO packaging platforms, such as TSMC's InFO_MS and InFO_oS, offer alternative solutions to TSMC's CoWoS (chip-on-wafer-on-substrate 2.5D Si interposer technology) for applications that seek cost/performance optimization in the mid- to high-end of the HPC market. TSMC developed 3D multi-stack (MUST) system integration technology and 3D MUST-in-MUST (3D-MiM) FO as the next generation wafer-level FO technology. [20] 3D-MiM technology utilizes a more simplified architecture that eliminates BGAs between packages for system-level performance, power, and form-factor (PPA) purposes. This technology also makes use of a modularized approach for both design and integration flow to improve design flexibility and integration efficiency. Known-good pre-stacked memory cubes and/or logic-memory cubes are fabricated by leveraging the established integrated FO technology platform (InFO) in tools, materials, design rules, and processes to shorten development cycle time and achieve cost effectiveness. Near memory processing is attractive for high bandwidth, low latency, and power saving. 3D-MiM, a next generation integrated FO technology, is proposed to be an alternative heterogeneous integration solution to FC-PoP for mobile and 3DIC stacking for HPC applications, to realize in-package near-memory computing.

Currently, TSMC InFO research and development (R&D) results have been highly enticing and promising for next generation FO-WLP. Near memory processing is attractive for high bandwidth, low latency, and power saving. 3D InFO_MiM is positioned by TSMC as an alternative heterogeneous integration solution for advanced mobile and HPC applications to meet 5G/AI-driven application demands for years to come.

TSMC recently developed a novel wafer-scale system integration solution, InFO_SoW (system-on-wafer), to integrate known-good chip arrays with power and thermal modules for HPC. [21] InFO_SoW eliminates the use of substrate and PCB by serving as the carrier itself. Close packed multiple chip arrays within a compact system enable the solution to reap wafer-scale benefits, such as low latency chip-to-chip communication, high bandwidth density, and low impedance, for greater computing performance and power efficiency. In addition to heterogeneous chip integration, its wafer-field processing capability has enabled a chiplet-based design for greater cost saving and design flexibility.

1.3.2.7 PTI

PTI is considered a new entrant in the FO landscape with FO-PLP. PTI FO-PLP was production-ready in 2018, and PTI commenced running volume production in 2018. PTI offers four technologies, called CHIEFS, CLIP, PiFO, and BF^2O: a chip-first face-up, a chip-last, a FO-PoP, and a chip-first face-down, respectively. Thanks to this diversity, PTI can address numerous applications. PTI has set an aggressive roadmap to compensate for its delay compared to companies that already have a long history in FO-WLP: small L/S (2 μm/2 μm) and high RDL count (3 RDL). PTI is betting high on FO-PLP technology. See the last paragraph of Section 1.1.1 for more details about PTI.

1.3.2.8 Samsung Electronics

As a reaction to TSMC's InFO, Samsung Electronics and Samsung Electro-Mechanics (SEMCO) announced in June 2016 that they were joining forces to staff a packaging project and launch a solution to compete for the Apple APE market. In 2016 $200 M was invested to change LCD assembly lines in Cheonan into IC packaging lines. In 2019, it was reported that Samsung Electronics acquired the FO-PLP business unit from Samsung Electro-Mechanics (SEMCO) to enhance its capability in the FO-WLP market. [16] The technology developed by SEMCO is a new type of FO: die are embedded in a substrate (PCB type) after being placed in cavities created in the substrate. This brings several potential advantages: it is easier to work with panels since equipment already exists to process substrates at panel level, while handling reconstituted panels made of mold compound is still challenging. It may also bring advantages in dealing with warpage since an organic substrate may be more rigid. Standard RDL made of dielectric material slit-coated on the panel surface and copper deposition constitute the interconnections. This is competitive with FO-WLP. Samsung FO-PLP has run production for its own smartwatch, but it is not a HD or UHD FO solution.

1.3.2.9 Huatian

Huatian developed embedded silicon fan-out (eSiFO) technology and is currently in HVM. The principle is to create cavities into Si wafers and to use them as carriers to embed thinned die. This would effectively replace mold compound. The subsequent RDL process is standard. The eSiFO technology is compatible with PoP architecture.

1.3.3 Supply Chain Overview

FO-WLP involves a simplification and consolidation of the packaging, assembly, and testing in a "middle-end" type of infrastructure, where the legacy is controlled by experienced OSAT companies in the semiconductor industry, but it requires collaboration at the design level. Figure 1.19 shows the global map of key companies involved in FO activities.

For the OSATs, eWLB pioneers (JCET, Amkor, and ASE) are still supplying consistent volumes of core FO, mainly core FO-WLP. Furthermore, some OSATs (Nepes, PTI, and ASE) are positioning FO-PLP to catch existing business by reducing packaging cost. They are also focused on large die size applications for the future. This attracts fabless customers such as MediaTek. It is noted that SPIL and Amkor Korea are running qualification lots for FO. Also, companies and institutes such as 3D Plus, Aurora semiconductor, CEA-Leti, Fraunhofer IZM (FIZM), and Institute of Microelectronics (IME Singapore) are all in collaboration to assess new FO capability.

Regarding foundries, it is only TSMC that is shipping FO-WLP in HVM. TSMC is the only supplier of APE FE and packaging for Apple's iPhones and smartwatches.

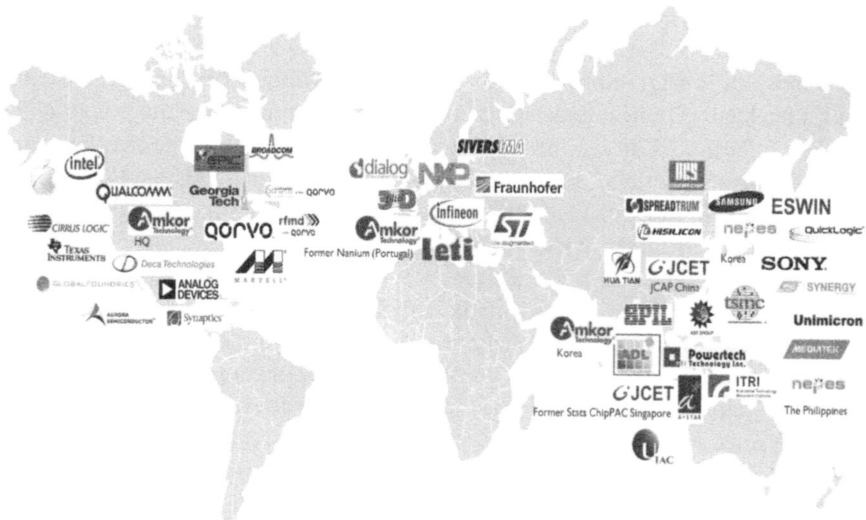

Figure 1.19 Global map of companies active in FO.

TSMC is expected to utilize InFO for HPC and 5G applications, tapping into a new pool of customers such as MediaTek, HiSilicon, and Nvidia for FO-WLP. GlobalFoundries collaborated with R&D institutes before assessing FO capability and now leverages OSAT capabilities.

Regarding integrated device manufacturers (IDM), Samsung Electronics (previously through SEMCO) has commercialized FO-PLP with APE/PMIC for the Galaxy watch (consumer). The focus is to enable HD FO for Samsung's internal smartphone and, if possible, to secure FE and packaging business for Apple's APE, which was lost to TSMC in 2015. NXP and Infineon outsourced a majority of their production to OSATs. They run R&D and low volume manufacturing (LVM) within internal units and are focused on the automotive market.

Deca has changed its business model and adopted a licensing model. The goal now is to do technology transfers of M-Series and AP to leading manufacturers. Fabless companies such as Qualcomm, Apple, MediaTek, HiSilicon, and Xilinx have been qualifying FO-WLP for megtrend-driven applications (5G, HPC, IoT). New companies in IoT and AiP have also emerged in recent years. For example, Sivers IMA, Synaptics, and Synergy have all been running LVM qualification lots since 2019.

1.3.4 Analysis of the Latest Developments in the Supply Chain

In this section, the supplier-customer relationships of leading companies within each business model, foundry, IDM, OSAT, and licenser, as shown in Figure 1.20, are explored. Not all suppliers are listed as the focus is on suppliers who have had more developments within the supply chain in recent years; therefore, this is a non-exhaustive list of companies. Customers are connected to suppliers in a more complex manner than purely existing customers, new customers, and speculated customers. Analyzing supplier-customer relationships by business model can help us better understand the impact on the dynamics of the business because the relationships have various facets. The structure of supplier-customer relationships is complex and reflects capability, resources, activities, and technology positioning.

As seen in Figure 1.20, TSMC is the leading foundry supplying FO-WLP. TSMC is not only positioning itself as an advanced foundry but also a high-end packaging house in BE. This unique business model will continue to lead the way in creating higher value and innovative breakthroughs. As such, FO-WLP will attract more end customers with new applications at the right cost.

InFO_oS (on substrate) technology is now being utilized for HPC running in LVM for MediaTek, Nephos, and Xilinx. Nvidia, AMD, and Qualcomm are believed to have expressed interest and have qualified test chips.

HiSilicon used to be TSMC's customer for UHD FO. However, the US enacted legislation in May 2020 that prevented foreign manufacturers of semiconductors that use US software and technology in their operations from shipping their products to Huawei unless they got a license from the US. TSMC stopped supplying new orders of UHD FO for Huawei. This essentially cut Huawei off from the supply of leading-edge chips, limiting Huawei's competitiveness. At the same time, TSMC lost a significant revenue stream from one of its major customers, Huawei.

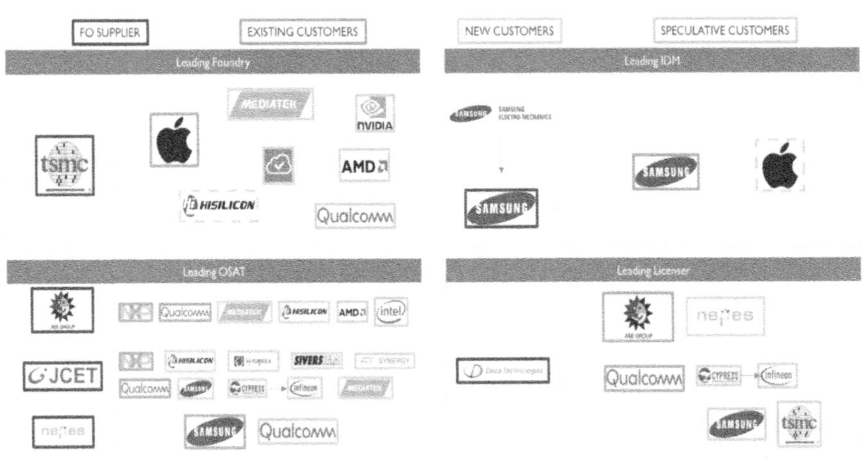

Figure 1.20 Leading supplier-customer relationship by business model (non-exhaustive list of companies).

As shown in Figure 1.20, Samsung Electronics is the leading IDM actively manufacturing FO-WLP internally. Samsung Electronics, being a leading IDM, had strong reasons and resources to invest in SEMCO for a FO-PLP solution and subsequently to reacquire it to expedite yield improvement and synergies with FE die internally. Consequently, Samsung Electronics' new model is the same as TSMC's model, which won the Apple APE die and packaging business in 2015.

Samsung has been instrumental in design, memory, logic, packaging, chipset assembly, and end product, so it can now drive innovation efficiently and decisively. Currently, HD FO-PLP is still in production for the Samsung Galaxy Watch APE and PMIC. Next, it will be for APE in smartphones. This is, however, easier said than done with new FO-PLP technology because the yield and reliability will prove to be another technical challenge when it comes to the larger die size and increased I/O counts required for mobile APE. Samsung will continue to innovate cost-effective HD FO in order to compete with TSMC for Apple's packaging and FE business. As of 2021, however, it will still be for internal customer use within Samsung.

As illustrated in Figure 1.20, the leading OSATs include the ASE Group, JCET, and Nepes, all of which have more developments in recent years. ASE's FOCoS is gaining traction for HPC. JCET is starting to enjoy the 5G boom. MediaTek is a confirmed ASE customer running LVM. These are all mainly for server applications. HiSilicon is looking to ASE for FO-WLP because TSMC's production is mostly dedicated to Apple. HiSilicon practices a dual or multi- vendor policy. However, because of similar legislation by the US, the ASE Group is expected to stop supplying HiSilicon.

JCET China is very HiSilicon-oriented. In fact, it is noted that an HD FO line was built in China supporting HiSilicon's HD FO request. HiSilicon is trying to enable domestic OSATs whenever possible due to the trade war. In a way this benefitted JCET. JCET Singapore used to enjoy large volumes from Qualcomm. Although the volume is not as significant as before, JCET Singapore has started to attract many new customers, such as Synergy, Synaptic, Sivers IMA, and NXP for 5G-driven applications. Also, there are new applications with existing customers for AiP and PMIC, also thanks to 5G. In summary, JCET is enjoying good business from existing PMICs and is qualifying for new applications with the start of the 5G boom.

In 2020, Nepes announced that it has spun off its FO-PLP business division into an independent subsidiary company called Nepes Laweh. A new independent management will be leading this company. This is a significant and rather bold move by Nepes to restructure and reinvent the new and old businesses in order to grow faster, especially after acquiring Deca's FO-PLP technology and line. Apart from Samsung, we believe it is Qualcomm driving this initiative. Moving forward, we are expecting Nepes to go full force on the development of advanced semiconductor process services with FO technology and achieve business competitiveness within core FO.

As shown in Figure 1.20, Deca repositioned itself as a leading industry licensor with its propriety technologies. With strong endorsement from Qualcomm, ASE, Nepes, and Cypress (acquired by Infineon), Deca has transformed from a manufacturer into a fully-fledged independent technology development and licensing company in 2020. Deca's manufacturing line in the Philippines has been sold to Nepes. This liberated Deca from the manufacturing business. Presently, Deca Technologies' focus is to enable applications in the mobile and consumer markets. Moving forward, it will target HPC and advanced networking. The targeted audience in the future could be Samsung or TSMC.

1.4 Fan-Out Panel-Level Packaging (FO-PLP)

1.4.1 Motivation and Challenges for FO-PLP

The demand for lower cost with higher performance has driven the semiconductor industry to develop innovative solutions. Similarly, the OSATs and their customers are always asking for ever-lower prices. One approach to reduce overall cost is to switch from wafer and strip level to a larger-size panel format to take advantage of efficiencies and economies of scale. Going from wafer to panel (for example 12-in. wafer to 18-in. × 24-in. panel) could enable cost reductions up to 50% if the technologies are ready and the yield is higher than 90%. Moving from wafer to panel format has major cost and productivity advantages due to higher efficiency and economies of scale. The usage area of a panel-shape carrier is higher than round wafer format. Panel level packaging process can produce more FO because of this high panel area usage ratio. The carrier area usage ratio of 8-in./12-in. wafers is < 85% due to the influence of edge incomplete packages, whereas carrier area usage of panels for FO can reach 95% with rectangular chips on rectangular panel carrier. Panel manufacturing has higher throughput due to reduced handling/transfer time compared to round wafer processing. Also, there is a huge decrease in the number of packages obtained per wafer because of increasing package size, which provides an incentive to move to larger panel carriers to boost profitability, as shown in Figure 1.21.

Certain criteria must be fulfilled and certain challenges overcome for the broad adoption of FO-PLP. These challenges are linked to large capex investment, standardization, multi-source availability, and most important, market availability to keep the panel line running. There are technical challenges too, such as warpage control, die placement accuracy, and fabrication of sub-10/10 μm line on large panels. Standardization of the panel size and assembly process is the biggest hurdle

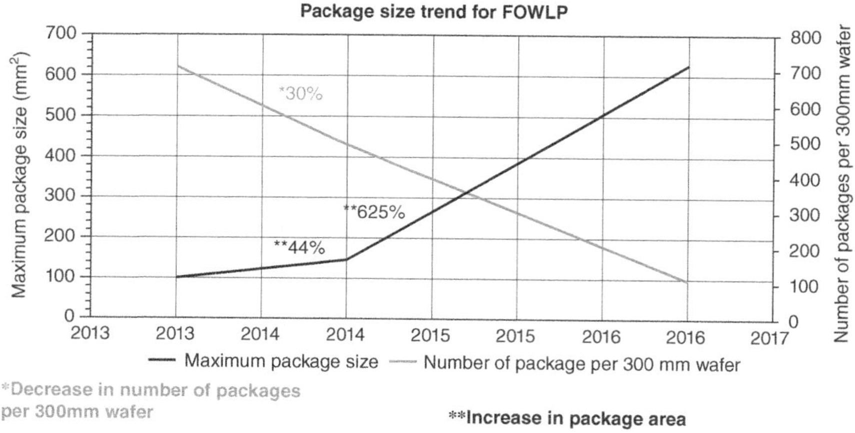

Figure 1.21 Package size trend for FOWLP showing decrease in number of packages per 300 mm reconstituted wafer.

for the adoption of FO-PLP. Each player is developing their own process using different panel sizes and infrastructure (PCB/LCD/WLP/PV/mix) catering to specific applications and customers. In this scenario, it is exceedingly difficult for end customers to multisource. Also, it is not profitable for equipment suppliers to design and manufacture equipment according to different customers' requirements. Warpage minimization of the panel is another big issue so that it can be processed using standard semiconductor processing tools. The larger the panel, the more complicated it gets because the loss of one panel represents a much higher value. Another challenge is the feasibility and reliability of new processes needed in case wafer solutions are difficult to adapt. For instance, spinning deposition of materials is more complicated for panels, and the use of sheets (for mold compound, for instance) or spray (for photoresist, for instance) could be a replacement.

1.4.2 FO-PLP Market and Applications

The applications for FO-PLP can be defined by segmenting the technical specifications required for RDL technology in terms of L/S: 2 μm for high-end applications such as CPU/GPU, FPGA; 5–8 μm for mid-end applications baseband, processors, power management module, and RFIC; and 10–15 μm for low-end applications such as mobile, consumers, RF, Wi-Fi, and power management. Currently FO-PLP in production targets low- and middle-end applications. Because of the technical challenges associated with fine L/S on panel, companies such as Nepes and PTI entered FO-PLP production targeting low-end applications (L/S > 10/10 μm). Samsung/Samsung Electromechanics (SEMCO) entered FO-PLP production targeting mid-end applications with an integrated APU for the Galaxy Watch. We believe the sweet spot for FO-PLP applications is larger package size (> 15 mm × 15 mm) and L/S ≤ 5/5 μm, and this is the goal toward which many companies such as PTI, ASE, and ESWIN, are working toward.

Panel will progressively take a share of production but still in a limited way. FO-PLP is expected to take off much later when big die applications are adopted. By 2025, we expect FO-PLP to account for 18% of the total fan-out packaging production by wafer (300 mm eq), as shown in Figure 1.22. The readiness of key companies for FO-PLP can be seen in Figure 1.23.

1.4.3 FO-PLP Supplier Overview

Panel-level manufacturing has the potential to leverage the knowledge and infrastructure of the WLP and the PCB / flat-panel display / photovoltaic industries. There are various factors that are driving the development of PLP and attracting various companies from the supply chain (including equipment and materials) to invest in panel infrastructure. On the one side, top fabless companies want OSATs to reduce the cost of high-density FO-WLP, and panel is seen as the key step to significantly reduce the package price. FO-PLP is on the roadmap of every big OSAT. On the other side, there are companies who are investing in and developing PLP capability and are really pushing for its adoption as part of their strategy. These companies are mainly driven by the success and publicity around FO-WLP and are those who:

- Missed the early FO-WLP (eWLB) wave (e.g. PTI);
- Are affected by the loss in substrate business and want a new business model while leveraging their experience in substrate manufacturing (e.g. SEMCO, Unimicron);

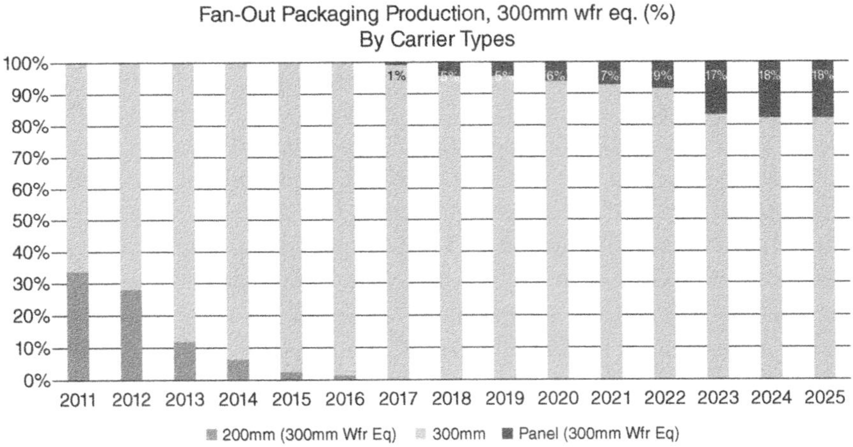

Figure 1.22 FO packaging production trend by carrier types.

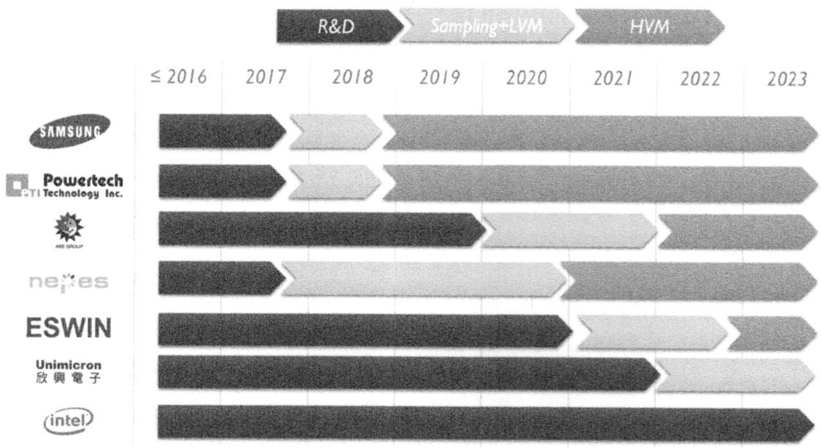

Figure 1.23 Readiness of key companies for FO-PLP production.

- Already have experience in panel processes such as LCD packaging and think they can leverage their experience to PLP (e.g. Nepes);
- Want to develop HD yet low-cost packaging to support their FE chip business (e.g. Samsung Electronics, Intel).

There are many companies developing FO-PLP technology, and after years of development, qualification, and sampling, many companies are going into HVM, such as Samsung, Nepes, and PTI. Nepes has been in low volume production since 2017. ASE in partnership with Deca Technologies is in an advanced stage of development and will go to volume production in 2022. Each player has its own business strategy and is working on its own FO-PLP technology (panel size, leveraging different infrastructure, etc.). For example, Nepes will initially focus on a coarse design (> 10/10 μm L/S) targeting automotive, sensors, and IoT application and will go into high-density design at a later stage, whereas the long-term aim of PTI and SEMCO is to target mid- and high-end applications that require 8/8 μm or lower L/S. ESWIN is potentially the first FO-PLP player in China to run LVM production by 2022. Unimicron is working on a different business model where they will manufacture the HD RDL, and further assembly will be done by a foundry, OSAT partner, or customer. On the other hand, prominent OSATs such as Amkor and JCET/STATS ChipPAC are currently in a wait-and-watch stage and are evaluating various options and will not go into volume production before 2022. Figure 1.24 shows the status of the key companies involved in FO-PLP.

Players	Production status	Technology	Panel manufacturing type	Panel size	*L/S minimal features (2018)	Driving applications
ASE	• Panel production ready • Target for volume production is 2022	• eWLB license (Chip-First Face-Down) • M-Series license (Chip-First Face-Up)	• PCB-based	• 300*300m^2 • 615*625m^2 (for M-series with Deca) (Yole estimation)	• 2μm L/S (for 300*300m^2 panels) • 10μm L/S (for M-series with Deca)	BB, PMU, RF
Samsung	• Production ready for low-end apps (PMIC, etc..) • Production for APE commenced in 2018	• SEMCO PLP	• PCB-based	• 510x415mm^2	8/14μm L/s	• RF SiP • BB, PMU, RF • APE + PMIC
Powertech Technology (PTI)	• LVM started in 2018 • New fab construction still in progress as of 2020	• CHIEFS (Chip-First) • CLIP (Chip-Last) • PiFO • BF^{2}O	• Mixed between LCD and PCB	• 510*515mm^2	• 5/5μm L/S expected	• BB, PMU, RF • APE • FPGA, CPU, GPU
nepes	• Production started in Q2 2017 • Production of M-series after 2019 acquisition	• VF-FOWLP	• LCD-based	• Gen 3 (~650x550mm^2) (Yole estimation)	• 10μm L/S	• BB, PMU, RF
Eswin	• Panel line ready in 2021 • Production to start in 2022	• eMFO (Embedded Molded Fan-Out, Chip-First) • ESWIN-FOPLP	• Mixed between LCD, semiconductor and PCB	• 510*515mm^2	• 5/5μm L/S for 2021; • 2/2μm L/S for 2022	• BB, PMU, RF • APE, • Logic, • SiP

Figure 1.24 Status of the key companies involved in FO-PLP.

Equipment readiness for PLP is not the bottleneck today. Tools are available in the market supporting various process steps in panel processing. However, the tools that support high-density panel packaging are especially expensive, so availability is not the issue, but the cost of the tool is. For some process steps (e.g. plating, physical vapor deposition [PVD], molding, die attach, and dicing), most of the tools are readily available in the market and can be adapted from the PCB, flat-panel display, or LCD industries and are likely to be reused to produce panels. However, some key process steps inherent to advanced packaging, such as lithography, necessitate the development of new upgraded tool capabilities to support fine L/S patterning on panels. Thick-resist lithography, panel-handling capabilities, exposure field size, and depth of focus have been developed by equipment suppliers in recent years. Equipment suppliers are adopting different strategies to enter the PLP business: by acquisition (e.g. Rudolph Technologies has developed tools suitable for PLP based on the knowledge they acquired through the acquisition of AZORES Flat Panel Display Panel Printer), by leveraging tool experience in other businesses and upgrading it (e.g. Evatech, Atotech, SCREEN), or by organically developing tools for PLP from scratch (e.g. ASM). Also, some tool suppliers have a strong position in the FO-WLP market. However, they are skeptical of the PLP business and thus are taking a wait-and-see approach. Ultratech, Applied Materials, and Lam Research are part of this group of companies. On the materials side, the trend for key materials suppliers is to use their current market-ready products for LCD and advanced packaging applications and optimize them to meet panel package requirements.

1.5 System Device Teardowns

1.5.1 Teardown of End-Systems with Fan-Out Packaging

FO technology has a very wide application space. From automotive to consumer, the market is large as described above. With many applications this packaging technology has brought many advantages, as described in Figure 1.25 below. Among them, we can list:

- Thermal dissipation;
- Simplified supply chain and manufacturing;
- Cost simplification;
- IC protection;
- Interconnection;
- High flexibility;
- Thin package;
- High BLR (board level reliability);

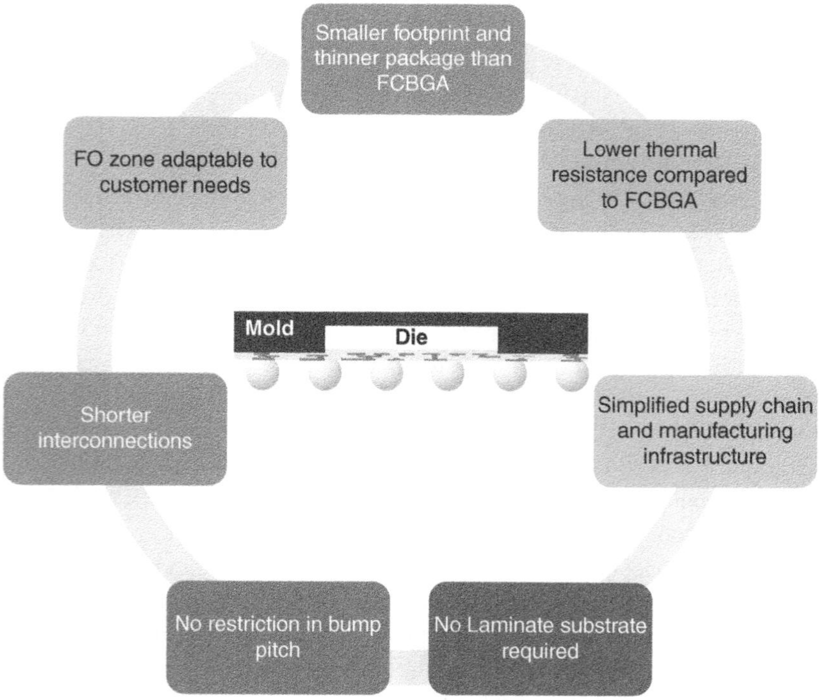

Figure 1.25 Key differentiators for fan-out packaging.

In automotive, the only system that uses FO technology is radar. This system is based on an MMIC die that transmits and receives the signal to sense the car's environment. Previously, the system was implemented in discrete mode on a dedicated RF board. In the MMIC implementation, several technologies are used on bulk silicon substrate, CMOS, and SiGe BiCMOS. As MMIC power is high, the diffused heat could rise rapidly. The main advantages of FO include thermal management using solder balls around the main power amplifier in the transmit path. Another advantage is the shorter interconnections for the receiver (Rx) and transmitter (Tx) path. Using shorter interconnections allows a reduction of parasitic signal. Two technologies have appeared in the market. RCP was developed by NXP in partnership with Nepes. Infineon was the first MMIC manufacturer to use this packaging technology. Now, as the technology has been licensed to several OSATs, newcomers, such as Calterah Semiconductor and MediaTek, are using the packaging technology for radar systems. For MediaTek, the small size and reduced cost of the system was also achieved with the use of FO technology by including the antenna system in the RDL.

These advantages can change for the consumer application. As consumer systems use low power, heat is not a concern in choosing the packaging. Specific factors here are more related to the cost and the size. Indeed, FO technology can provide a small form factor and thinner packaging while keeping to a relatively low-cost solution. Since 2011, the number of references on the market using FO technology has exponentially grown year over year as shown in Figure 1.26. The graph counts the number of component reference models per year. For example, since 2018, Qualcomm has released a huge volume of PMICs with the reference model PM8150. This reference appears in 2018, 2019, and 2020 and counts for one in each year.

Other companies have tried to integrate the packaging technology into their components, but no one else has really succeeded like Qualcomm. The first device from Qualcomm was a PMIC that used eWLB to spread out the die interconnections to fit the ball pitch of the PCB board. At the time, the proportion of the die on the component was close to 70%. Now, depending on the final application, the proportion of the die in the packaging can be close to 90%, as shown in Figure 1.27. The tendency is clearly toward die densification in the packaging and not toward real fan out of the interconnects over a larger area than the die size. In this case, the goal of the FO has become only for protection. Indeed, today FO-WLP technology allows the packaging of any die on 12-in. wafer or panel and provides side wall protection for any die size. In this way, the final OEMs can provide a packaging for a high technology node (0.18 μm), based on 8-in. wafer, on a 12-in. wafer with molding all around the die. This not only provides protection but also reduces the cost of the packaging compared to a WLP.

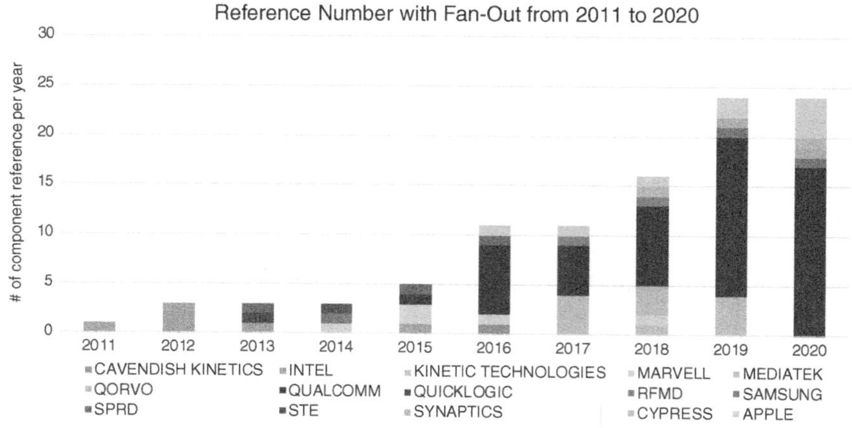

Figure 1.26 Fan-out packaging evolution by reference model from 2011.

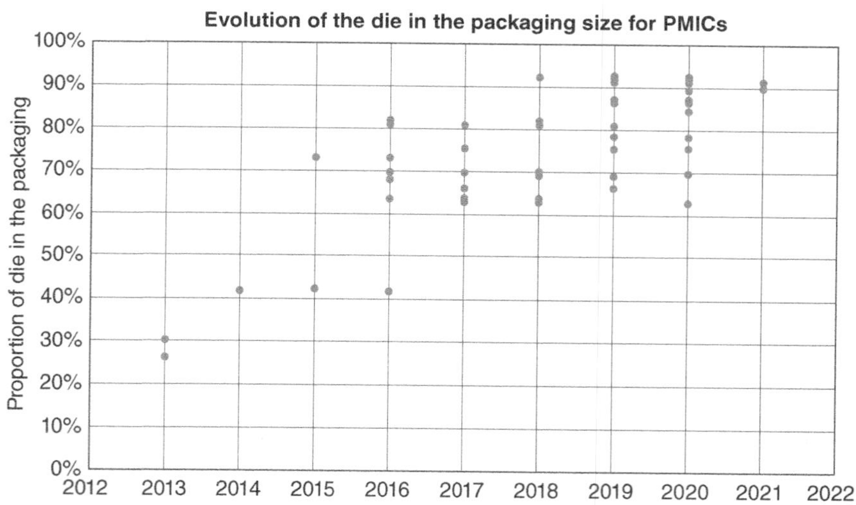

Figure 1.27 Evolution of die proportion for FO packaging from 2013 until 2021.

Now, practically all the PMICs from Qualcomm are using FO technology from ASE to provide 5-side wall protection for the ICs. Only the backside of the die is not covered with molding, but it could be covered with backside laminate and is offered as an option in the M-Series packaging process. One other application of FO in the consumer world is for application processors. As shown in Figure 1.28, the application processor (AP) percentage is small in the market but as the components are much larger than a typical PMIC, the opportunity is large.

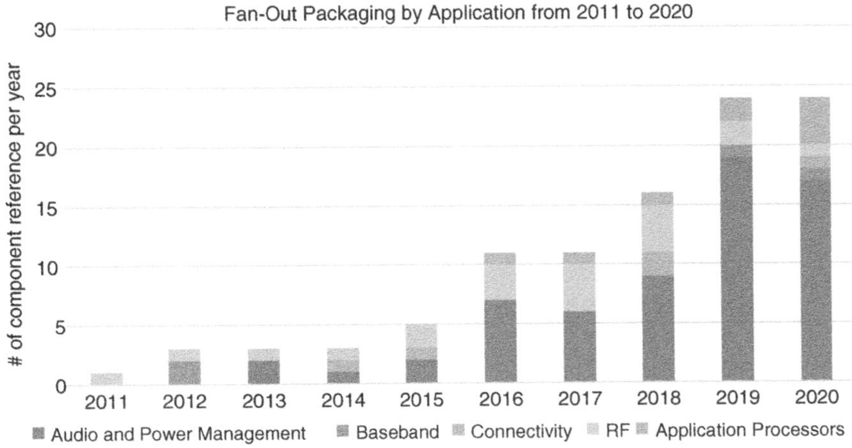

Figure 1.28 Fan-out packaging evolution by application from 2011.

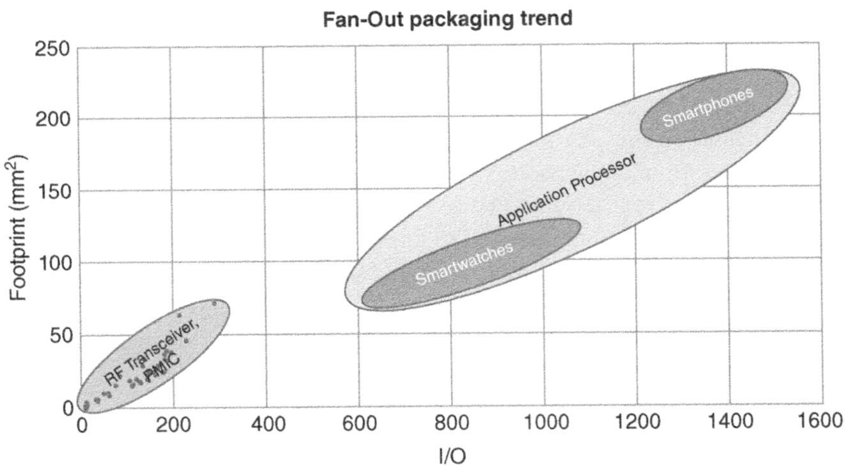

Figure 1.29 Fan-out packaging bump density trend.

Finally, based on a database of consumer components since 2011, a trend in the bump density and the component size can be found and is illustrated in Figure 1.29.

Two main regions can be identified. One is dedicated to small components such as the RF transceiver in the past and PMIC now, with a low number of I/Os and small area mainly using FO for side wall protection. The other, a more spread-out region, is dedicated to APE. In this region, two smaller areas can be defined. One is for smartwatch components with a medium number of I/Os. The other is for the APE of smartphones with a very high level of bump density as the component performs multiple tasks and requires a very high number of IOs.

1.5.2 Technology Comparison

In this section, a comparison by application will be made for each FO technology found in the market, from radar to the SiP. eWLB technology will be presented in several applications as it is the most popular FO packaging technology today. To start with the radar MMIC, a comparison between RCP technology will be made with eWLB from Infineon and its application to the AiP system from STATS ChipPAC for MediaTek.

1.5.2.1 Radar IC: eWLB vs. RCP

The first implementation of RCP technology, chip-first face-down configuration as developed by NXP (formerly Freescale) had an issue that every FO manufacturer has faced, die shifting. Freescale had an interesting solution with the implementation of a copper frame that hardens the structure and prevents a large die shift during the molding. Moreover, the frame was grounded in the package to enhance the isolation of the MMIC. The eWLB implementation is like the RCP technology with a chip-first face-down configuration, but the molding needs to be controlled to avoid any die shifting. As mentioned before, the form factor was not the prime driver in this kind of application. The thermomechanical enhancement was the main concern and the criterion of choice for the FO technology. Indeed, by looking at the die relative to the packaging size, only 25% of the packaging was occupied by the MMIC. All the rest was for grounding, thermal dissipation, BLR, and short interconnections with the PCB substrate. Today, as the MMIC die technology evolves, the tone is not the same. As shown in Figure 1.30, the form factor became one of the main criteria.

The first generation of MMIC was made of three 6 mm x 6 mm components with different die functions integrated in the SiGe BiCMOS process. The last generation by Infineon is larger, but a view of the bump side shows a very small area with RDL compared to the previous generation. Moreover, Infineon managed to get particularly good thermomechanical behavior with solder bumps directly under the die. Finally, advanced RFCMOS technology with eWLB packaging can offer a very small form factor and allows a radar sensor in the parking sensor application. MediaTek managed to fit an integrated antenna system into the same packaging as the MMIC die to fit in a device with a form factor close to the ultrasonic sensor available on the market. To do so, STATS ChipPAC added a very large area of copper RDLs to provide thermal dissipation and antenna isolation. In this configuration, no interconnect is used in the RF path. The antennas are directly connected to the die I/O.

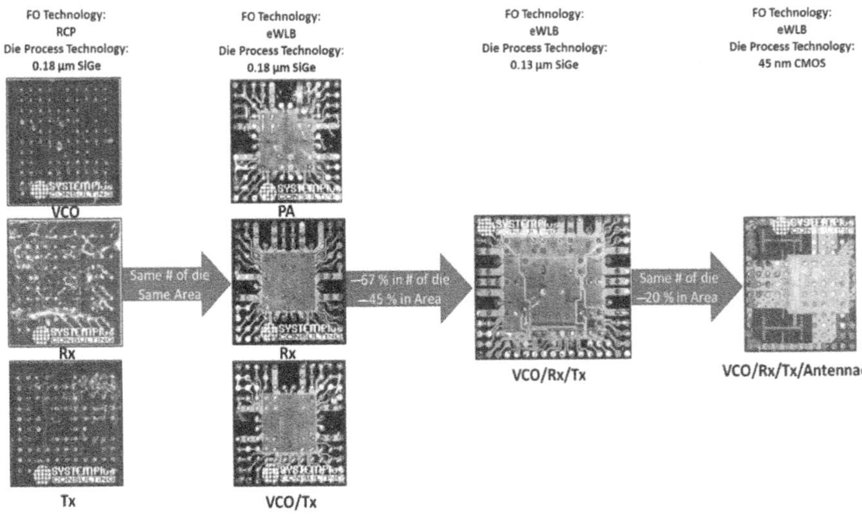

Figure 1.30 MMIC FO packaging technology comparison.

1.5.2.2 MCM/SiP: RCP-SiP vs. eWLB

Both previous technologies can be also used for SiP design. Two designs can be found in the market for industrial and consumer purposes. Both use FO packaging technology for the form factor. For the industrial application, NXP and Nepes developed the RCP into a version that uses a PCB via frame as a replacement for the copper frame used to prevent die shifting; in this case, the via frame is also used to provide interconnects between the bottom RDL and a memory package enabling a PoP configuration.

In very small packages, the components feature a MCU, a PMIC, and a flash memory along with several passive components placed in SMD configuration. The die process is similar to the RCP packaging used for the radar MMIC. For the eWLB, the application of SiP in the market is limited to the consumer world, with two die and good yield losses. A device-like connectivity FE module can be found in the smartwatch application where the low-profile packaging is required. Cypress is the first company to provide such a solution in high volume manufacturing. The device integrated an MCU with a Bluetooth SoC in the same package.

Today, the number of die is limited because of the potential for severe die shifting with several die on a wafer. For a higher number of die, the AP technology developed by Deca Technologies is a promising solution. This technique allows a modification of the routing depending on the die shifting by using a laser direct imaging (LDI) system. It allowed Deca Technologies to develop their own FO packaging technology, called M-Series. In the case of Cypress's SiP, both technologies—eWLB

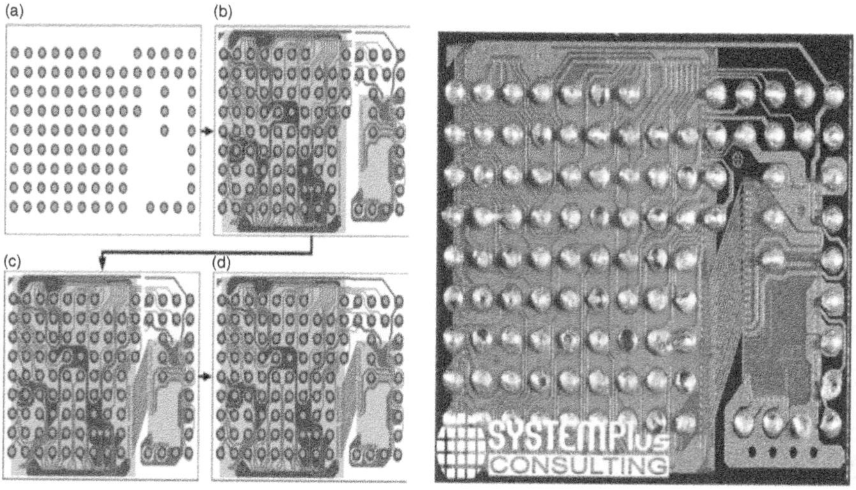

Figure 1.31 Adaptive patterning applied on Cypress SiP project and Cypress CY8C68237FM-BLE bottom side.

and M-Series—were in competition to produce the packaging of the system, as shown in Figure 1.31. Finally, it was STATS ChipPAC who made the SiP using eWLB technology. STATS ChipPAC managed to control the die shifting and realized a routing between the MCU and the SoC with the smallest state-of-the-art line/space value.

1.5.2.3 PMIC: eWLB vs. M-Series

For now, the FO market is moving toward M-Series. Indeed, anywhere Qualcomm uses FO-WLP, M-Series technology has been implemented. Qualcomm introduced FO packaging technology in smartphones in 2015. This was initially based on eWLB from Nanium and STATS ChipPAC. eWLB enabled side wall protection for these die (5 sides including backside). As it is a chip-first face-down process, the fifth side, which is the backside of the die, received protection with no additional process step. This was before the growth of the M-Series.

M-Series totally captured the FO market dedicated to eWLB since it uses a chip-first face-up configuration with active die side protection, AP lithography, and it can provide 6-side wall protection with an optional backside lamination step.

As described in Figure 1.32, the technology uses copper studs and overmold on the die face before the RDL manufacturing process. In this way, the die is completely protected with a molding layer on the top and it provides a very flat surface for RDL manufacturing. Moreover, by using AP technology in the process flow,

Figure 1.32 FO packaging of PMIC components: eWLB (left) and M-Series (right).

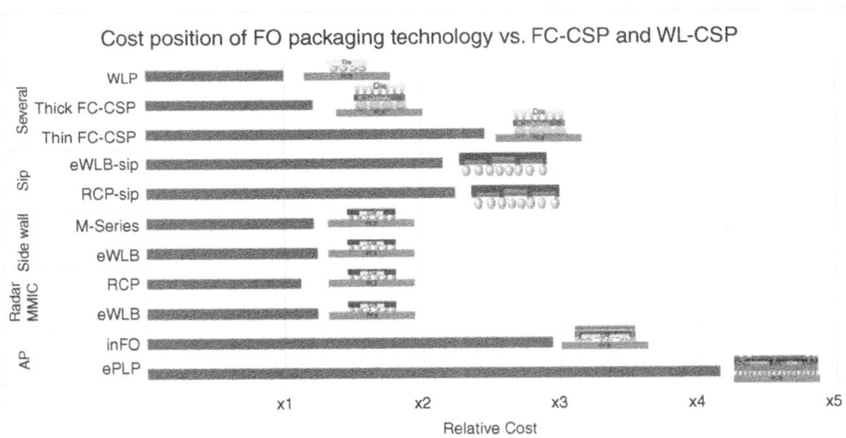

Figure 1.33 Cost comparison of FO packaging technology available in the market.

the package manufacturer can correct for die shifting without any redesign of the mask set using laser direct imaging (LDI) to compensate.

1.5.3 Cost Comparison

Each packaging type has been simulated to evaluate the cost of the FO packaging. Each step has been integrated describing the process flow with all the equipment and related consumables. In this way, the costlier steps of each process can be identified, and a comparison can be made. As shown in Figure 1.33, the relative cost of each FO packaging type can be obtained with reference to WLP.

For simple FO packaging, such as eWLB, RCP or M-Series, the difference compared to WLP is small. FO shows a 20% cost increase compared to a WLP. For all of them, the pick and place step is the costliest. The need for accurate placement of the die before molding is crucial. Only RCP is costlier because of the use of a metal frame to handle die shifting.

FO technology for integration in devices like a radar MMIC, side wall protection shows a relative cost in the same range as thick FC-CSP. One of the key differentiations is the BLR, which is obviously higher for FC-CSP compared to FO packaging. Yet M-Series technology, with its molding protection and hardening at the face of the die, allows the BLR of the technology to be closer to thick FC-CSP and makes a difference compared to eWLB.

Of course, for SiP versions of the FO technology, the relative cost is not the same. FO technology allows a multi-chip packaging level compared to a WLP, which clearly results in a cost increase. But the costlier step is not the same as for single chip FO packaging. Indeed, the use of a metal frame structure to enable PoP in this case creates a process step that makes up 10% of the total cost of the packaging. Also, as the surface to cover with molding is larger, the steps are costlier. All combined, the relative cost is almost twice that of single chip FO packaging.

Finally, by comparing the FO technology for thin packaging of an APE, the cost of the FO packaging from TSMC or SEMCO is three or four times more than WLP. Compared to a thin FC-CSP, the cost of these APE FO are only 20–70% higher. Between TSMC's InFO and a thin FC-CSP, the cost is not the main differentiator. Again, BLR is also one of the key specifications. As the cost is almost the same, for a better BLR clients will choose thin FC-CSP. Since InFO allows thinner packaging and the possibility to have integrated passive devices (IPD) in a face-to-face configuration, the choice would be InFO technology for performance. Moreover, if the ePLP solution from Samsung is considered, the estimated cost will be higher than InFO. Here the step that increases the cost is not related to the PoP enabler as with the copper pillar in InFO, but to the molding. Indeed, the amount and the type of the molding is quite different compared to other FO packaging technologies. Used at panel level, the required molding needs to be fluid enough to overmold the die and the via frame. Also, it needs to be reliable enough to be processed for backside RDL manufacturing.

1.6 Conclusion

FO is not new and has been in HVM since 2007. Recent technological progress is pushing the frontiers of what it can do, and FO is profoundly changing the industry's perception of its potential. This is especially so after TSMC's InFO HD FO technology was chosen to package the APE (A10) of the iPhone 7 in 2016. TSMC's InFO has pushed the high-end smartphone into a whole new higher-end market segment, defined as HD FO.

FO has a stronger appeal than conventional FC-CSP because of its thinner package dimension, increased I/O density, and lower thermal resistance. Additionally, the fanning out of the electrical I/O and the multi-die nature has produced significant scaling benefits for 2D, 2.5D, and 3D integration. The 2.5D FO structure as a replacement of the 2.5D Si interposer in the high-end applications domain has been featured and commercialized in the market. More evolutions of 3D FO-PoP are expected to result in a thinner profile, with more memory capacity and memory bandwidth. In essence, FO is becoming more dynamic than ever, offering solutions ranging from low-end packaging technology up to high-performing and cost-effective integration platforms.

Valued at $1256 M in 2019, the FO market value is set to reach $3046 M by 2025 at 15.9% CAGR. Total production volume of FO is projected to grow from 1703 kwspy in 2019 at a 12.3% CAGR, reaching 3419 kwspy in 2025. For FO, the mobile and consumer end-market dominates with overall revenue of $741 M in 2019 while it is projected to reach $1498 M by 2025 at 12.5% CAGR. Doing equally well is the telecom and infrastructure end-market, which is growing at an astounding 20.2% CAGR from $504 M in 2019 to $1523 M in 2025. Automotive and mobility is expected to grow by 13.8% CAGR from $11 M in 2019 to $24 in 2025. Lastly, the medical end market is expected to shrink with little adoption.

TSMC is all-in on FO for a new pool of customers. InFO_oS and InFO_MS are now being utilized in HPC for MediaTek and Xilinx. Samsung Electronics is hungry to recapture Apple's Si business with HD FO. Samsung Electronics, being a leading IDM, obviously had strong resources to initially invest in SEMCO for an FO-PLP solution and subsequently reacquire it to expedite yield improvements and synergies with FE die internally. Consequently, Samsung Electronics' new model is the same as TSMC's model which won the Apple APE die and packaging business deal in 2015. Samsung could enable HD FO-PLP in its own smartphone devices to self-endorse its performance, cost, and reliability to Apple.

ASE's FOCoS is gaining traction for HPC. HiSilicon and MediaTek are the confirmed new customers running LVM. JCET is starting to enjoy the 5G boom. HiSilicon is trying to enable domestic OSATs whenever possible due to the trade war. This, in a way, benefitted JCET China. JCET Singapore is enjoying good business from existing core PMICs and is qualifying for new applications with the 5G boom, which started in 2019.

Nepes has committed to a strong FO strategy by investing heavily in 2019 and 2020. In 2019, Nepes Laweh invested close to $35 M in acquiring Deca's M-Series and Philippines operations. Then in 2020, a new FO-WLP spin-off company called Nepes Laweh was established. Nepes Laweh started to lay down the groundwork for a new facility in Korea, targeting production by end of 2020. Nepes Laweh is expecting ~$330 M in revenue by 2024, which is about 40% of Nepes total forecast revenue in 2024.

With strong endorsement from Qualcomm, ASE, and Nepes, Deca transformed from a manufacturer into a fully-fledged independent technology development and licensing company in 2020. Deca's business model now includes technology transfers to leading manufacturing companies (foundries, OSATs, semiconductor companies) and license agreements to over 40 issued and pending patents. This includes associated know-how in M-Series and AP and AP design systems in partnership with leading EDA providers combined with proven high-volume real-time manufacturing design.

Panel makers entering the FO game are struggling to achieve high-end products with good L/S and high reliability, though this is the sweet spot for FO-PLP. For FO-PLP companies, the challenge is technical and most likely will require some time to be addressed to ensure a good yield. Existing FO companies have different views. For them to scale-up to panel is more of a financial issue, since new equipment targeting the same performance as wafer manufacturing is required. ASE seems aggressive, with two panel lines been set up (300 mm × 300 mm for fine L/S and 600 mm × 600 mm for M-Series). They seem to be gearing up for a big battle in the core FO market and are planning an entry for HD FO. Given the technical challenges that adversely affect the yield, the FO-PLP that is going into HVM production will support a relatively simple design: > 10/10 μm L/S, < 10 mm × 10 mm package size, maximum 2-layer RDL. With the maturation of the technology and experience gained, FO-PLP will be adopted for HD design with < 10/10 μm L/S, multi-layer RDL, > 15 mm × 15 mm package size and multi-die SiP integration.

Today, the main player using the M-Series from Deca Technologies and ASE is Qualcomm. The die size as a percentage of the packaging area is now close to 90% for PMIC. The FO is not used for spreading the interconnection but for side wall protection and cost reduction. In RF applications, the market has evolved toward connectivity and SiP applications.

The APE has started to grow and represents a very large market potential in the next few years. Regarding trends in density, three patterns have been observed. One for PMIC with low density, small footprint, and side wall protection. The second for smartwatches, with medium footprint, medium bump pitch, and low z-height profile. The third for smartphones with HD, large footprint, and low z-height profile with bump pitch and size flexibility.

The goal of using FO has changed in most of the applications. For MMIC, the predominant factor is no longer thermomechanical behavior. It has now become integration and cost reduction. For SiP, cost reduction is now critical. For the PMIC, side wall protection is the main concern. With AP on the other hand, z-profile and high bump density is still the main concern. Compared to all the technology available for WLP, FO is still costlier due to additional processes such as pick and place and molding affecting the cost. Depending on the specification required by the customers, the technology could be more profitable than WLP in term of flexibility, simplicity of design, thermomechanical behavior, and in some case cost reduction.

References

1 Keser, B., Amrine, C., Fay, O. et al. (2007). The redistributed chip package: a breakthrough for advanced packaging. *Proceedings of the Electronic Components and Technology Conference,* IEEE.

2 Brunnbauer, M., Fugut, E., Beer, G., and Meyer, T. (2006). Embedded Wafer level ball grid array (eWLB). *Proceedings of the Electronic Packaging and Technology Conference*, IEEE.

3 https://www.infineon.com/cms/en/about-infineon/press/market-news/2007/INFCOM200711-013.html.

4 https://www.infineon.com/cms/en/about-infineon/press/market-news/2008/INFCOM200808-084.html.

5 https://evertiq.com/news/16928.

6 https://newsroom.intel.com/news-releases/intel-completes-acquisition-of-infineons-wireless-solutions-business/#gs.xj0vd4.

7 https://www.globenewswire.com/news-release/2017/03/15/1078919/0/en/STATS-ChipPAC-Achieves-1-5-Billion-Unit-Milestone-in-Fan-out-Wafer-Level-Packaging-Shipments.html?culture=en-us.

8 https://www.cnbc.com/2012/10/10/nanium-passes-production-milestone-200-million-ewlb-components-shipped-in-less-than-two-years.html.

9 https://ir.amkor.com/news-releases/news-release-details/amkor-technology-acquire-nanium.

10 Scanlan, C., Olson, T., Rogers, B. et al. (2012). Adaptive patterning for panelized packaging. *Proceedings of the IWLPC*, SMTA.

11 https://www.prnewswire.com/news-releases/cypress-subsidiary-deca-technologies-to-receive-60-million-investment-from-ase-300259281.html.

12 www.nepes.co.kr/en/ir/overview.php?idx=707&bgu=view.

13 Zwenger, C. and Huemoeller, R. (2015). Silicon wafer integrated fan-out technology. *IMAPS 11th International Conference on Device Packaging.*

14 Liu, C., Chen, S., Kuo, F. et al. (2012). High-performance integrated fan-out wafer level packaging (InFO-WLP): technology and system integration. *Proceeding of the International Electron Devices Meeting*, IEEE.

15 https://www.samsungsem.com/global/newsroom/news/view.do?id=5.

16 https://english.etnews.com/20190423200002.

17 www.pti.com.tw/ptiweb/Pressrelease/Powertech%20Technology%20Inc.%20Hsinchu%20Science%20Park%20Plant%20III%20Groundbreaking%20Press%20Release.pdf.

18 Wang, C., Tang, T., Lin, C. et al. (2018). InFO_AiP technology for high performance and compact 5G millimeter wave system integration. *Proceedings of the Electronic Components and Technology Conference (ECTC),* IEEE.

19 Tsai, C., Hsieh, J., Lin, W. et al. (2015). High performance passive devices for millimeter wave system integration on integrated fan-out (InFO) wafer level packaging technology. *Proceedings of the International Electron Devices Meeting (IEDM)*, IEEE.

20 Su, A., Ku, T., Tsai, C. et al. (2019). 3D-MiM (MUST-in-MUST) technology for advanced system integration. *Proceedings of the Electronic Components and Technology Conference (ECTC)*, IEEE.

21 Chun, S., Kuo, T., Tsai, H. et al. (2020). InFO_SoW (system-on-wafer) for high performance computing. *Proceeding of the Electronic Components and Technology Conference (ECTC)*, IEEE.

2

Cost Comparison of FO-WLP with Other Technologies

Amy Palesko Lujan

SavanSys Solutions, LLC, Austin, TX, USA

2.1 Introduction

When selecting a packaging technology, there are numerous factors to consider. Many have to do with the design itself: there are thermal and electrical considerations, size requirements, yield concerns, and more. Cost is one of these many factors. It is not more important than any other requirement, but it is a key consideration. If every technical requirement can be met but the cost is prohibitive, the design cycle must start over.

In this chapter, the cost of fan-out wafer-level packaging (FO-WLP) is analyzed. The cost of FO-WLP itself is discussed, including variations on the process and comparisons between FO-WLP and other technologies are included.

The scope of the comparisons in this chapter is cost only. When comparing two technologies, the author will not go into detail about the technical benefits of the different technologies. If two (or more) packages are compared, the assumption is that all can meet the design requirements. For technical detail regarding different technologies offered, please refer to the other chapters in this publication.

2.2 Activity-Based Cost Modeling

All cost comparisons in this chapter are carried out using activity-based cost modeling. This is a detailed, bottom-up approach to cost. With activity-based cost modeling, a process flow is divided into a series of activities, and the total cost of each activity is calculated. The cost of each activity is determined by analyzing the following attributes:

- Direct labor: This is determined by the time required to perform the activity multiplied by the percentage of an operator required for that activity;

Embedded and Fan-Out Wafer and Panel Level Packaging Technologies for Advanced Application Spaces: High Performance Compute and System-in-Package, First Edition. Edited by Beth Keser and Steffen Kröhnert.

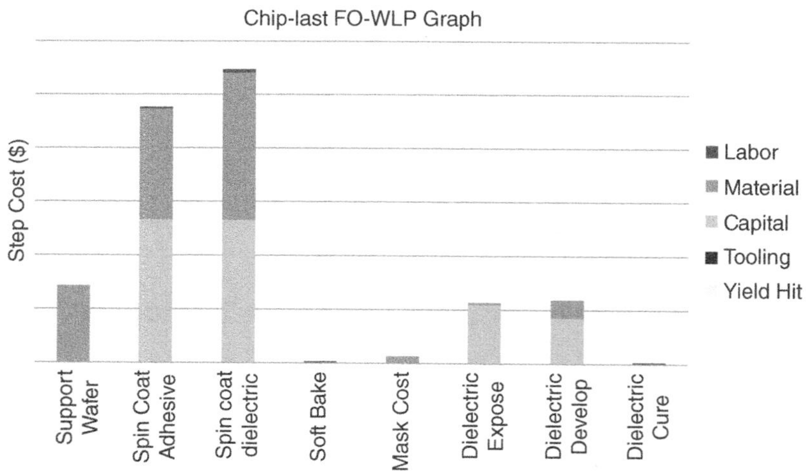

Figure 2.1 Activity-based cost modeling output.

- Material: Both consumable and permanent material costs are included;
- Capital: This is calculated by allocating the equipment depreciation cost based on how long the product uses that equipment;
- Tooling cost: This will be any nonrecurring engineering or fixtures;
- Yield hit: For assembly steps, this will be defects per million opportunities; for fabrication steps, this will be defects per square centimeter. Even though these defects are not visible until later in the process when testing is done, they add cost when they are created.

The result of using activity-based cost modeling is a detailed understanding of the direct costs of each activity in a process. The graph in Figure 2.1 shows the results of the first few steps in a die-last FO-WLP process.

There are other costs to consider, which make up the difference between direct cost and price. Figure 2.2 provides an overview of the different types of cost.

Component	Description	
Direct Cost	Measured Cost — May be done at the activity level or at the factory level	Cost models are used to estimate this directly
Indirect Cost	Factory cost that is not directly associated with an activity. Support, quality, manufacturing engineering, utilities, plant, etc.	While all 4 of These vary widely, their total is driven to a level of consistency by the market. They are usually applied on top of direct cost as a % per manufacturing object
Overhead	Company cost that needs to covered. Typically G&A, marketing, engineering, etc.	
Profit Margin	Usually a percentage on total cost	
Risk Factor	A higher than ususal margin allocated to new technologies	

Figure 2.2 Types of cost.

Indirect costs, overhead, profit margin, and risk factor must ultimately be considered when it comes to the final price of a product, but they are not part of the scope of this chapter.

2.3 Cost Analysis of FO-WLP Variations

Before carrying out a comparison with other technologies, it is important to understand the process and cost drivers of FO-WLP itself. There are three major variations of the process. The process flow for die-first face-down FO-WLP is shown in Figure 2.3. The three light gray boxes reflect the processing that is done on the chip to be placed; the darker boxes represent FO-WLP steps.

There are variations that can exist within the process shown. Some fan-out packages may not require underbump metallization (UBM); others may require a backside lamination process. Furthermore, different approaches may be used to create the redistribution layer (RDL). All cost comparisons in this chapter are based on

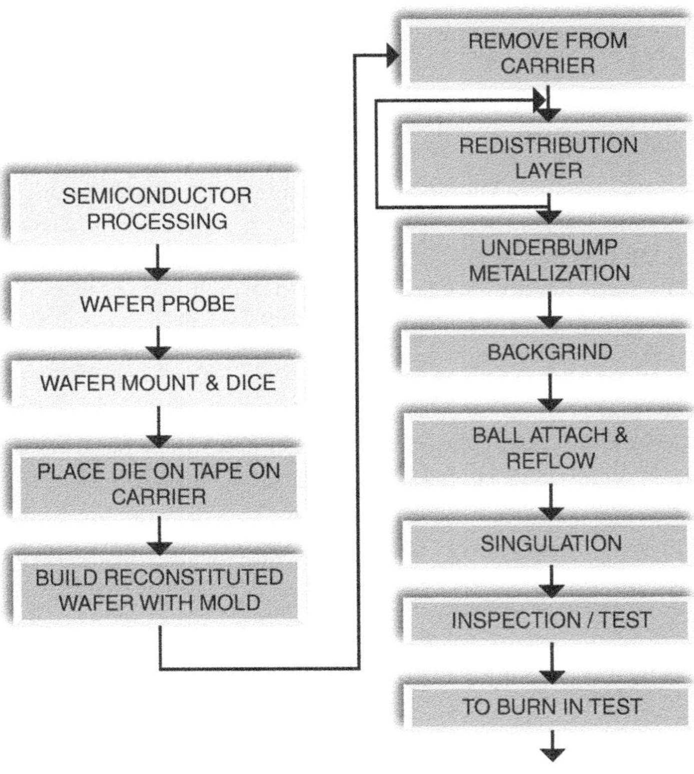

Figure 2.3 Typical die-first face-down FO-WLP process flow.

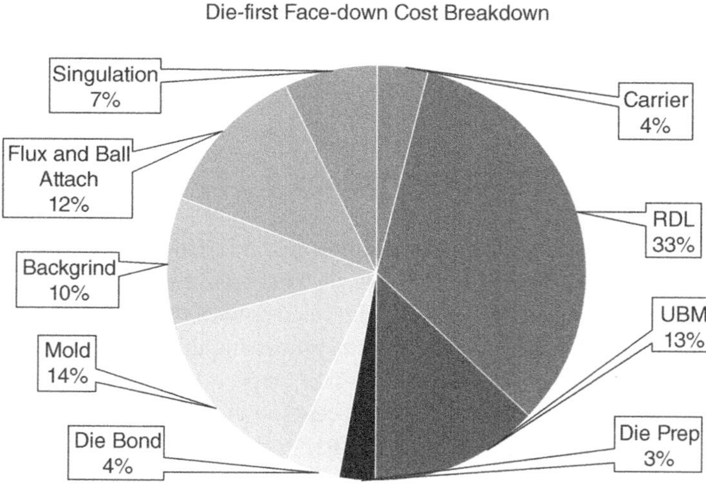

Figure 2.4 Die-first face-down cost breakdown by process segment.

a copper-polymer RDL that uses photoimageable dielectric, but there are other options such as a dual damascene process (more expensive), or RDLs that use a maskless process [1].

2.3.1 Process Segment Costs

The cost of a die-first face-down package is broken down into process segments in Figure 2.4. Each process segment is discussed briefly to further identify where the package costs are coming from. Note that various metrology steps are rolled up as part of each process segment instead of captured in their own category. For example, there are warpage measurements and die drift measurements rolled up as part of the category of mold steps.

2.3.1.1 Die Preparation

This refers to the cost of processing the incoming wafer that will be placed in the package. In the case of die-first face-down, this is the cost to backside grind and singulate the wafer. This category will also include the cost to bump the incoming wafer if the fan-out packaging type requires bumps such as die-first face-up FO-WLP.

2.3.1.2 Carrier

In a die-first face-down process, the activities captured in the carrier process segment are simple. The carrier itself is inexpensive and reusable, and the first activity is to apply tape on the carrier, onto which the die will be placed. The second activity rolled up in this process segment is the cost of removing the reconstituted wafer (die surrounded by mold), since the mold is thick enough to support the die as they continue through the process.

The main types of cost in the carrier process are the equipment costs of taping the carrier and later debonding the reconstituted wafer. The material costs of the simple carrier and tape are negligible in comparison.

2.3.1.3 Die Bond

The die bond portion of this process is not a true bonding. It is a simple placement of the die, face-down, onto the tape on the carrier. Since the die is face-down, there is no connection that needs to be made during bonding.

The main cost driver is throughput, and the main type of cost is capital. The longer each die placement takes, the longer the equipment is in use, driving up the capital cost associated with this activity.

2.3.1.4 Mold

This segment represents a wafer level compression molding process. The mold is applied in a thick layer to act as a carrier going forward until the end of the process when the mold is ground down to the desired final thickness.

The main type of cost associated with this process segment is the material cost of the mold. While wafer level compression molding equipment is not inexpensive, the capital costs associated with this step are less than a quarter of the material cost of the mold compound itself.

2.3.1.5 Backgrinding

This activity refers to the steps when the thick mold that has been serving as a carrier is ground away. This occurs after all interconnect processing is done, immediately before flux and ball attach and singulation.

The costs associated with this step are capital. The more mold that must be ground away, the longer the equipment will be in use.

2.3.1.6 RDL

Unlike the other process segments, this one comprises more than a few steps. In some fan-out processes, the RDL accounts for more than half the steps in the entire process flow. For the style of RDL modeled here, a photoimageable dielectric is applied, patterned, and then a seed layer is deposited. Next, photoresist is applied to define the area where the copper is plated up before the resist is stripped and the seed layer etched. Lastly, a final photoimageable dielectric layer is applied and patterned to complete a single layer of RDL.

Material and capital are the major cost contributors to the RDL process. A spin-coater is required for applying the dielectrics and photoresist; a sputter tool is required for the seed layer; a plating line is required for metallization; a stepper is assumed to be used for exposure, although the cost would go down slightly if an aligner could be used while still meeting the design requirements. These are all expensive pieces of equipment.

The highest material costs are the dielectric, photoresist, and metals. The total material cost is considered, even if that material is not left on the final product. For example, the cost model takes into account the total photoresist that must be dispensed to coat the wafer to the target thickness, even if 90% of that material is spun off [2].

2.3.1.7 UBM

The creation of UBM is a series of patterning and plating steps, similar to the RDL but where there can be many RDL layers, there is only one thick Cu UBM layer. The cost is driven by both material and equipment.

2.3.1.8 Flux and Ball Attach

This is the activity of attaching the solder balls before singulation to prepare for the eventual placement of the package onto a substrate.

The major cost driver is the material cost of the solder balls themselves, although the capital costs are not insubstantial, totaling about half of the material costs.

2.3.1.9 Singulation

This refers to the singulation process, of which there are many variations. Modeled here is a laser scribe followed by mechanical saw, which is a cost-effective approach. The cost of this step will depend directly on the size of the package; the smaller the package, the longer dicing will take.

While there are some material costs associated with dicing, the cost is dominated by the capital considerations of how long the equipment is being used.

2.3.2 FO-WLP Variations

Die-first face-down is the process discussed in the previous section.

Die-first face-up is the second variation of FO-WLP. Because this chapter focuses on FO-WLP costs, this section will only address differences in the process flow that affect cost. Other chapters in this publication address the process flow variations in greater detail as it relates to actual processing and technical considerations.

The two items that differentiate the die-first face-up process from the die-first face-down process from a cost perspective are:

1. The incoming die must have bumps, copper pillars, or copper studs before it is placed (with those bumps/pillars/studs facing up). After mold application, the mold is immediately ground down to reveal these studs before RDL processing begins.

2. It is assumed that adaptive patterning is used to account for die shift due to mold application [3]. Adaptive patterning carries an added cost but also a favorable yield impact.

Die-last is the third variation of FO-WLP. The items that differentiate the die-last process from a cost perspective are:

1. The die must have bumps or copper pillars for placement onto the RDL.
2. A robust carrier and debond process must be used.
3. The die are placed onto a fully formed RDL, instead of having the RDL built up around them after placement. This means that die can be placed on known good RDL locations, which brings a yield benefit to this process.
4. Die-last uses the same mold compound as the die-first processes, but that alone is not sufficient. Capillary underfill (CUF) must also be applied to protect the bump or copper pillar interconnects.

An example package design was run through the cost model for all three FO-WLP variations. The design is a 3 mm × 3 mm die in a 6 mm × 6 mm package with a single RDL. The die cost category includes the cost of the wafer itself (300 mm, mature node) and the cost of processing that wafer for placement into the package. The results are shown in Table 2.1, expressed as numbers relative to the die-first face-down package.

The total direct costs are close, with die-first face-down a little more cost-effective than the other two processes. The categories of steps with different direct costs between the three technologies are discussed below.

Table 2.1 Comparison of three FO-WLP variations.

	Die-first face-down	Die-first face-up	Die-last
Carrier	1.00	1.00	2.86
RDL	1.00	1.00	1.00
UBM	1.00	1.00	1.00
Die cost and preparation	1.00	1.06	1.06
Die bond	1.00	1.00	1.82
Mold/mold + CUF	1.00	0.96	1.04
Backgrind/post-mold grind	1.00	0.76	0.76
Flux and ball attach	1.00	1.00	1.00
Package singulation	1.00	1.00	1.00
Scrap	1.00	0.61	0.14
Total	1.00	1.06	1.06

2.3.2.1 Carrier

The die-last process requires a most robust and expensive carrier process. In the die-first processes, it is primarily mold that is placed upon the carrier, along with the die; in the die-last process, the RDL itself is built up on the carrier. Debonding the almost fully processed wafer, with the RDL directly on the temporary carrier, makes the die-last carrier process the most expensive.

2.3.2.2 Die Cost and Preparation

This is the cost of the incoming silicon wafer itself, as well as the cost to process that wafer. For all FO-WLP options, this includes the cost to dice the wafer for placement in the package. For die-first face-up and die-last, this also includes the cost to add bumps or copper pillars. The cost to bump the wafer is why this category is more expensive than the die preparation associated with the die-first face-down process.

2.3.2.3 Die Bond

Die are placed without making any connections in the die-first processes. This is a relatively low-cost process because no materials are required as part of the die placement, and the throughput is high since the RDL has yet to be formed [4]. In contrast, a full die bond must occur as the bumped die are placed in the die-last process. The die-last cost model includes the material cost of flux, and die placement takes longer since the RDL is already formed and the bumped die must be placed accurately.

2.3.2.4 Mold/Mold + CUF

The die-first face-down package has the highest cost in this category because the mold must be thick enough to act as a support for the wafer throughout processing. The die-first face-up process requires less mold, but this category includes the cost of adaptive patterning, which means the cost in this category is almost the same as in the die-first face-down case, despite the requirement for less mold. For the die-last process, the assumption is again that less mold is needed because the mold does not have to act as a support. On the other hand, this category includes the addition of capillary underfill material for the die-last process. With the added cost of CUF and the reduced amount of mold compound required, the total cost is slightly higher than the die-first face-down scenario.

2.3.2.5 Backgrind/Post-mold Grind

This activity occurs in different parts of the process flow depending on which FO-WLP variation is used, which is why it has two names. The thick mold compound is ground away near the end of the die-first process; the mold is on the backside of the wafer, so this process is called backgrinding. In the die-first face-up process, the mold is ground away immediately after application to reveal the copper studs. In the die-last process, the mold is applied near the end of the process (after RDL formation and die bonding), and grinding occurs soon after application. Since grinding occurs shortly after molding in the die-first face-up and die-last processes, this is called post-mold grind.

As stated in the previous section, the die-first face-down process is assumed to require the thickest application of mold, so the grinding of this thick mold takes longer than the post-mold grinds that occur in the other two processes. This is why the die-first face-down process has the highest cost in this category.

2.3.2.6 Scrap

The scrap cost is different for all three FO-WLP variations. The scrap cost is the highest for the die-first face-down option. This is because the chips are already embedded in the mold before any RDL processing is done, and no adaptive patterning is carried out. This means that die shift may result in some yield loss between the die and RDL. Furthermore, all defects introduced by the RDL process, which involves many photolithographic steps, will cause the loss of those die. In die-first face-up, the added cost of adaptive patterning means better yields because the RDL pattern has been adjusted to match the die shift, but any defects introduced during RDL processing will still cause the loss of some die. The die-last option has the best yield because the assumption is that the die will only be placed on known good RDL locations, which means defects introduced by the RDL process will not cause the loss of any die.

As the number of RDLs are increased, the relative relationship between die-last and die-first face-down remains stable. Multiple RDLs result in more defects, but since the die in the die-last process is assumed to be placed on known good RDL locations, only carrier and RDL processing costs are lost to RDL defects.

On the other hand, the relative relationship between die-first face-up and face-down does change with RDL count. As stated, the main yield difference between these two technologies is that mold shift does not cause the loss of die in die-first face-up. The scrap cost associated with mold shift is stable; it hits the face-down process and does not hit the face-up process. As the RDLs increase, the scrap cost associated with RDL count increases for both technologies, making the die-shift yield savings less significant for die-first face-up. As shown in the chart, the relative scrap cost for die-first face-up is 0.61 of the scrap cost of die-first face-down. With two RDLs, this increases to 0.73. With three RDLs, it is 0.79.

2.4 Cost of FO-WLP versus Wire Bond and Flip Chip

The first sections of this chapter broke down the cost contributions of different segments of the FO-WLP process and discussed how these costs change depending on the type of fan-out being used. This section compares the cost of different packages built with FO-WLP technology to that same package built using more established technologies.

Wire bond is the most mature of the three technologies evaluated. For both wire bond and flip chip packaging, there are two distinct processes: substrate creation and assembly. These are discrete parts of the processes and may occur at two entirely separate locations. This contrasts with FO-WLP, which is essentially both substrate fabrication and assembly all in one process.

Flip chip and wire bond substrates are created with printed circuit board (PCB) technology. There are many process options for creating these substrates. The build-up process may be additive, semi-additive (SAP), or modified semi-additive (mSAP), depending on line and space (L/S) requirements [5]. There are different core and build-up materials available, and layers can be added to increase the complexity of the substrate as needed. There may be mechanical vias or laser vias in the substrate, and there are many surface finish options. After the substrate for a flip chip or wire bond package is created on a panel, it is often singulated into strips to go into assembly processing for flip chip chip-scale packaging (FC-CSP). Singulation into single substrates for flip chip ball grid array (FC-BGA) is another process variation, but the modeling in this chapter assumes all flip chip and wire bond assemblies are processed in strips for FC-CSP and wirebond CSP.

The assembly process is straightforward. For a wire bond package, the die is placed on the substrate, and then wires are used to create the connection between the die and the substrate. There are many options for the type of wire bonding process to use, as well as the type of wire. For a flip chip package, a die with bumps or copper pillars is placed onto the substrate, and the connection is generally made via a standard mass reflow process or a thermocompression bonding process.

Once the die is connected to the substrate via wires or bumps, several things may happen; molding may be applied, a lid may be attached, passive components may be placed, and so forth. Lastly, solder balls are attached to the bottom of the substrate and the packages are singulated.

The comparisons in this chapter focus on the direct costs for each technology, although factories for flip chip and wire bond assembly, substrate fabrication, and fan-out packaging will have different overhead and indirect cost considerations. Furthermore, other differences, such as the capital expenditures associated with a newer fan-out product line versus a fully depreciated wire bond line, are not considered. The goal is to make the comparisons as similar as possible.

For the first analysis, a die-first face-up fan-out package is compared to a wire bond package that uses a simple, two-layer substrate with a core. In one scenario, the wire bond assembly uses copper wires; in the second scenario, palladium coated copper (PdCu) wires are used. A range of die sizes are tested. The die sizes range from 3 mm × 3 mm to 13 mm × 13 mm. For the fan-out

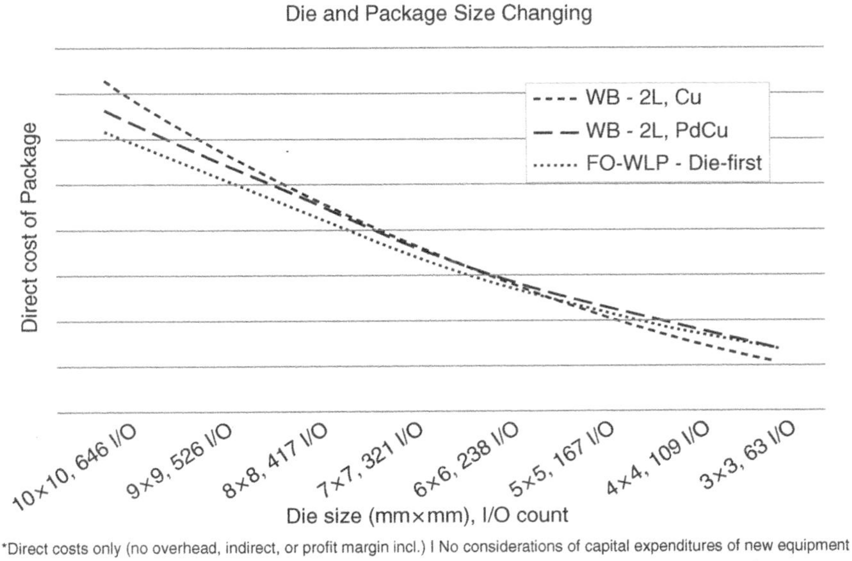

Figure 2.5 Wire bond versus FO-WLP.

scenario, the associated package size is 1 mm greater on each side than the die size (3 mm × 3 mm die in a 4 mm × 4 mm fan-out package). For the wire bond scenario, the package is 2 mm greater on each side (3 mm × 3 mm die in a 5 mm × 5 mm wire bond package).

The results are shown in Figure 2.5. There are two crossover points visible in the chart. The first crossover point is between the die-first fan-out design and the wire bond design that uses copper wire. Up until the die size surpasses 5 mm × 5 mm, FO-WLP is cost-effective. Wire bond is traditionally a cost-effective technology, and if the wire bond package size were the same as the FO-WLP package size, wire bond would be the cost-effective option for each design tested. However, given the fact that additional space must be left for the wire bonds between the die and the package, it is difficult for a 6 mm × 6 mm wire bond package to compete with a 5 mm × 5 mm fan-out package; each millimeter makes a difference at such a small size.

The second crossover point is when the fan-out package becomes more cost-effective than the wire bond design that uses PdCu wires. PdCu wires carry a slightly higher material cost than copper, and that wire bond design does not become noticeably more cost-effective than the equivalent fan-out package until an 8 mm × 8 mm die size is surpassed.

The takeaway from this analysis is that FO-WLP may be more cost-effective than wire bonding technology if the package size must be reduced beyond the capabilities of the wire bonding process. In general, if the mature wire bonding

Table 2.2 Package design for two 11 mm × 6 mm Die (n/a here means not applicable).

Pkg size (mm)	4L wire bond package	1-2-1 wire bond package	2 RDL die-first FO-WLP
16 × 16	1.03		
15 × 15	1.00	1.17	1.40
14 × 14	n/a	n/a	1.17
13 × 13	n/a	n/a	1.01

Table 2.3 Package design for two 3 mm × 3 mm die.

Pkg size (mm)	4L wire bond package	2 RDL die-first FO-WLP
5 × 10	1.00	1.25
4 × 9	n/a	0.94

process can support a design, there is rarely a compelling cost reason to move to FO-WLP.

Figure 2.5 evaluated the cost of placing a single die in a package that is only slightly larger than the die size. Tables 2.2 and 2.3 evaluate how the cost of a two-die package changes with different wire bond and FO-WLP design assumptions.

The first design is a larger package into which two 11 mm × 6 mm die are placed. Two different complexities of wire bond substrate are tested, and a two-layer RDL is assumed for the FO-WLP scenario. Both wire bond substrates have the same number of layers—an inner layer pair and an outer layer pair—but the materials and processes used on the outside layers are different. The 4L option has a core, then a layer of prepreg with copper foil on top on bottom, and only contains through hole vias. The 1-2-1 has a core, a layer of ABF material on the top and bottom, and allows for greater routing complexity by supporting both laser vias and through hole vias.

Both wire bond scenarios assume copper wire. The solder ball size is 0.4 mm for both technologies for the two larger package sizes; for the two smaller fan-out packages, a 0.35 mm solder ball is assumed in order to support the same number of I/Os as in the large package. The 15 mm × 15 mm package scenario with a 4L substrate is the basis for the relative cost comparison.

The smallest wire bond package is assumed to be a 15 mm × 15 mm package, to allow for room for the wires. That restriction does not exist for FO-WLP, which allows the package size to shrink. This reduction in package size is what allows FO-WLP to be cost competitive, even with 2 RDLs. The smallest fan-out package is cost competitive with the 16 mm × 16 mm 4L wire bond package. The 14 mm × 14 mm fan-out package is cost competitive with the 15 mm × 15 mm wire bond scenario if a more complex 1-2-1 substrate is required to support the connections.

The next design places two 3 mm × 3 mm die into wire bond and fan-out packages. The number of RDLs required is one of the key cost drivers for any FO-WLP design. Table 2.3 shows that even with 2 RDLs, shrinking the package by 1 mm on each side is enough to make this two-die design cost-effective with FO-WLP.

These examples reinforce that package size is a major cost driver for both technologies. The ability to shrink the package size by just 1 mm or 2 mm can make FO-WLP, a newer technology, more cost-efficient than the established wire bond process.

Next, FO-WLP is compared to flip chip packaging in a few different ways to understand the design features that drive the cost of each technology. In the first comparison, shown in Figure 2.6, a 5 mm × 5 mm die is placed into different package sizes, ranging from 6 mm × 6 mm to 14 mm × 14 mm. The cost of the flip chip package considers the cost of adding bumps to the die (a cost the die-first FO-WLP design doesn't require).

Three different substrate options are shown for the flip chip package. The simplest is a 2L substrate, which uses a single core; the next option utilizes a coreless

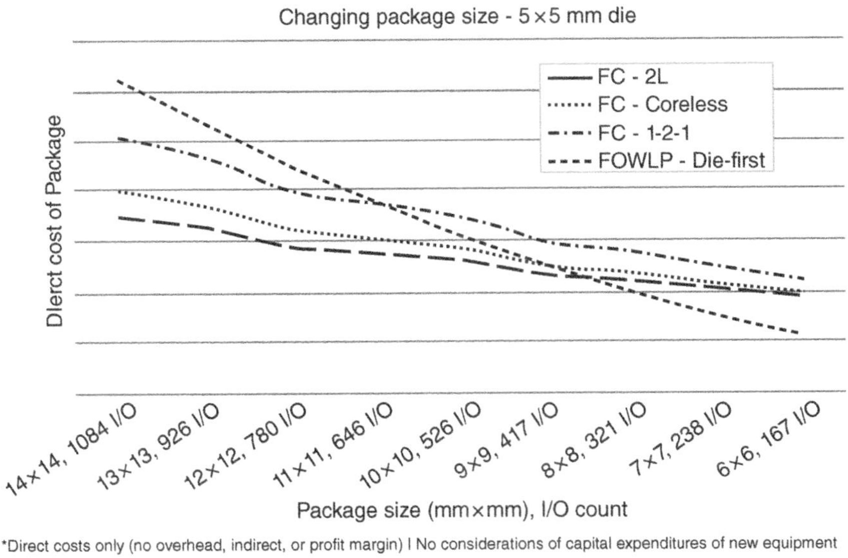

Figure 2.6 Flip chip versus FO-WLP—changing package size.

process that results in a coreless substrate with two layers; the most complex substrate is a 1-2-1, which has a core layer pair and a build-up layer pair. The crossover points for the three flip chip package options with the fan-out package occur at different points. This crossover is the point at which the flip chip option becomes cost-effective compared to FO-WLP.

Note that the fan-out line is smooth, while the flip chip lines appear wavy. This is because the relationship between package size and direct cost is not entirely linear for flip chip packaging. This is because of the need to consider the panel size, strip size, and package size for a flip chip package. Instead of only losing the space between packages and at the edge of a wafer, as is the case for calculating the number of die from a wafer, the space between packages, the edge of the strip, and the strip-to-strip spacing must all be considered when calculating the number of packages per panel. The addition of spacing as it relates to strips is why the relationship is not smoothly linear for the flip chip packages.

The flip chip package with the simplest substrate becomes cost-effective when the 5 mm × 5 mm die is being packaged in approximately an 8 mm × 8 mm package or larger. The coreless option is cost-effective when the 5 mm × 5 mm die is packaged in a 9 mm × 9 mm package, and the more complex substrate option becomes cost-effective only for package sizes around 11 mm × 11 mm or greater.

The takeaway from Figure 2.6 is that FO-WLP is more cost-effective than flip chip technology when dealing with smaller packages. At these smaller package sizes, more packages can be processed on a single 300 mm wafer and benefit from the wafer-level activities, as compared to the number of packages that can be processed from a single substrate panel. Package size is a key cost driver for both technologies.

The chart in Figure 2.7 holds the package size steady and changes the die size. The fan-out price is nearly static, even becoming slightly lower as the die grows. This is because the amount of mold needed is smaller as the die gets closer to the package size. In contrast, as the die size increases, the flip chip cost increases steadily. This is because of the cost required to bump the die before it is placed in the package. The cost to bump a wafer is stable. The larger the die, the more of that cost is amortized to that die. As larger die are placed into the 14 mm × 14 mm flip chip package, the cost to bump those die becomes a barrier to cost-effectiveness when compared to FO-WLP. The fact that die-first fan-out does not require bumps means die size is not a major cost driver for this type of FO-WLP.

In conclusion, when comparing FO-WLP to the more mature options of wire bond and flip chip packaging, there are three key takeaways regarding cost.

1. FO-WLP is more likely to be cost-effective at smaller package sizes.
2. The cost to add bumps to a die that will be placed in a flip chip package is not insignificant, which makes die size a key cost driver for the flip chip process. On the other hand, die size has very little impact on the cost of a FO-WLP package built using a die-first face-down process.

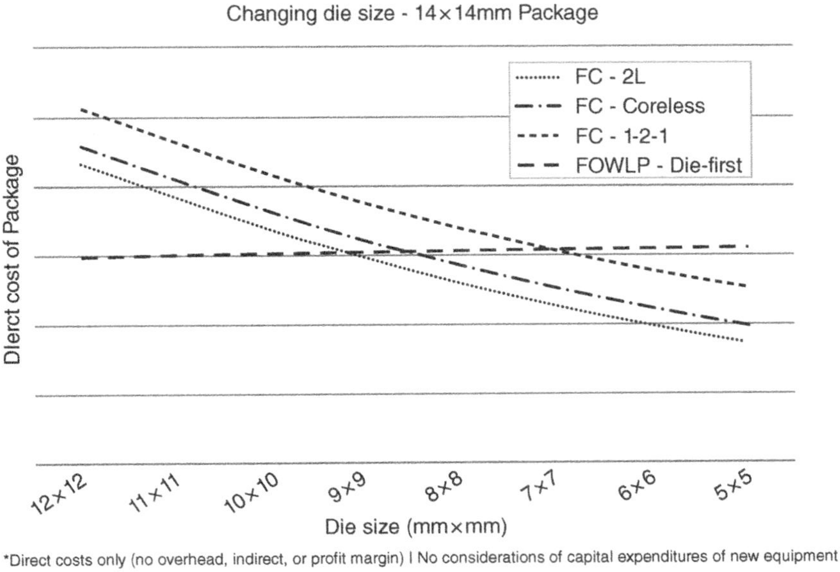

Figure 2.7 Flip chip versus FO—changing die size.

3. Package size is a major cost driver for all three technologies. If FO-WLP allows for even a small reduction in package size, a design may be cost-effective with FO-WLP even when compared to the mature, low-cost wire bonding method.

2.5 Package-on-Package Cost Analysis

The previous sections focused on placing one die in a package, or two die side by side. Moving forward in complexity, the next type of package to evaluate is one where die are stacked. This occurs when the package cannot be shrunk further in length or width and so vertical real estate must be used.

There are many ways to stack die. The most mature method is package-on-package (PoP) stacking. There are other stacking processes that involve putting vias in the die and stacking them directly on top of one another (3D) or placing die on an interposer and then placing that interposer on a substrate (commonly called 2.5D) [6].

The PoP process is set apart from the more recent die stacking trends because it is exactly what it sounds like—two packages being stacked, not simply two die. Original PoP packages typically involved a wire bonded die for both the top and the bottom packages [7]. The bottom die and wires were covered with mold to protect them, while the sides of the board with the pads were left exposed for the solder balls of a second package to be placed on top.

The PoP process is described here. A die is placed in the bottom package with flip chip packaging, the entire package is molded, and then vias are drilled in the mold

Table 2.4 Flip chip and die-last FO-WLP with TMVs—top level costs.

	FC with TMVs	Die-last fan-out with TMVs
2L top substrate cost	1.00	1.00
Bottom substrate + assembly cost	1.00	0.93
Die preparation cost	1.00	1.00
Total cost	**1.00**	**0.96**

and filled with solder. These are called through-mold vias (TMVs), and these allow another package, with solder balls on the bottom of the substrate, to be placed on top of the first package.

The scope of the cost analysis includes the bumping of the flip chip die, the creation of the bottom substrate, the flip chip assembly process (including TMVs), and the creation of a top substrate with solder balls attached. The cost of the die being assembled onto the top substrate is out of scope. Two different FO-WLP variations are included in the analysis: die-last and die-first face-up.

Table 2.4 compares a TMV PoP flip chip package with a die-last fan-out package that also uses TMVs. This is the closest comparison that can be constructed—they both use CUF, are overmolded with a standard mold compound, have TMVs, and require a bumped die. A 3-2-3 mSAP substrate is selected for the flip chip scenario; this refers to a substrate with one core layer and three build-up layers built with mSAP technology. A 2-layer RDL is modeled for the fan-out scenario.

The package is 14 mm × 14 mm with 900 I/Os; the die is 10 mm × 10 mm, and there are 300 TMVs. A 2L 14 mm × 14 mm substrate is placed on top of each bottom package. The cost of the top substrate includes the creation of the 2L substrate itself, the attachment of solder balls for placement onto the bottom substrate, and singulation of the 2L substrate.

As a reminder, all costs are direct costs only; no overhead, profit margin, or indirect costs are applied. All numbers are expressed relative to the flip chip with TMVs scenario. Of the three categories, the only one different among these scenarios is bottom substrate and assembly. For flip chip, this refers to the two separate processes of building a 3-2-3 substrate and then doing flip chip assembly with TMVs. For die-last FO-WLP, there is only one packaging process, representing both substrate and assembly. For the selected design, the costs are similar for substrate and assembly across both technologies.

Table 2.5 breaks down the activities that exist in both substrate and assembly processes.

In comparing activities from the two technologies in similar buckets, they are all slightly different, though many are close. Each is addressed below, with the differences in cost discussed.

Table 2.5 Flip chip and die-last FO-WLP with TMVs—detail costs.

	FC with TMVs	Die-last fan-out with TMVs
FC: 3-2-3 substrate cost FO-WLP: carrier +2 RDLs cost	1.00	0.79
Die bond cost	1.00	0.97
CUF and mold cost	1.00	1.07
Ball attach cost	1.00	1.10
Singulation cost	1.00	0.89
TMV cost	1.00	0.98
Total cost	**1.00**	**0.93**

2.5.1.1 Substrate/RDLs

This compares the largest number of steps between the two processes. This bucket encompasses the entire 3-2-3 mSAP substrate creation process for the flip chip package, and the creation of two RDLs for the fan-out package, including the cost of the carrier (and the subsequent cost to debond from the carrier). One reason FO-WLP has the advantage here is that only two RDLs are necessary. It is assumed that two layers of routing are enough for the fan-out package due to the ability to achieve finer L/S with the FO-WLP RDL process than with PCB-based substrate processes [8]. The mSAP process, modeled here, achieves finer L/S than subtractive but not enough to reduce the substrate to only two layers. That is not to say that one layer of PCB routing is equivalent in cost to one layer of RDL in FO-WLP; the costs are not the same. But the ability to have fewer FO-WLP RDLs is beneficial regardless of cost equivalencies.

2.5.1.2 Die Bond

The cost of die placement onto the completed RDL or the completed substrate is virtually the same. Placement time per die is relatively equal across the two technologies, and the amount of flux required is relatively equal. But the activity is done at the wafer level for FO-WLP and at the strip level for flip chip, so there are different equipment costs and wafer versus strip throughputs to consider. These different assumptions result in a small difference in cost totals for this category.

2.5.1.3 CUF and Mold Cost

This category captures multiple activities. The first activity is the application of capillary underfill immediately after the die are placed. The second activity is the application of the mold compound to the desired thickness. The costs of the CUF and mold compound activities are primarily driven by the design features of chip size, package size, and mold height. This means they are approximately the same for both packages. The main reason the costs are slightly different is because of the different assumptions related to throughput and equipment costs for handling wafers versus handling strips.

2.5.1.4 Ball Attach

Similar to the case with the die bond category, the cost difference here is due to processing at the wafer level versus the strip level. The actual cost of the solder balls, which dominates the total cost in this category, is the same—an equal number of the same solder balls are being placed on the 14 mm × 14 mm package regardless of whether it is flip chip or FO-WLP. The cost and throughput of doing ball drop for an entire wafer is the reason ball attach is slightly more for FO-WLP.

2.5.1.5 Singulation

The time to singulate is based on the package size, which is the same for both technologies. The cost for singulating the 14 mm × 14 mm packages in the FO-WLP scenario is lower because there is only a single wafer mount before the entire wafer is diced. This contrasts with having to mount and singulate each strip in the flip chip process. Singulation is only a small portion of the total cost for both packages, and the difference in this category is not a major contributor to the flip chip or FO-WLP totals.

2.5.1.6 TMV

Similar to die bond and ball attach, the cost is close between the two technologies in this category. The time to drill the holes in the mold and the material cost of dropping in the solder balls are approximately equal across both technologies. The next comparison in this section will introduce the option of tall copper pillars being used in place of TMVs.

This example assumed that the limitations on total package height were such that having a second package on top of the first package was not problematic and that the electrical requirements could be met with solder as the through-package via material. In cases where more rigorous design features must be met, large copper pillars [9] may be used in place of solder ball vias, and an RDL can be constructed directly on top of the bottom package, removing the need for a second substrate. With an RDL on top to handle routing, a second die can be placed directly on top of the first package.

For this type of fan-out PoP construction, die-first face-up processing is assumed. The incoming die have copper pillars, are placed with the studs facing up on a carrier, and then large copper pillars are built up. Mold is applied, the mold is ground down to reveal both the large copper pillars and the copper pillars on the die, and the bottom package fan-out process continues to RDL creation, solder ball attach, debonding, and singulation.

In Table 2.6, this option of die-first face-up with copper pillars and a top RDL is added to the comparison. All numbers are still relative to the flip chip with TMVs scenario, and all costs continue to be direct cost only.

With the addition of the large copper pillars, the die-first face-up fan-out cost in the substrate and assembly bucket is higher than the other scenarios. Additionally, the cost of creating a 1-layer RDL is more than the cost of a 2L top substrate.

Table 2.7 adds the die-first face-up fan-out package with copper pillars to the comparison of similar groups of activities in the substrate and assembly portions of the processes. A few categories are the same for both FO-WLP options relative to flip chip with TMVs, but a few are different. The categories that have changed between die-last fan-out with TMVs and die-first face-up fan-out with large copper pillars are addressed below.

Table 2.6 Flip chip, die-last FO-WLP with TMVs, FO-WLP with copper pillars—top level costs.

	Flip chip with TMVs	**Die-last fan-out with TMVs**	**Die-first face-up fan-out with large copper pillars**
2L top substrate cost Top RDL cost	1.00	1.00	1.16
Bottom substrate + assembly cost	1.00	0.93	1.03
Die preparation cost	1.00	1.00	1.00
Total cost	**1.00**	**0.96**	**1.04**

Table 2.7 Flip chip, die-last FO-WLP with TMVs, FO-WLP with copper pillars—detail costs.

	Flip chip with TMVs	**Die-last Fan-out with TMVs**	**Die-first face-up fan-out with large copper pillars**
FC: 3-2-3 substrate cost FO-WLP: carrier +2 RDLs cost	1.00	0.79	0.79
Die bond cost	1.00	0.97	0.40
CUF and mold cost	1.00	1.07	1.37
Ball attach cost	1.00	1.10	1.10
Singulation cost	1.00	0.89	0.89
TMV/large copper pillars cost	1.00	0.98	1.67
Total cost	**1.00**	0.93	1.03

2.5.1.7 Die Bond

This is less in the die-first face-up process because the die are being placed face-up on a carrier and no connection is being made; placement is fast, and no flux is required.

2.5.1.8 CUF and Mold Cost

CUF is not required with die-first face-up processing, which represents a material cost savings. However, an additional activity is required—the mold must be ground down to reveal the copper pillars. This additional activity is why this category is somewhat more expensive for the die-first face-up process.

2.5.1.9 TMV/Large Copper Pillars

The creation of large copper pillars to serve as connections between the top die and bottom interconnect is more expensive than the creation of through mold vias. The TMV process requires drilling through mold, which is an activity not needed in the die-first face-up process with large copper pillars. However, plating lines are expensive, and the time required on this expensive equipment to plate up large copper pillars is substantial. Furthermore, this category includes the photolithography required to define the area where the copper pillars will be, whereas lithography steps are not needed in the TMV scenarios. Between the plating line and lithography steps, creating large copper pillars is more expensive than drilling vias in mold and dropping in solder.

The key takeaways for package-on-package designs are:

1. Wire bond, flip chip, and FO-WLP technologies all support different versions of PoP.
2. Newer versions of PoP such as FO-WLP may be cost-effective, even compared to PoP designs that use more mature technology, if the number of necessary routing layers can be reduced.
3. PoP package costs will tend to increase as size constraints and/or design requirements increase.

2.6 Conclusions

In this chapter, the cost of FO-WLP was broken down into process segments, variations on the process were analyzed, and FO-WLP was compared against other technologies.

Within the FO-WLP process, the RDL represents the largest cost contribution. This is true regardless of the variation of FO-WLP. Among the three variations of die-first face-up, die first-face down, and die-last, die-first face-down has the lowest processing cost but the highest yield loss.

From a design perspective, package size is the major cost driver for fan-out packages. In a comparison between FO-WLP and wire bond, the mature wire bond process will almost always generate the package with the lowest cost, but if FO-WLP allows the package size to be reduced, there are situations in which FO-WLP may be more cost-effective than wire bonding.

FO-WLP is more likely to be cost-effective compared to flip chip packaging at smaller package sizes. Die-first face-down is particularly cost-effective compared to flip chip packaging because the flip chip package must account for the cost of bumping the incoming die, and this is not a cost that die-first face-down fan-out shares.

PoP is a more complex structure supported by multiple technologies, including FO-WLP. The decision regarding whether to use wire bond, flip chip, or some type of fan-out depends on the design requirements. Analysis showed that a die-last PoP design with TMVs can be cost competitive with the more mature flip chip PoP design with TMVs, particularly if the number of routing layers can be reduced. Switching to a die-first face-up fan-out design that uses copper pillars increases the cost but will enable designs that a TMV structure cannot support.

References

1 Lapedus, M. (2019). Lithography challenges for fan-out. *Semiconductor Engineering*.

2 US Army Research Laboratory (2015). Optimization of thick negative photoresist for fabrication of interdigitated capacitor structures. Report ARL-TR-7258.

3 Scanlan, C., Olson, T., Robers, B. et al. (2012). Adaptive patterning for panelized packaging. Proceedings of SMTA's *International Wafer-Level Packaging Conference*, San Jose, CA, USA (November 6–7, 2013).

4 Strothmann, T. (2016). Optimizing equipment selection for diverse fan-out process flow. *Semicon Taiwan*, Taipei, Taiwan (September 7–9, 2016).

5 Dunn, T. (2018). Additive electronics: PCB scale to IC scale. *PCB007 Magazine* (September), pp. 12–16.

6 Waidhas, B. and Keser, B. (2020). 2.xD and 3D package architectures and challenges. *3D & Systems Summit*, Dresden, Germany (January 27–29, 2020).

7 Eslampour, H., Lee, S., Park, S.S. et al. (2010). Comparison of advanced PoP package configurations. *Electronic Components and Technology Conference*, Las Vegas, NV, USA (June 1–3. 2010).

8 Lin, E. and Kao, K. (2019). Fine-line patterning for SLPs calls for high-resolution photoresist. *DuPont Featured Solutions* (August 11, 2019).

9 Tao, M., Prabhu, A., Agrawal, Akash et al. (2016). Package-on-package interconnect for fan-out wafer level packages. Proceedings of SMTA's *International Wafer-Level Packaging Conference*, San Jose, CA, USA (October 18–20, 2016).

3

Integrated Fan-Out (InFO) for Mobile Computing

Doug C.H. Yu, John Yeh, Kuo-Chung Yee, and Chih Hang Tung

Taiwan Semiconductor Manufacturing Company (TSMC), Hsinchu, Taiwana

3.1 Introduction

Packaging technology plays a unique and vital role in the microelectronics industry. A microelectronics package physically protects the chip(s) inside the package while providing connections to the outside world, whether electrical and/or optical, wired or wirelessly. With market drivers from personal and home computing, to mobile computing, to high performance computing (HPC) and artificial intelligence/machine learning (AI/ML), microelectronic packaging embraces revolutionary challenges inside and out. Inside a package, system on chip (SoC) drives exponential growth of transistor count and memory capacity in a limited system footprint. Mobile and new HPC systems have extreme performance, power, thermal, and form factor requirements. These revolutionary challenges have been driving the microelectronics industry to search for new packaging and system integration solutions.

Wafer-level packaging (WLP) and wafer-level system integration (WLSI) have emerged as an attractive system integration solution in both mobile and HPC applications because of their cost competitiveness with the smallest form factors and structural simplicity. By leveraging wafer-level processes, new and innovative WLSI platforms have been successfully established in mass production for years, which in turn, is now propelling computing hardware to move beyond what was conceivable by using SoC with conventional packages alone. Innovative system integration is a paradigm shift and game changer. Now, the system package level scaling trend is unstoppable in not only boosting logic chips to More Moore (MM) transistor density/count scaling but also devouring non-CMOS chips for More-than-Moore (MtM) heterogeneous functionality into one single and powerful component.

Embedded and Fan-Out Wafer and Panel Level Packaging Technologies for Advanced Application Spaces: High Performance Compute and System-in-Package, First Edition. Edited by Beth Keser and Steffen Kröhnert.

In this chapter, integrated fan-out (InFO) for mobile computing is discussed. In a subsequent chapter, InFO for HPC computing will be introduced. WLSI is an innovative system integration platform. New integration schemes under the WLSI umbrella are evolving quickly. In this context, it is advisable, for the readers, to always look into the latest publications for updated information.

3.2 Fan-In Wafer-Level Packaging

Traditionally, there are two different variations of WLP. One is fan-in wafer-level chip scale package (WLP) where ball grid array (BGA) solder balls are placed directly on the silicon chip. The other is fan-out wafer level package (FO-WLP) where the BGA ball numbers exceed the silicon chip area, and an extension of the chip area is made with epoxy mold compound (EMC) to accommodate the additional BGA balls. Figure 3.1 shows both fan-in and fan-out WLP. As there is no additional interposer or substrate, the package dimension is minimal with relatively simple processes and attractive costs. For FO-WLP, there are two different technology approaches. One is wafer-scale processes based on wafer process technologies. The other is panel-scale processes using large rectangular panels

(a) (b)

Figure 3.1 Two wafer level package technology platforms, (a) fan-in and (b) fan-out, covering a wide variety of system integration needs from low pin count, small chip size, single chip component up to multiple chips, large component size, and high pin count system integration solutions.

inherited from the organic substrate and PCB industries. This chapter will focus on wafer-level fan-out processes.

WLP is one of the most simplified packaging technologies. Technologically, it is an extension of wafer-level processes since all process steps are done at wafer form. After wafers are finished with a surface passivation protection, the wafer is then electroplated with a redistribution layer (RDL, usually copper), coated with photosensitive polymer dielectric materials, plated with under bump metallization (UBM), and finally BGA solder balls are placed on the wafer to finish the WLP package. If all BGA balls are placed within the silicon chip, fan-in WLP technology is used. The wafer is then diced, and individual chips are ready for surface mount onto printed circuit boards (PCBs). In WLP, silicon chips are left without protection from backside and sidewalls. Some upgraded variations of WLP have added protection on the backside and sidewalls using FO-WLP processing resulting in added process costs. The advantages of such a simple packaging scheme are, obviously, cost, form factor, and process throughput time.

3.2.1 Dielectric and Redistribution Layers (RDL)

Depending on design complexity and requirement, one RDL is sufficient in most of the WLP packaging. When required, a second RDL with copper (Cu) UBM can be used, but at an added cost. Adding RDL layers provides additional mechanical buffer and improves the mechanical reliability of the component, which in turn, enables larger chip size to be made using WLP. Typical dielectric materials, such as poly(bisbenzoxazole) (PBO) or polyimide (PI), are used as dielectric layers. They are photo imaged to form via and trace structures using photolithography. Thicker dielectric layers give rise to better mechanical buffering to protect both solder balls and silicon chips from mechanical stress induced by the coefficient of thermal expansion (CTE) mismatch between the PCB and the silicon (Si) chip. However, a thicker dielectric has a more difficult time forming small vias and also induces higher residual stresses. The Cu RDL is created by physical vapor deposition (PVD) of the seed metals (typically titanium and Cu) and then electroplating Cu. A second dielectric layer is deposited followed by a via opening for a UBM and solder BGA placement. See details in Figure 3.2.

3.2.2 Under Bump Metallization (UBM)

A UBM layer is used to protect the RDL and silicon chip. Cu is used as a UBM layer to reduce the overall process cost. Additional UBM metals, such as electroplated Ni or electroless laminated alloy, such as ENEPIG, are also used. UBM is one of the main cost contributors in WLP and needs to be chosen carefully. Total mask layers for building WLP can run from four to six masks all the way down to two masks, determined by chip design requirements and cost, as shown in Figure 3.2.

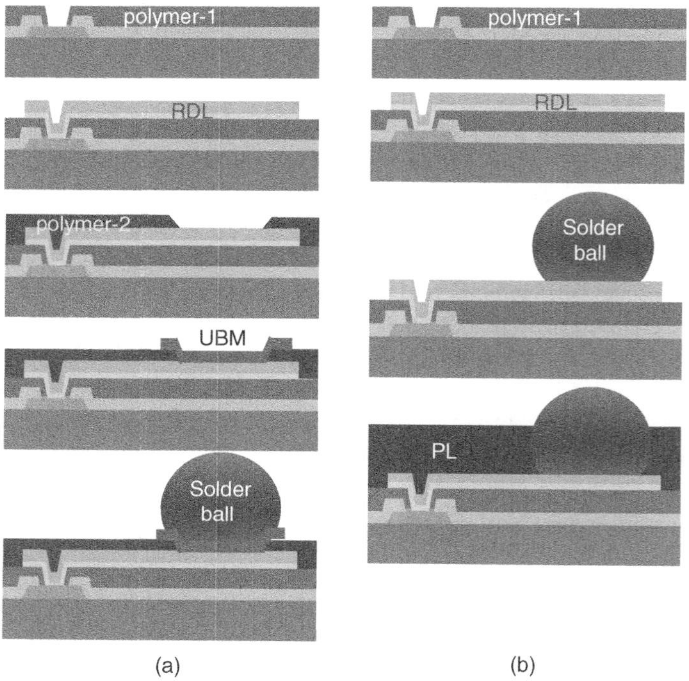

Figure 3.2 (a) Traditional four-mask WLP with RDL and UBM. (b) Two-mask WLP effectively reduces WLP cost and at the same time improves reliability performance. Protective Layer (PL) coating also helps to improve reliability.

3.2.3 Reliability and Challenges

Board-level reliability is the main challenge for WLP due to the thermal expansion coefficient (CTE) mismatch between silicon and PCB-induced mechanical shear stress during temperature cycle or thermal shock tests. As the shear stress is chip size dependent, the WLP chip size is often capped by this limitation. Traditionally, the WLP chip size is smaller than 7 mm × 7 mm to 8 mm × 8 mm. Larger chips give rise to fatigue failures of the BGA solder balls.

3.2.4 Large Die WLP

There are different approaches of achieving large die size WLP [1]. Reducing shear stress by buffering or redistributing the distance-from-neutral-point (DNP) are most common. Increasing the dielectric layer thickness, increasing the Cu RDL or UBM thickness, reducing UBM pad size, or increasing BGA ball height can all increase the overall chip stand-off when mounted on a PCB and improve the large

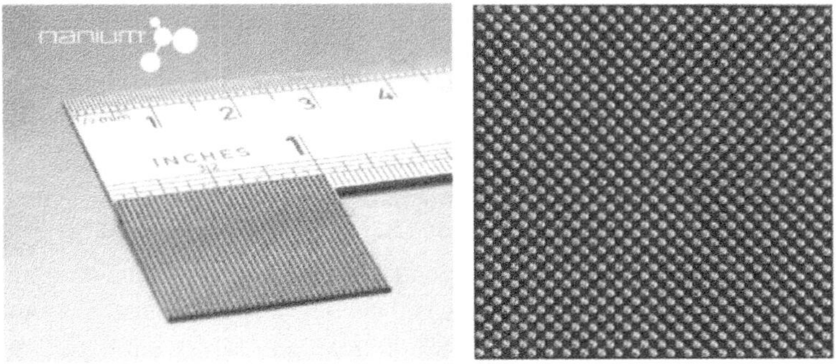

Figure 3.3 An ultra-large fan-in WLP component [1].

die WLP thermal cycle life. Using more flexible structures, such as ductile solder materials, compliant dielectric layers, or redesigning the Cu RDL and UBM pad shape also improve the shear reliability. Direct mechanical support of the BGA solder collar using an additional polymer coating layer after BGA formation also helps. Large die size WLP, as large as 1 in. × 1 in. (25.4 mm × 25.4 mm) are feasible, as shown in Figure 3.3, depending on the process complexity, reliability requirement, and process cost consideration. However, as stated before, 8 mm × 8 mm is the normal maximum considered in WLP design and typically WLP in high-volume manufacturing are less than 6 mm × 6 mm for margin in board-level reliability.

3.3 Fan-Out Wafer-Level System Integration

As implied by its name, early FO-WLP technology was fundamentally similar to fan-in with the main difference in its silicon body being extended to accommodate the additional IO and BGA balls using epoxy mold compound. This difference, however, had given rise to technological challenges that have made fan-out WLP adoption more difficult and much slower than fan-in. In a typical fan-out process, silicon wafers are diced into individual chips, and known good die (KGD) are picked and placed on a template or carrier, in wafer form or panel form, and then overmolded with epoxy molding compound (EMC). RDL, UBM, and BGA are then fabricated on the Si die and molding compound before the finished components are singulated to form individual packages. As component size is enlarged by molding compound, the area ratio of silicon chip to final component area with compound is called fan-out ratio, and thus the technology is named fan-out. Needless to say, the RDL has to cross the interface between the silicon chip and molding compound for BGA solder ball interconnections. Since the RDL replaces the organic substrate in flip-chip packaging, FO-WLP is cost-competitive compared to traditional flip chip packages. The FO-WLP is also thinner than flip chip packaging because of the lack of substrate. When FO-WLP was first introduced, it was a simple, low-density, one-RDL, single-chip, size-limited packaging solution such as eWLB [2]. This can be called the first-generation (Gen 1) fan-out package, which was only meant to accommodate excessive BGA balls. For a process flow schematic of Gen 1 chip-first FO-WLP, see Chapter 2, Figure 2.3. While driven by the recent mobile computing system packaging and scaling needs, FO-WLP has transformed enormously and become far more versatile, with high-density, multi-RDL, multiple chips, large body size, and 3D stackable system integration solutions. As this is a completely different platform, it can be called the second-generation (Gen 2) FO-WLP. The first fan-out technology proposed in the Gen 2 group is InFO [3]. InFO was shown to have advantages in performance, thermal, power, and form factor with competitive cost for mobile applications [4]. Fan-out is now readily competitive to other system packaging and system integration technologies, such as multi-chip-module (MCM), package-on-package (PoP), silicon interposer, or 3DIC.

3.3.1 Chip-First vs. Chip-Last

There are currently two integration schemes for FO-WLP: chip-first and chip-last. Chip-first scheme builds a reconstituted wafer by placing all KGD onto a carrier first, either face-down or face-up, depending on process design. This is followed by a wafer-level or panel-level molding process to embed all KGD within an EMC and then forming the RDL layer within the package and UBM for interconnect to the external system. Typical examples are eWLB [2] and InFO [4], \ SESUB is slightly different since the carrier is a substrate with cavities for the die, which become part of the package and is not removable and reusable like an eWLB and InFO carrier [5]. Since the Cu RDL metal is deposited directly onto silicon chip connection pads, there is no solder reflow for chip to RDL joining as in flip chip packaging. Chip-first tends to produce a thinner package profile without additional height due to interconnect bump joints and underfill. A key process challenge for chip-first is die-shift during molding. Extensive studies have been done to ensure minimization of the die-shift during molding and compensating the shift in post molding RDL process using lithography and alignment. Compensation of die shift can also be done through carefully characterized pick and place compensation factors, which place the die farther out than their target location and then mold cure shrinkage moves them to their final location. Figure 3.4 shows an example of die shift measurement across a 300 mm wafer.

Alternatively, chip-last scheme builds a reconstituted wafer by forming the RDL layer first on a carrier, followed by bonding all KGD onto the RDL layer by flip

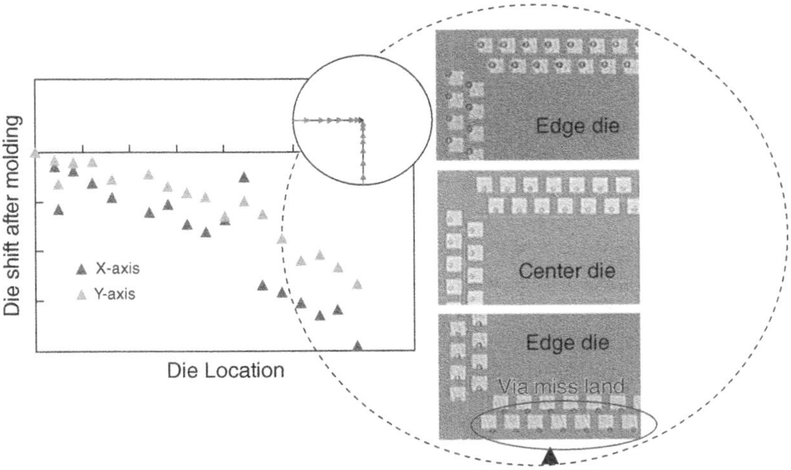

Figure 3.4 Die shift measured on 300 mm wafer. In general, center die tend to shift less than edge die, depending on compound dispensing, chip distribution, and curing condition.

chip bonding and underfilling. Then the wafer is molded to embed all the KGD within the EMC. Finally a UBM is formed for BGA solder ball attach. Examples can be found such as silicon-less integrated module (SLIM) [6], silicon wafer integrated fan-out technology (SWIFT) [7], and FOCLP [8]. Since chip mounting onto premade RDL requires high-accuracy placement and permanent metallic joints, chip-last requires solder joining structure, which limits both joining pitch density and results in taller package height. In chip-first and chip-last alike, high yield is a must in order to make the technology manufacturing feasible because the silicon chips embedded in fan-out package are often the costlier part of the process. Any minute yield lost represents a substantial added cost and lost profit margin.

3.3.2 Molding and Planarization

Molding compound provides structural integrity and protection for the silicon chips and RDL, ensures the mechanical and environmental reliability of the components, and provides electrical interconnection to the other components on the PCB. For wafer form fan-out, molding is done at wafer level, where a wafer-form mold cavity is used. Both injection transfer molding and compression molding have been developed. Conventional injection transfer molding is challenging here as the molding compound flow pattern is greatly affected by the flow path, chip layout pattern, chip-to-chip spacing, chip height variation, and edge die. Because of less physical barriers, the mold flow along wafer edge is faster than the wafer center, and voids happen at the vicinity of flow end. Also, the size and shape of the filler in the molding compound need to be selected to avoid filler trapping between the chip-to-chip spacing, which causes voids and other defects. However, due to its maturity and lower cost, injection transfer molding remains a molding option for wafer-level molding. An alternative wafer-molding technology is compression molding. In this case, liquid molding compound is dispensed on wafers at a predetermined amount followed by mechanical head compression to push the compound into the mold cavity. The liquid molding compound used in this case needs to have excellent flow and gap-filling capabilities. In the chip-first face-up case, a mold compound grinding and planarization process needs to be added to ensure proper pad exposure for RDL deposition and interconnection. The molding and post-molding planarization are extremely critical in warpage control as well as fine line/space (L/S) RDL formation. In wafer-level molding, the EMC material properties such as CTE, Young's modulus, curing-induced shrinkage and warpage, and glass transition temperature (T_g) are essential to the mechanical stability and reliability of the fan-out package. On the other hand, viscosity, filler material, and filler size are not only closely related to the CTE, Young's modulus, and T_g but also determine the workability and molding quality.

To enable fine L/S RDL, the wafer warpage and surface planarization are crucial. In a chip-first scheme, the wafer planarization depends on whether the die is placed face-down or face-up. For a chip face-down scheme, the wafer planarization is controlled by the carrier surface, while for a face-up scheme, the wafer planarization is typically controlled by a wafer level EMC grinding process. In a chip-last scheme, as chips are flip chip bonded to a preformed RDL, the molded wafer planarization is conducted on the backside in order to create vertical via interconnection and backside RDL. There are two potential defects in surface planarization that can affect the quality of fine L/S RDL. For chip-first face-down FO-WLP, one is the chip/EMC edge recess caused by the carrier release layer when using chip-first face-down scheme, such as the case in eWLB. The other defect is the EMC cavity caused by the large-size filler breakage during the EMC grinding.

For panel-level molding, the process becomes more challenging because of limited industry experience in handling large molding area and complex dispense pattern design with long flow distance [9]. Different dispensing techniques have been developed to overcome these issues. Homogeneous granular dispensing can eliminate post-mold flow mark or knit lines by using either sieve or vibration. Sheet lamination molding is another technique in which preformed blanket compound sheets are melted and flow around chips for encapsulation. Different molding techniques have different challenges and require careful evaluation from a process integration point of view. Figure 3.5 shows a very large panel molding. Uniformity in temperature, stress, and curing chemistry is a challenge when panel sizes become large.

Figure 3.5 Panel-level molding requires large panel handling capability with precise control of molding thickness uniformity and warpage across the panel [9].

3.3.3 Redistribution Layer (RDL)

The RDL L/S determines the routing density and number of RDL layers for a given chip size, and I/O pin count and RDL L/S are critical for the final package yield and cost. In a chip-first scheme, the RDL is built on the EMC using wafer-level process tools typically used for bump interconnect formation and WLP. The L/S capability is determined by the type of lithographic tools (aligner, stepper), photo resist (PR) materials, the molded wafer warpage, and chip-to-chip coplanarity within the molded wafer. In chip-last scheme, depending on the L/S requirements, there are two schemes to form the RDL layer currently adopted by leading assembly houses. The first RDL scheme adopts wafer-bumping tools and dielectric layers to provide an RDL capability of L/S = 2–5 μm, thus named SWIFT [7]. The second RDL scheme adopts silicon BEOL tools and dielectrics materials to provide an RDL capability of L/S in the submicron range, like those used for silicon interposer. The silicon is removed in a later stage, and thus the scheme is called SLIM [7].

RDL is one of the important cost-determining processes in fan-out WLP (see Chapter 2 for details). This is where wafer level and substrate-like panel-level processes are distinguished from each other. In wafer-level processes, the RDL is made with dielectric layer deposition and lithographic patterning followed by sputtered seed layers, mask and litho-patterning, and Cu electrochemical plating (ECP). Additional RDL layers can be added with the processes repeating. The processes are fully compatible with wafer processes with relatively relaxed pitch and line dimensions compared to Cu back-end-of-line (BEOL). Line width and spacing are anywhere from 10/10 μm line width and spacing down to submicrometer range, limited by dielectric materials, thickness, and contact/via aspect ratio. The RDL pitch density is near to Cu BEOL top metal layers pitch density, which has enabled and extended FO-WLP to high-performance mobile application

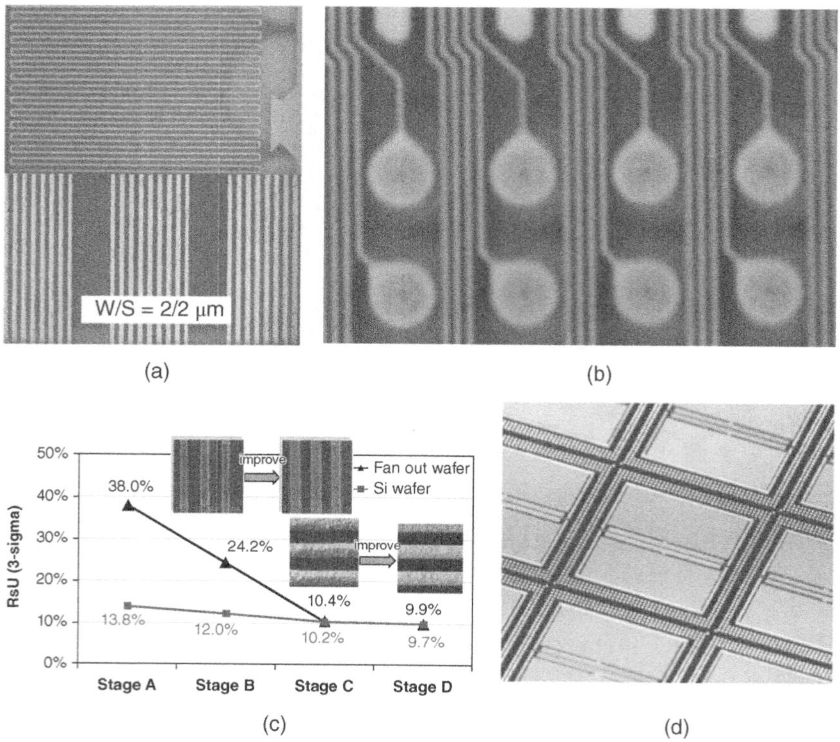

Figure 3.6 Fine pitch Cu lines in wafer-level processes enable lateral die partition using InFO platform. (a) 2/2 μm L/S high-density RDL. (b) Fine pitch RDL routing between through interconnect vias (TIV). (c) InFO wafer scale RDL electrical resistance (RsU) uniformity equivalent to silicon wafer BEOL processes has been demonstrated at production scale. (d) Logic die partition and lateral integration can be achieved using InFO fine pitch RDL.

domains, such as logic and memory lateral integration and high-performance SoC logic chip partition. A detailed discussion is available in the InFO for HPC chapter (See Chapter 4). Thick Cu is often needed for power, ground and integrated passive devices (IPD), such as antenna, inductor, and capacitors. In summary, the wafer-level Cu RDL dimension is scalable with minimum effort when pitch shrinkage and density increase are needed over time. Figure 3.6 shows examples of fine pitch fan-out RDL with uniformity and performance comparable to wafer BEOL Cu metallization.

Panel-level RDL, however, is made with a different technology. Evolving from organic substrate and PCB processes, panel Cu RDL is formed using a semi-additive process (SAP). The SAP uses a thin electroless copper seed layer, which can be etched after pattern plating. Lateral etching undercut is inevitable. The resist height needs to be slightly higher than copper pattern plating. These conditions, plus line and space widths, define the etch channel aspect ratio. Lateral etching at the electroless-copper/plated-copper interface is worse because this interface etches faster. It is this undesirable effect that limits the SAP linewidth to about 8–10 μm [10]. For this reason, panel-level Cu RDL L/S is limited to about 15–10 μm. However, new recessed Cu line processes are being developed, which promise finer lines [11]. Contrary to traditional Cu lines standing and protruding on flat dielectric film surface, recessed Cu lines are embedded in a dielectric film with the sidewalls and the bottom covered by the film. Trenches for copper lines in the dielectric are formed by laser ablation or by mechanical imprint, as shown in Figure 3.7a. A metal seed layer is then applied to the entire surface, and the trenches are filled with copper by an electroplating process. The unwanted surface copper is then removed by etching, chemical-mechanical polishing (CMP) or a combination of chemical/mechanical planarization steps. The result is shown in Figure 3.7b. Panel level RDL also needs to overcome tool size limitations when the panel size becomes larger. Large panel handling and warpage during the FO-WLP processes are a challenge, particularly for fine pitch RDL uniformity control across the whole panel.

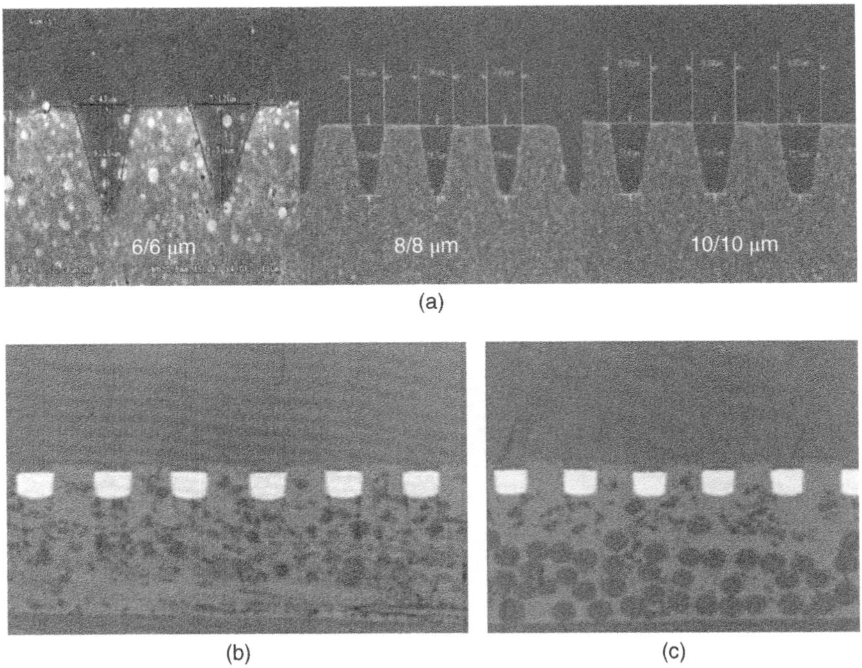

Figure 3.7 Fine pitch Cu lines using embedding recessed Cu line technology. (a) Dielectric recess trenches with different size and spacing. (b) 8/8 μm L/S Cu embedded in dielectric. (c) 7/7 μm L/S [11].

3.3.4 Through Via and Vertical Interconnection

Like traditional flip chip packages, FO-WLP can be stacked using vertical vias to provide an electrical path between top and bottom packages. Such vertical via interconnections are called TIV (through interconnect via), TPV (through package via), or TMV (through mold via) depending on the scheme and methods of vertical via formation. In a chip-first scheme, the vertical via interconnection is formed after EMC by a laser ablation process and followed by a metal filling. For a metal filling, one option is using solder reflow. On the other hand, in a chip-last scheme, the vertical via interconnection is formed after RDL formation by a lithographic process, followed by an electrical chemical plating (ECP). Typical FO-WLP via pitch is from below 150 μm [12] up to 350 μm [5], which is close to the number of BGA. A 150 μm pitch TIV is close to silicon interposer TSV pitch density, which is typically at about 100 μm. Fine pitch TIV enables and extends FO-WLP vertical PoP stacking for high-performance system integration such as high bandwidth logic and memory stacking [13]. Figure 3.8 shows examples of such high-performance FO_PoP stacking using high-density TIV. To fully utilize the advantages of fine pitch TIV, a backside RDL layer can be added, which restores BGA area array on the backside of FO package instead of being limited to the peripheral, an advantage that cannot be achieved by traditional FC_PoP without adding additional substrate layers.

Figure 3.8 A high-density InFO_PoP memory-on-logic stacking using TIV with pitch density higher than BGA. Both fine pitch TIV and backside RDL enables InFO_PoP for high-performance system integration [12].

3.4 Integrated Passive Devices (IPDs)

3.4.1 High Q-Factor 3D Solenoid Inductor

For RF system design, a high Q-factor inductor is highly desirable as it helps to improve system performance. This drives the need for integration of active chips and passive components in a package. A high Q-factor (quality factor) 3D solenoid inductor formed by RDL and copper via in the molding compound on InFO package has been fabricated. The 3D solenoid inductor with three turns achieves a Q-factor of 59 at 3 GHz and 0.68x lower resistance and 1.6x higher Q-factor than a 2D spiral inductor at the same inductance. The coupling between two closely placed inductors was also demonstrated and it was found that the coupling in orthogonal placement is 12 dB lower than in parallel placement [14]. The 3D inductor has 0.68x lower resistance and 1.6x higher Q-factor than those of the 2D spiral inductor under the same inductance. The 3D inductor with thick metal in the low loss molding compound can result in a higher Q-factor as compared to the 2D spiral inductor using thin metal, as shown in Figure 3.9.

3.4.2 Antenna in Package (AiP) and 5G Communication

The technology development for high-performance and low-power mmWave RF system integration has drawn a lot of attention from the rise of IEEE 802.11ad 60GHz Wi-Fi and 5G mobile communication. From a system integration point of

	2D sprial inductor	3D solenoid inductor
Structure	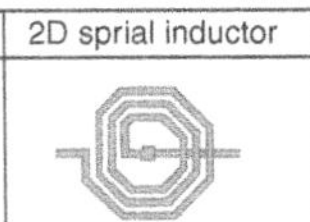	
Inductance	3.87 nH (1x)	3.9 nH (1x)
Resistance	1.98 Ω (1x)	1.35 Ω (0.68x)
Q-factor	36.8 (1x)	59 (1.6x)

(a)

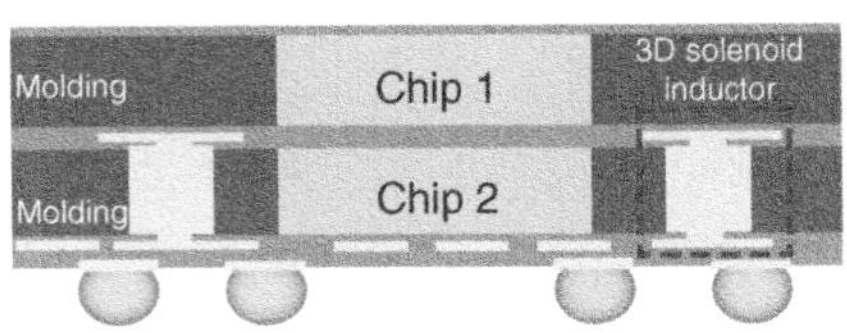

(b)

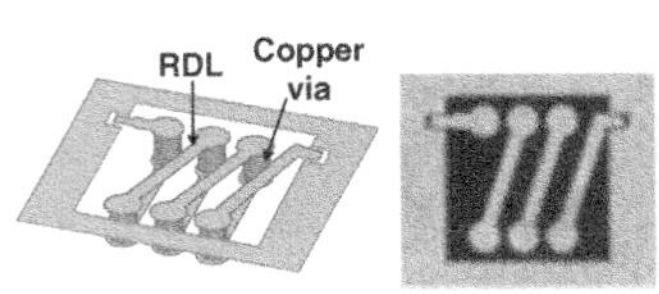

(c)

Figure 3.9 (a) Performance comparison between 2D spiral inductor and the 3D solenoid inductor on InFO package. (b) InFO structure for RF system integration. (c) 3D solenoid inductor on InFO chip fabricated [14].

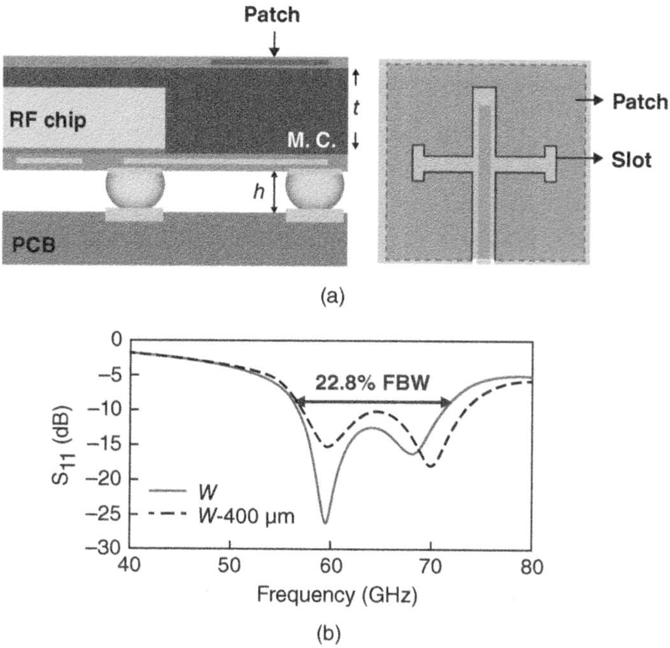

Figure 3.10 (a) Slot-coupled patch antenna on InFO_AiP in InFO package. (b) The S11 of the wideband patch antenna [15].

view, low transmission loss from the RF chip to antenna and integration process tolerance in mmWave application are required to obtain high system performance.

InFO_AiP technology, with low loss chip-to antenna interconnect and wideband slot-coupled patch antenna, is demonstrated for low power, high performance, and compact 5G millimeter wave (mmW) system integration. The low loss chip-to-antenna interconnect is related to the low metal surface roughness of the RDL and a smooth interconnect transition between chip and package in InFO technology. The InFO RDL with low metal roughness results in transmission loss of 0.3 dB/mm, which is lower than the loss of Cu trace in organic substrates, and the smooth interconnect transition with low discontinuity structure reduces the interconnect loss by 0.78 dB at 60 GHz. A wideband slot-coupled patch antenna has been successfully demonstrated with 22.8% fractional bandwidth (FBW) (56.6–71.2 GHz) and over antenna gain of 3 dB in the operating band, as shown in Figure 3.10 [15].

3.4.3 Passive Devices for Millimeter Wave System Integration

High-performance passive devices for mmW systems, including inductor, ring resonator, power combiner, coupler, balun, transmission line, and antenna, are realized using InFO WLP technology. The inductors have a Q factor over 40; the power combiner, coupler, and balun show lower transmission loss than on-chip passives; the antenna has efficiency of over 60%. These devices on InFO enable low noise and power mmW system for mobile communication and IoT applications [16].

Figure 3.11a shows the proposed mmW system architecture based on the InFO technology. It is composed of a 65 nm CMOS RF integrated circuit (RFIC), dielectric, molding compound, RDL, and passive devices. The 65 nm CMOS RFIC is surrounded by molding compound; the dielectric covers the molding compound and CMOS RFIC; RDLs are distributed on the dielectric and connected to RFIC I/O pads; and the passive devices designed above RFIC or molding compound are linked to RFIC interconnects through RDL and bumpless interconnection. Figure 3.11b shows the optical images of the realized passive devices on the RDLs, including (a) inductor, (b) ring resonator, (c) antenna, (d) power combiner, (e) coupler, (f) balun, (g) coplanar waveguide (CPW), and (h) microstrip line, hereafter called InFO passives.

Figure 3.12a shows a 300 mm InFO wafer with all the passives integrated. The performance of the InFO inductor, ring resonator, power combiner, coupler, and balun are designed and measured, respectively. For the inductor, three one-turn coils in rectangular shape with different opening size are designed by using RDL. One of the inductors with a 100 μm opening has a Q-factor of over 20 from 5 to 49 GHz, as shown in Figure 3.12b, and these three inductors have high Q peak of 45, 40, and 36 with respect to 300, 460, and 600 pH, respectively. For the resonator realized in microstrip-line ring type, the measurement shows the resonances at 32 and 64 GHz, correctly predicted by the 3D EM simulation, as shown in Figure 3.12c.

The power combiner with a measured 3.5 dB loss outperforms the on-chip one by 1.8 dB, a 34% improvement; the coupler with a measured 4.1 dB loss outperforms the on-chip one by 1 dB, a 20% improvement; and the balun with a simulated 4.1 dB loss outperforms the on-chip one by 2.3 dB, a 36% improvement [16]. The frequency responses of the power combiner, coupler, and balun are calculated from 50 to 80 GHz and shown in Figure 3.13. The InFO passives can provide performance better than on-chip passives, aiming at low power and low noise mmW system.

Co-planar waveguide (CPW) and microstrip lines are two common RF transmission line structures. A CPW transmission line is designed to 50 Ω and with a length of 4 mm. It is measured from DC to 67 GHz and compared with simulation, as shown in Figure 3.9. The transmission loss is below 1.8 dB, and the return loss is above 10 dB up to 67 GHz. In addition, a microstrip line is designed to 50 Ω, and its transmission loss and return loss were measured. The transmission loss for CPW line and microstrip line are 0.35 and 0.34 dB/mm, respectively, at 60 GHz. Compared with those on the BEOL of CMOS, the CPW and microstrip lines on InFO provide 30 and 32% improvement in loss reduction [17, 18].

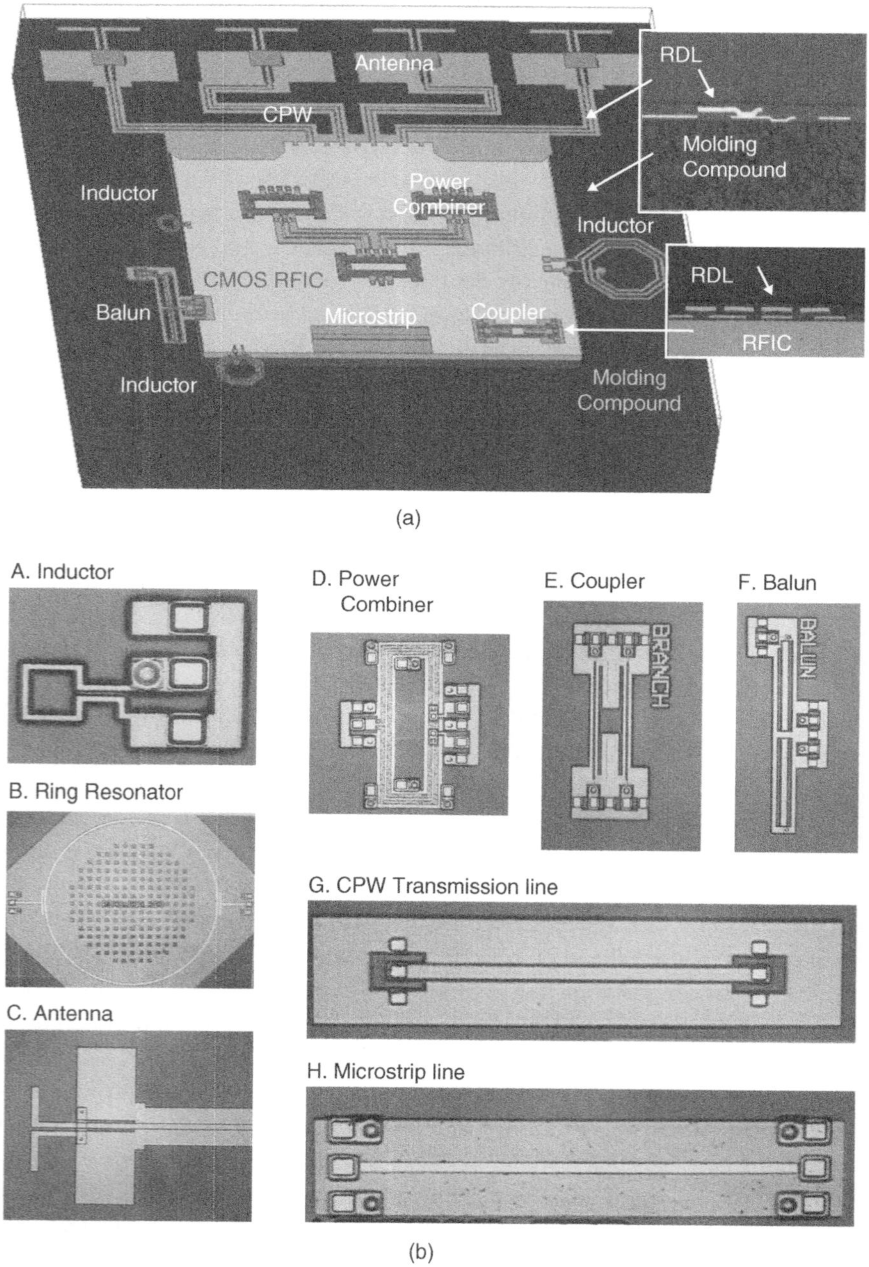

Figure 3.11 (a) Schematic and cross section images of millimeter wave system circuit, including RFIC, passive devices, transmission lines, and antennas on InFO structure. (b) Optical photos of realized passive devices: (A) inductor, (B) ring resonator, (C) antenna, (D) power combiner, (E) coupler, (F) balun, (G) CPW, and (H) microstrip [16].

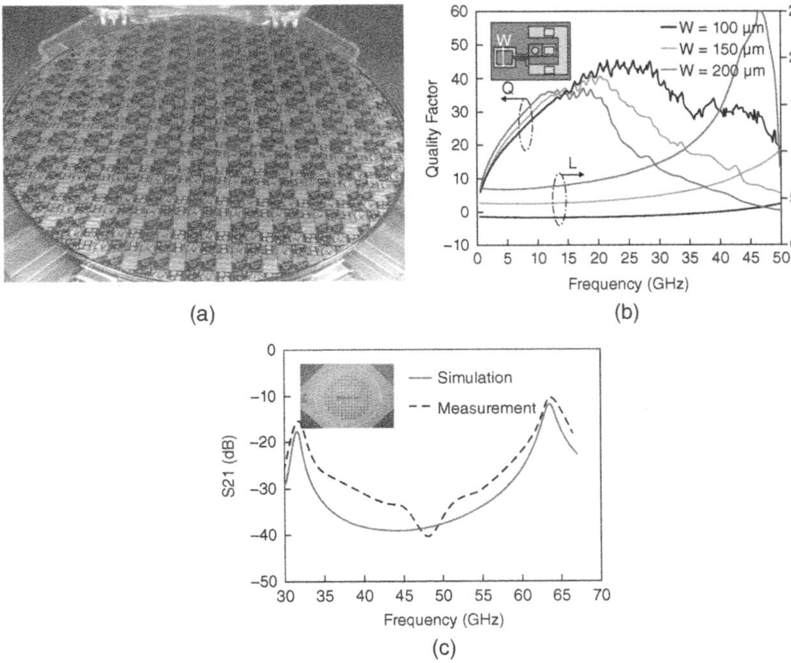

Figure 3.12 (a) 12″ InFO wafer for millimeter wave system integration incorporating silicon die, RF passive devices, transmission lines, and antenna. (b) The measurement of L and Q of the inductors with W = 100, 150, and 200 μm implemented by InFO RDLs. (c) The measurement and simulation of the transmission loss for the ring resonator [16].

3.5 Power, Performance, Form Factor, and Cost

To demonstrate the full advantages of a fan-out WLSI package, an InFO_PoP together with currently available mobile computing system integration solutions such as, flip chip PoP (FC_PoP) and state-of-the-art 3DIC with TSV are benchmarked. Key performance indicators compared including signal and power integrity, thickness form factor, heat dissipation, junction leakage, SoC power cap, memory bandwidth and cost. InFO, which was especially developed for mobile devices and HPC chiplet system integration, are used as the fan-out representative example. InFO is readily adaptive to 3D stacking using conventional package-on-package (PoP) architecture. The new InFO_PoP is benchmarked with alternative 3D stacking technologies, including existing flip-chip PoP (FC_PoP), flip-chip high memory bandwidth PoP (FC-HMB_PoP), and the 3D chip stacking using TSV (3DIC). Figure 3.14 shows the schematic cross section of the four different logic and memory stacking structures in this study [13].

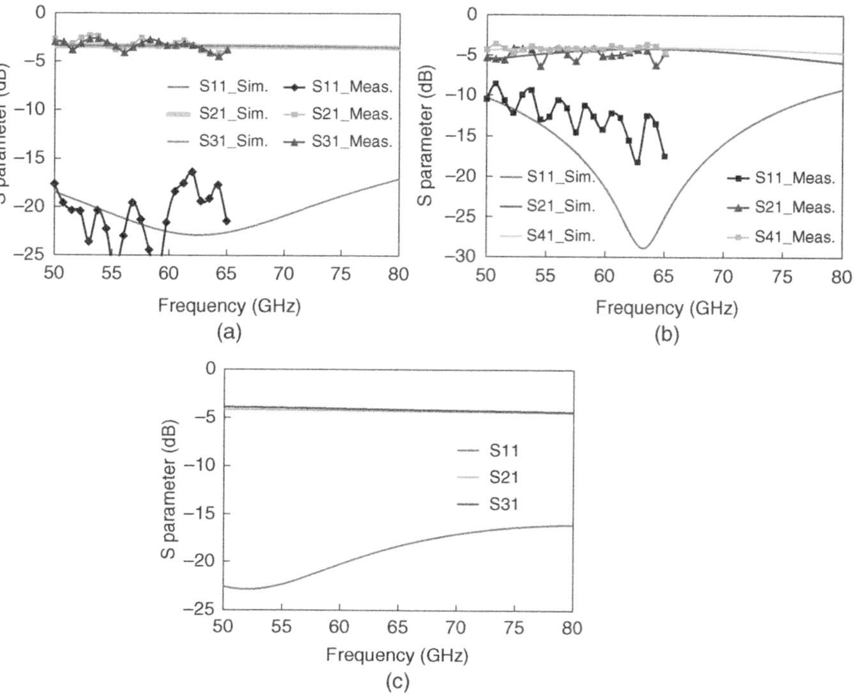

Figure 3.13 (a) Measurement and simulation of the transmission loss and return loss for the InFO power combiner. (b) Measurement and simulation of the transmission loss and return loss for the InFO coupler. (c) Simulation of the transmission loss and return loss for the InFO balun [16].

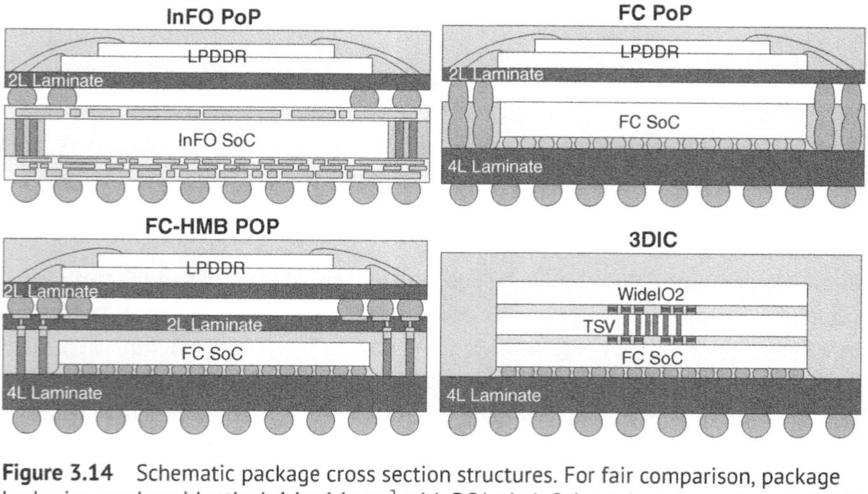

Figure 3.14 Schematic package cross section structures. For fair comparison, package body size used are identical, $14 \times 14\,\text{mm}^2$ with BGA pitch 0.4 mm (not drawn to scale) [13].

3.5.1 Signal and Power Integrity

By routing optimization, SoC signal integrity in InFO_PoP can achieve 20% better than FC_PoP because of its ultra-fine RDL capability. Additional ground lines can be added to enhance shielding when using commodity LPDDR DRAM with fixed pitch, as seen in Figure 3.15.

FC_PoP can support only limited peripheral BGA balls for topside wirebond memory package. InFO_PoP resolves this issue by providing backside RDL and transforming peripheral to areal BGA array and simultaneously improving signal integrity and power consumption. FC-HMB_PoP can also provide BGA array by adding another substrate between packages but at the penalties of adding package height and additional bill-of-materials (BoM), while still limited by organic substrate routing dimensions.

InFO_PoP SoC power integrity, power delivery network (PDN) is improved because of efficient SoC-to-PCB power/ground routing, as seen in Figure 3.16. The improvement originates from close distance between silicon-to-PCB ground planes by eliminating organic substrate and bumps and by thickness reduction within InFO RDL power and ground layers.

Signal Integrity of Bottom PKG (Eye Height)	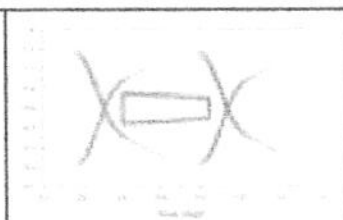			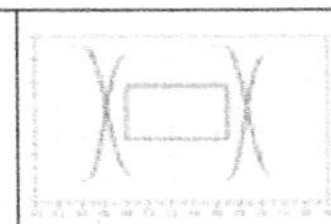
	1.20x	1x	0.91x	--
Memory I/O Power Consumption	0.998x	1x	1.003x	0.82x
Data I/O Count per PKG	64	64	64	512
Package Type	InFO_PoP	FC_PoP	FC-HMB_PoP	3DIC

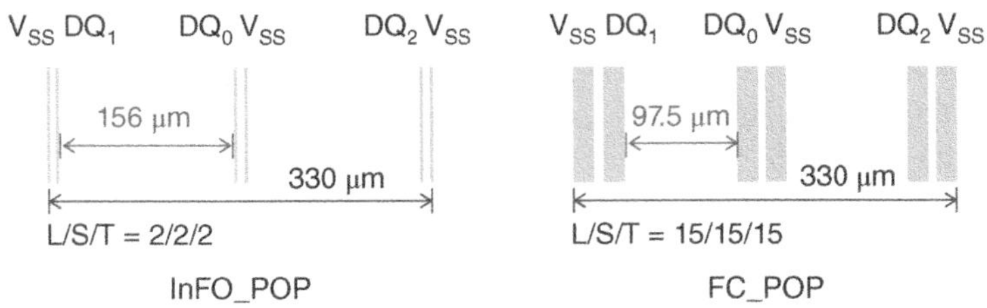

Figure 3.15 Package signal integrity comparison among InFO_PoP (Line/Space/Thickness = 2/2/2 μm), FC_PoP (L/S/T = 15/15/15 μm), FC-HMB_PoP (L/S/T = 15/15/15 μm, and 3DIC. InFO_PoP is 20% better than flip chip PoPs (DQ_0). For all three PoPs using LPDDR3, IO speed is 1.6 Gb/s and bandwidth 12.8 GB/s. For wide IO2 3DIC, IO speed is 0.8 Gb/s and bandwidth 51.2 GB/s. Package signal integrity is calculated based on the single layer interconnect from SoC Al pad to LPDDR BGA [13].

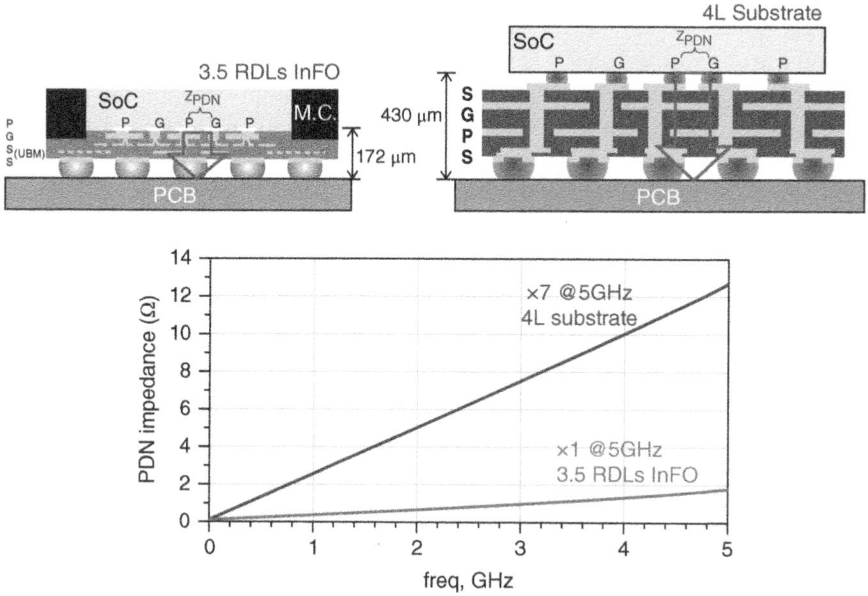

Figure 3.16 InFO to PCB direct power/ground routing significantly improve PDN (power delivery network) impedance for seven times at 5 GHz, compared to traditional FC_PoP and 3DIC using substrates [13].

3.5.2 Heat Dissipation and Thermal Performance

Figure 3.17 compares the steady-state thermal performance and junction temperature. InFO_PoP shows lowest junction temperature, $T_{j,max}$ and thermal resistance, θ_{JA}. It mounts logic chip(s) directly onto printed circuit board (PCB) without organic substrate and greatly reduces junction-to-environment thermal resistance. Due to low junction temperature achieved at the same power level, InFO_PoP device junction leakage current is substantially lower than the other three schemes.

Figure 3.18 shows that under the same operating power level, junction leakage of InFO_PoP is only 66% of the conventional FC_PoP and is the lowest among the four 3D integration schemes. Low junction leakage leads to power efficiency and long battery life for mobile computing. Low θ_{JA} and $T_{j,max}$ and low junction leakage allow designers to push for higher computing capability. Figure 3.19 shows maximum allowable power for the four configurations considered. An average of 19–23% more SoC power is allowed when using InFO_PoP as compared to FC_PoP. This allows better design flexibility and functions integration in SoC or SiP (system-in-package).

In transient thermal behavior, the time constant, τ, is defined as $(T_\tau - T_0)/(T_{steady} - T_0) = 1\text{–}1/e$ (where e is Euler's number). Figure 3.20 compares the

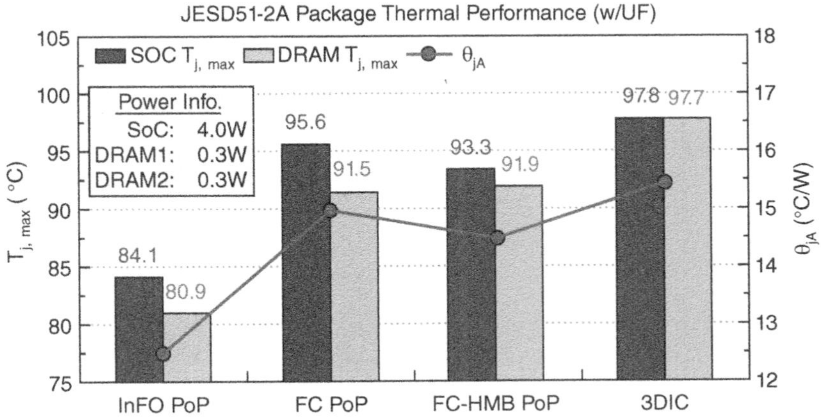

Figure 3.17 Package steady-state thermal performance comparison among InFO_PoP, FC_PoP, FC-HMB_PoP, and 3DIC at the same system power level. InFO_PoP has the lowest junction temperature and best heat dissipation performance [13].

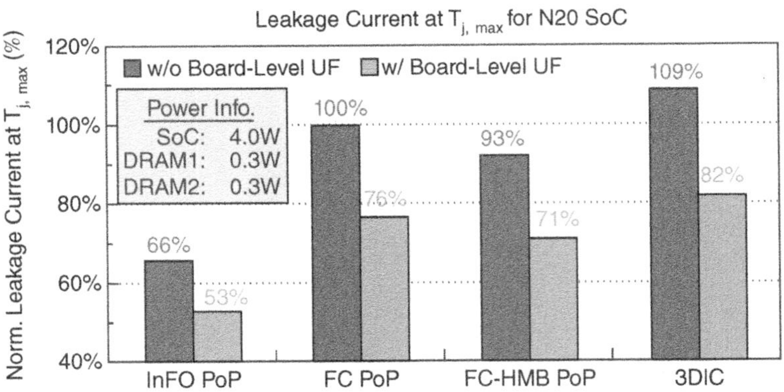

Figure 3.18 Junction leakage current at $T_{j,max}$ using advanced node, 20 nm HKMG, in four different schemes. Due to its low junction temperature, InFO_PoP achieves 33% lower junction leakage current compared to FC_PoP [13].

relative time constants of the four package schemes both with and without PCB. When mounted on PCB, InFO_PoP outperforms FC_PoP by more than 10%. FC-HMB_PoP, however, shows the best transient thermal behavior owing to its multiple substrates plus molding, which act as heat reservoirs and delay the time required to heat/cool the system. However, this solution incurs BoM and thickness penalties. It is noted that the thermal performance of 3DIC (TSV) is among the worst, as shown in Figures 3.17–3.20. InFO_PoP has more sensitive thermal transient property because of its thin package structure and minimal thermal mass.

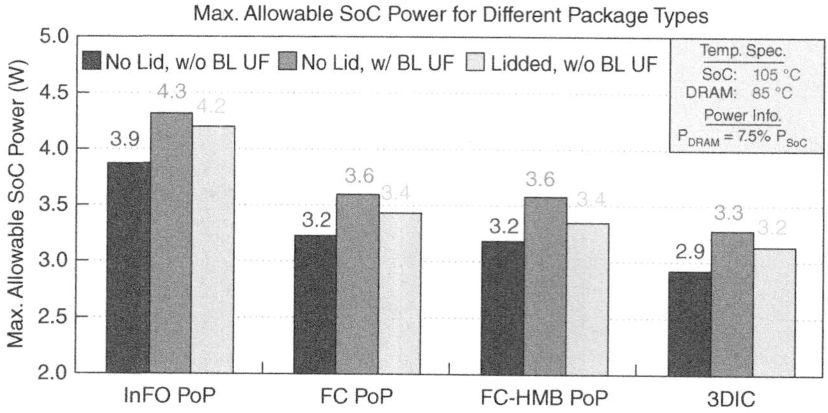

Figure 3.19 Maximum allowable SoC power. InFO_PoP demonstrates highest maximum allowed power for SoC chip, a 20–23% more than FC_PoP in package with different underfill and heat spreader configurations [13].

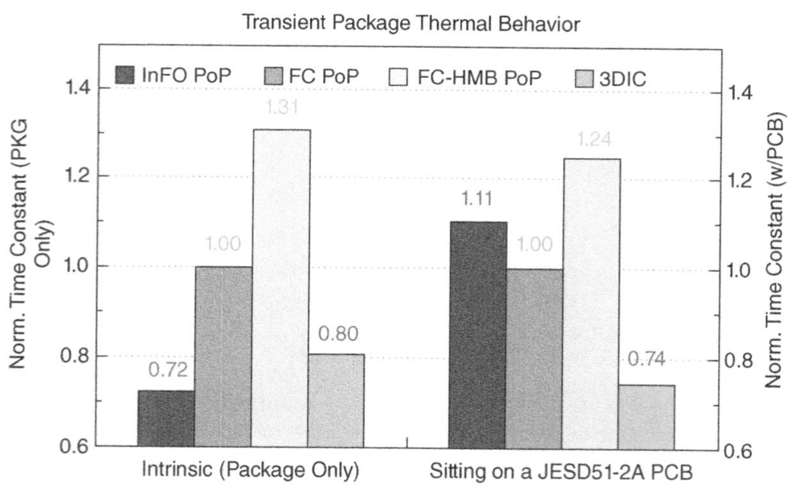

Figure 3.20 Time constant, τ, is defined as the time at which the temperature change reaches 1 − 1/e (~63.2%) of its overall temperature difference between the initial state and the final steady state under the influence of a step heating or cooling. Relative time constants of different package schemes with and without PCB. The higher the time constant is, the better the package transient heat dissipation will be. FC-HMB_PoP has the highest τ among the four due to its largest effective thermal mass [13].

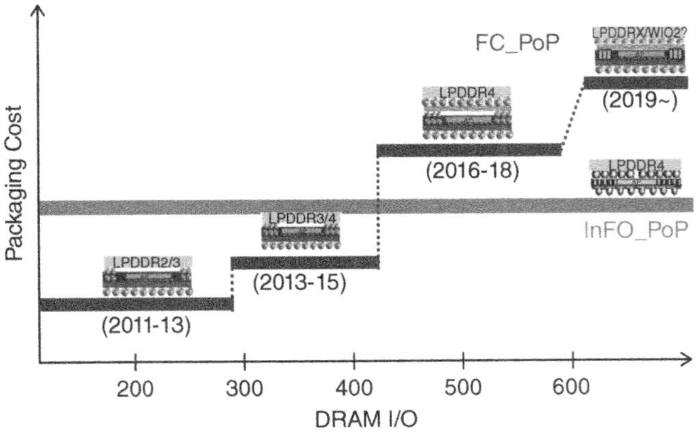

Figure 3.21 FC_PoP package cost increases as the DRAM I/O count pushing up with ever more mobile computing bandwidth demands. As the I/O count increases, the InFO_PoP cost remains flat because of its innovative slim packaging structure, fine pitch RDL, and fine pitch TIV. Both are wafer-level processes [13].

3.5.3 Form Factor and Thickness

For total package thickness, InFO_PoP achieves ≤0.8 mm because of the lack of substrate and bump interconnects. An FC_PoP total thickness is currently at about 1.0 mm. FC-HMB_PoP is expected to be thicker than FC_PoP as one more substrate is used. The total thickness of a 3DIC is expected to be between 0.8 and 1.0 mm depending on the substrate used. In comparison, InFO_PoP enables mobile smartphone or tablet by further reducing total thickness.

3.5.4 Cycle Time to Market and Cost

InFO-PoP leverages WLSI tools, materials, process control, process know-how, yield learning and unique package-on-wafer (PoW) in DRAM package stack to overall shorten customer cycle time to market for market competitiveness. Figure 3.21 shows the cost comparison between FC_PoP and InFO_PoP. While FC_PoP cost trends up to meet DRAM bandwidth increases, InFO_PoP remains relatively constant because of its highly scalable RDL and TIV structure. InFO_PoP is the smallest, lightest, simplest, and most cost-effective packaging structure among the four.

3.6 Summary

WLP has transformed from a traditional size-limited large-pitch component packaging technology into a cost-effective system integration platform for both mobile and high-performance computing, thanks to the adoption of a WLSI platform that leverages front-end wafer processes [19]. Integrated passives (IPD) and heterogeneous integration on both lateral and vertical directions, by adopting fine pitch RDL and TIV, have extended WLP from single die package into multi-chip system integration domains. Both fan-in WLP and fan-out FO-WLP are now gradually replacing traditional flip chip-based packaging technologies in the cost competitive markets of the Internet of Things (IoT), wearables, mobile computing, and high-performance data centers.

References

1 Chatinho, V., Cardoso, A., Campos, J., and Geraldes, J. (2015). Development of very large fan-in WLP/ WLCSP for volume production. *IEEE Electronic Components & Technology Conference*: 1096–1101.

2 Brunnbauer, M., Fürgüt, E., Beer, G., and Meyer, T. (2006). Embedded wafer level ball grid array (eWLB). *IEEE Electronics Packaging Technology Conference*: 1–5.

3 Yu, Doug C.H. (2013). Innovative wafer-based interconnect enabling system integration and semiconductor paradigm shift. *IITC Plenary presentation*, Kyoto, Japan.

4 Yu, Doug C.H. (2014). Wafer level system integration for SiP. *IEEE IEDM*, 27.1.1–27.1.4, pp. 626–629.

5 Fujioka, H. (2015). SESUB technology-embedded package. *SiP Global Summit 2015 - Embedded and Wafer Level Package Technology Forum*, SEMICON Taiwan, September.

6 Yoo, M. (2015). Alternative 3D—TSV-less interposer. SiP Global Summit 2015 – System Integration by 3D-SiP, SEMICON Taiwan, September.

7 Huemoeller, R. and Zwenger, C. (2015). Silicon wafer integrated fan-out technology. *Chip Scale Review*, March/April.

8 Chen, S., Wang, S., Lee, C., and Hunt, J. (2015). Low cost chip last fanout package using coreless substrate. *11th International Conference and Exhibition on Device Packaging*, Fountain Hills, AZ, USA, March.

9 Braun, T., Raatz, S., Voges, S. et al. (2015). Compression molding for large area fan-out wafer/panel level packaging. *SEMI Packaging Technology Seminar*, June.

10 Dietz, K.H. (2008). Challenges and limitations of subtractive processing in PWB and substrate fabrication. *Microsystems, Packaging, Assembly & Circuits Technology Conference, 2008. IMPACT 2008. 3rd International*, 231–234.

11 Hu, D.C. (2013). Panel level embedded technology. *System in a Package Global Summit 2013*, SEMI Taiwan.

12 Yu, Doug C.H. (2014). From silicon to system, from mobile to cloud computing, and from 'today' to 'next big things. 2014 SEMI Taiwan, SIP Keynote, September 3.

13 Yu, D.C.H. (2015). A new integration technology platform: integrated fan-out wafer-level-packaging for Mobile applications. *IEEE VLSI-T*: T46–T47.

14 Tang, Tzu-Chun, Lin, Chia-Chia, Hsu, Che-Wei et al. (2019). High Q-factor 3D solenoid inductor on InFO package for RF system integration. *2019 Electrical Design of Advanced Packaging and Systems (EDAPS)*, 16–18 December, KAOHSIUNG, Taiwan.

15 Wang, C.-T., Tang, T.-C., Lin, C.-W. (2018). InFO_AiP technology for high performance and compact 5G millimeter wave system integration. *IEEE 68th Electronic Components and Technology Conference*, pp. 202–207.

16 Tsai, Chung-Hao, Hsieh, Jeng-Shien, Lin, Wei-Heng et al. (2015). High performance passive devices for millimeter wave system integration on integrated fan-out (InFO) wafer level packaging technology. *IEEE IEDM*, pp. 632–635.

17 Hsu, Che-Wei, Tsai, Chung-Hao, Hsieh, Jeng-Shien et al. (2017). High performance chip-partitioned millimeter wave passive devices on smooth and fine pitch InFO RDL. *IEEE 67th Electronic Components and Technology Conference*, pp. 254–259.

18 Chen, S.M., Huang, L.H., Yeh, J.H. et al. (2013). *IEEE VLSI-T*, pp. T46–T47.

19 Yu, Doug C.H. (2014). invited paper, New system-in-package (SiP) integration technologies. *IEEE Custom Integrated Circuits Conference, IEEE CICC*, pp. 1–6.

4

Integrated Fan-Out (InFO) for High Performance Computing

Doug C.H. Yu, John Yeh, Kuo-Chung Yee, and Chih Hang Tung

Taiwan Semiconductor Manufacturing Company (TSMC), Hsinchu, Taiwan

4.1 Introduction

Gordon Moore, in an article published in 1965, suggested that transistor count doubles every 18–24 months on an integrated chip. The prediction, later known as Moore's law, guided semiconductor technology progression for five decades in high performance computing (HPC) systems, including central processing unit (CPU), graphics processing unit (GPU), field-programmable gate array (FPGA), mobile processors, and some purpose-built special accelerators [1]. Despite the continuous buildup of technological challenges and cost escalation, Moore's law remains healthy and strong thanks to the wafer-level system integration (WLSI) platforms introduced by foundries [2]. Today, the biennial system level transistor count doubling trend is not only sustained but in some cases accelerates more aggressively than Moore's law prediction, as shown in Figure 4.1. Complementary to transistor count scaling, this new system level scaling based on WLSI platforms has helped to boost system performance, power, package form factor (area), and cost (PPAC) in a similar way as Moore's law did in the last five decades.

As Moore's Law evolves from transistor scaling alone (Moore's law 1.0) onto transistor plus system-level scaling (Moore's law 2.0), it is important to learn how such enormous systems evolved continuously with sustainable performance gain and cost benefits. There are a few notable innovations. Firstly, the new wafer-level system scaling is largely leveraging established wafer processes in terms of capability and capacity. Other than a few new processes, such as through silicon via (TSV) and chip stacking, many design houses' and foundries' existing technologies can be shared and reused, including intellectual property (IP) and electronic design automation (EDA) tools. Secondly, as transistor counts continue to grow exponentially, increasingly diversified IP circuit blocks demand circuit

Embedded and Fan-Out Wafer and Panel Level Packaging Technologies for Advanced Application Spaces: High Performance Compute and System-in-Package, First Edition. Edited by Beth Keser and Steffen Kröhnert.

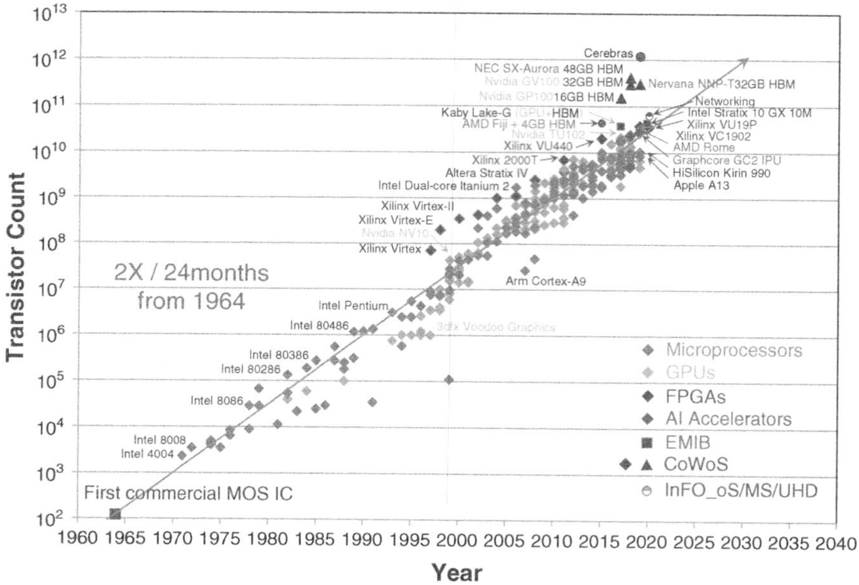

Figure 4.1 Moore's law depicts transistor count doubling every 24 months. This remains true for HPC system-on-chip (SoC), including microprocessors, GPU, FPGA, and other application-specific accelerators. As computing power exponentially grows for over 50 years, SoC chip sizes grow beyond lithographic reticle sizes (26 mm × 32 mm) and system-on-package (SoP) emerges as an effective way to sustain Moore's Law (2.0) beyond its SoC limitation, as shown in RED points [3].

and transistor level optimization in order to gain best scaling benefits. Despite great efforts made by design houses and foundries, co-optimization among different functional blocks based on one specific transistor node technology has become formidable, if not impossible. An emerging option is to partition a large SoC chip into isolated IP functions, each manufactured separately using appropriate technologies, and re-assembled with SoC-like, or monolithic-like, integration processes, using either 2D, 2.5D, or 3D packaging structures. Partitioned smaller chips also gain yield and wafer die-count advantages, thus optimizing not only performance but also cost [3].

Contact gate pitch is used as an index for Moore's law SoC scaling to enable performance, power, area, and value. A new index 3DID (3D interconnect density) metric has been proposed for Moore's law 2.0, system on package (SoP) scaling [2]. 3DID is defined as the highest chip-to-chip horizontal interconnect density (# lines/mm) multiplied by the highest vertical interconnect density (#bonds/mm^2) between chips within a package or system. Chip scaling has been gauged by contact gate pitch scaling. Similarly, 3DID scaling can be used for measuring system performance, power, and form factor values for SoP scaling. Figure 4.2 shows

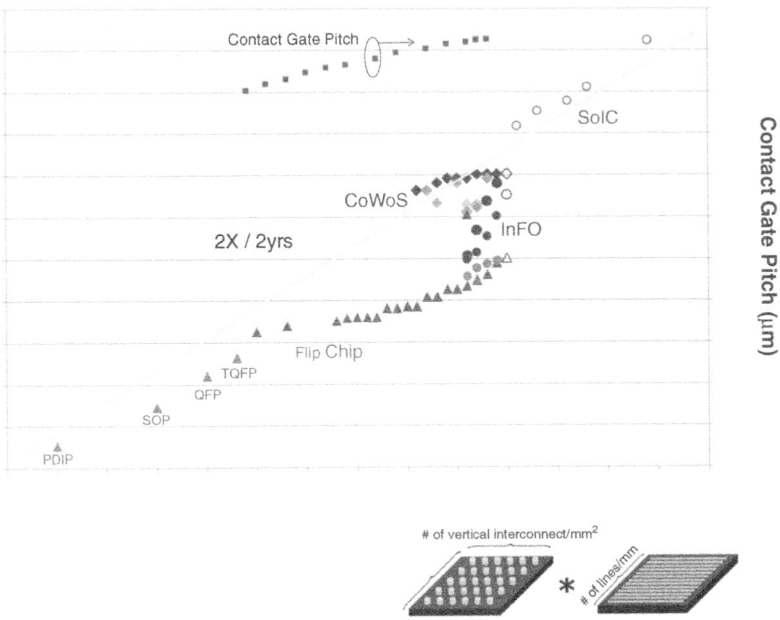

Figure 4.2 New scaling index. 3DID is the product of vertical bond density and horizontal line pitch. 2×/2 years 3DID cadency is expected with packaging technology advancement.

how 3DID has been improved effectively by innovative packaging technologies such as chip-on-wafer-on-substrate (CoWoS), a leadership chip-last Si interposer, and integrated fan-out (InFO), an innovative chip-first redistribution layer (RDL) interposer, in recent years [4]. 3D WLSI is further enhanced with the introduction of system on integrated chip (SOIC), a disruptive front-end 3D integrated circuit (3DIC) [5]. The holistic integration of frontend 3D SoIC plus backend 3D (CoWoS, InFO) can support the 2×/2 years 3DID cadency in the coming decades.

4.2 3DFabric and System-on-Integrated-Chip (SoIC)

As system integration technologies evolve quickly with different platforms mixed and matched in increasingly complex application scenarios, it is necessary to reexam the existing technologies regarding fan-out, interposer, silicon interconnection bridges, chip embedding, and their associated process flow and sequences. For foundry-enabled technologies, all these are under the umbrella of 3D/2.5D/2D wafer-level integration, and thus named 3DFabric™ for its enablement of fine pitch intertwined chip connection and a unified fabrication philosophy leveraging existing processes, tools, and technology platforms. Under 3DFabric, all processes with chips embedding first before interconnection are called InFO, as the fundamental manufacturing process started with chip embedding first, followed by creating fan out interconnection. All processes that start with making a redistribution interconnection (RDL) followed by chip placement onto premade RDL are called CoWoS since the RDL are deposited before chip placement, regardless of whether a silicon interposer is being used. This new naming system, despite the slight difference from generally accepted nomenclature, truly reflects the process essence and is able to point to their future development paths. InFO, CoWoS, system on wafer (SoW) and system-on-integrated-substrate (SoIS), form a universal WLSI technology family that will drive the industry to meet the future system scaling needs faced by increasingly more challenging and diversified computing requirements. Figure 4.3 shows simplified family trees of the foundry's 3DFabric [5].

In 3DFabric, chip-first is named InFO and chip-last is named CoWoS. In production, however, the two do not need to be independent. There are cases where one may need both chip embedding first for some chips followed by chip-last for additional chips. The possibility of combining the two allows system architects to design and implement an optimized design with ample options in local interconnect density, chip shape and sizes (including height), different process thermal budgets for individual chips, final power/energy solutions, and of course, yield and cost consideration along with supply chain and ecosystem management. This

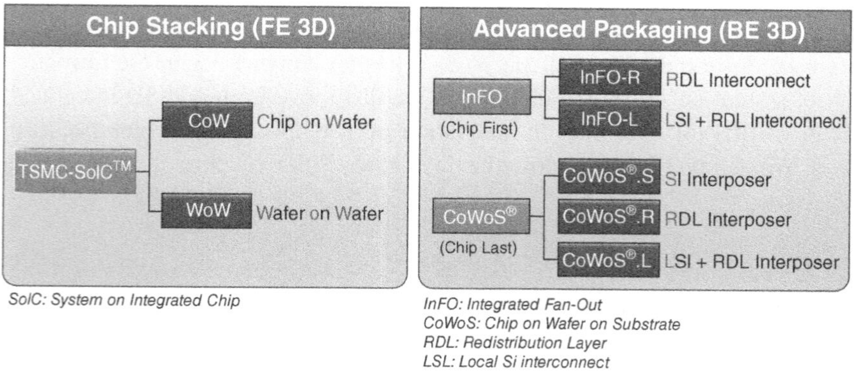

Figure 4.3 TSMC 3DFabric™ wafer-level system integration platforms. Front-end 3D (FE 3D) is called SoIC, which includes chip-on-wafer (CoW) and wafer-on-wafer (WoW) stacking. Back-end 3D includes redefined InFO (chip-first) and CoWoS (chip-last). InFO has R (with RDL) and L (local silicon interconnection or LSI) varieties. CoWoS has R (with RDL), L (with LSI), and S (with Si interposer) varieties.

flexibility becomes essential, as we discuss more complicated integration scenarios later in this chapter.

SoIC, as the name implies, stacks various active and passive chips using chip-on-wafer or wafer-on-wafer 3D stacking. Chips made by different technologies separately with different materials and bond pitch density are tested to establish the known good die (KGD) before SoIC stacking and integration. SoIC utilizes front-end wafer processes with pitch density on par with copper (Cu) back-end-of-line (BEOL) enabling ultimate chip-to-chip interconnection density with SoC or monolithic equivalent density and performance. An SoIC integrated chip not only looks like but also behaves like an SoC chip in every aspect in terms of electrical performance and mechanical integrity and can be subsequently assembled using conventional packages or new WLSI platforms, such as CoWoS or InFO.

4.3 CoWoS-R, CoWoS-S, and CoWoS-L

CoWoS represents a chip-last process flow, in which interconnections are formed before chip placement. According to design and system requirement considerations, interconnection density, bandwidth, and latency requirements will vary with application, and thus different interconnection options are available as shown in Figure 4.4. CoWoS-S is the traditional CoWoS with silicon interposer (represented by the last "S") as the chip-to-chip interconnection. Ultra-high interconnection density and mature design and manufacturing technology are the main advantages of CoWoS-S. Silicon (Si) interposer can also prefabricate

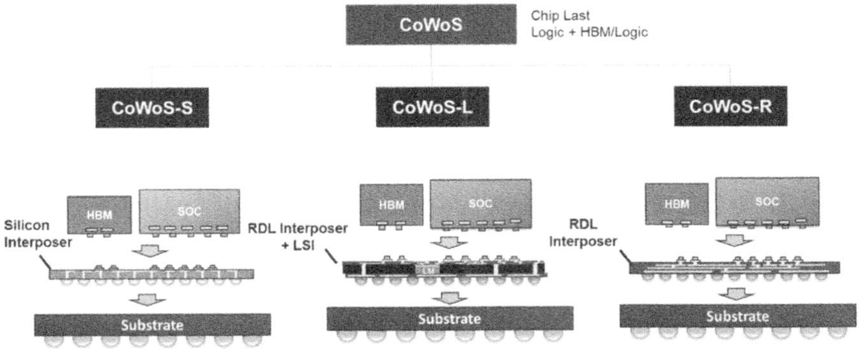

Figure 4.4 CoWoS chip-last wafer-level system integration platforms with various technology options. S represents a silicon interposer as the main interconnection. R represents RDL as the interposer for the main interconnection. L represents a local silicon interconnect (LSI) with small pieces of silicon to boost local interconnection densities.

passives and active circuitries, thus becoming an active interposer, to boost the system performance. Many of these additives cannot be done by using discrete components without performance or cost penalties. CoWoS-R, where R represents RDL, is the platform where the silicon interposer is replaced by prefabricated RDL for its cost, form factor, and power and signal integrity benefits, although the pitch density and bandwidth density may be slightly compromised. In most systems, interconnection density varies from chip to chip and location to location. Thus, the third variation of CoWoS is CoWoS-L, where L denotes local silicon interconnection (LSI). These local and small silicon interconnect pieces are embedded into RDL layers where high-density connections are needed and cannot be achieved by RDL alone. CoWoS-L enables local high-density interconnection eliminating the need for a large silicon interposer and thus balances the design complexity, high-speed performance, and manufacturing cost. The LSI silicon has a low profile and is a low parasitic discrete component. With deep trench capacitor (DTC) inside the LSI chip, CoWoS-L can boost SoC speed (iso-power) or reduce power consumption (iso-speed) by eliminating voltage droop against conventional FC-substrate with multilayer ceramic capacitor (MLCC). More discussion on this will be presented in the InFO-3DMS/CoWoS-L section.

4.4 InFO-L and InFO-R

InFO is a chip-first integration technology where chips are embedded before interconnecting RDL are manufactured. As interconnections are made using wafer-level processes, InFO does not require bulky chip-to-chip solder joining and their associated underfills. This improves not only form factor but also power and thermal integrity. Like in chip-last CoWoS, InFO-L uses LSI, and InFO-R uses RDL. InFO-R is the traditional InFO platform where chip-first embedding and then RDL are made on top of the chips for interconnection. InFO-L is made by adding LSI inside RDL layers.

4.5 Info Ultra-High-Density Interconnect (InFO-UHD)

One of the most important differences between a wafer-level process (WLP) and a panel-level process (PLP) is the interconnection RDL geometry and pitch density. By definition, they are different because they are made by different technologies, tools, and materials and aimed for different markets and applications. The cost of ownership (CoO) of WLP and PLP are different. It is always possible to improve PLP RDL pitch density and sizes by employing similar tools used by WLP but with the penalties of reducing the CoO gaps between the two. Many mature WLP processes are not readily available for PLP due to its fundamental geometry and process philosophy differences. Not all WLP processes can be copied into PLP, even with proper tools.

To further boost InFO-R pitch density, a fine pitch, multi-layer embedded copper (Cu) dual damascene RDL with stacked vias on a 300 mm wafer using a single lithography dielectric film is demonstrated. Each RDL layer is composed of sub-5 μm microvias and a 2 μm/1 μm line/space (L/S) escape routing using a Cu dual damascene process. A liquid photoimageable dielectric film was used for the fabrication of microvias and RDL trenches using a UV lithography tool. To achieve a good total thickness variation (TTV) control within the thin dielectric film, a chemical mechanical polishing (CMP) process was applied to remove the plated Cu overburden and seed metal from the dielectric surface while maintaining a smooth planarization surface to minimize the electrical transmission loss when system chips running at a high frequency [6].

Table 4.1 shows the comparisons among various fine pitch, high-density, Cu trace interconnection technologies, namely BEOL, embedded Cu trace, and semi-additive process (SAP), respectively, when all are processed at wafer level for InFO-UHD. A wafer-level platform rather than panel-level platform was used due to both technology maturity and tools and materials readiness in volume production requiring exceptionally fine pitch and multi-layer RDL interconnections. SAP has long been a mainstream technology to form copper trace of packages in IC packaging applications. However, issues of degrading adhesion strength between copper line and dielectric film and copper trace undercut due to seed metal etching lead to reliability concerns as the L/S scales down to 3 μm/3 μm and beyond [7]. Recently embedded Cu trace technology using UV laser to form fine pitch Cu traces was applied to organic substrates and panel-based fan out packages for low cost [8–11]. Compared to SAP technology, the embedded Cu

Table 4.1 Variant fine pitch RDL technologies [6].

Attributes	Silicon interposer	Embedded Cu trace	SAP Cu trace
Technology	BEOL	polymer with conformal seed metal	PR+ polymer with bottom seed metal
Platform	wafer	wafer	wafer
Maximum routing/ (L/S: μm)	~1300 (0.4/0.4)	~330 (2/1)	~500 (1/1)
Cost	high	low	low
RF insertion loss at high frequency	medium high	low	medium high

trace technology not only provides a better adhesion between Cu trace and dielectrics material but also eliminates undercut and sidewall etching issues, which may deteriorate the transmission line loss at a high frequency. The feasibility of a 2 μm/2 μm of fine embedded Cu trace on a 150 mm Si wafer has been demonstrated [12]. The embedded Cu trench was formed by a lithography process first on a cured Ajinomoto buildup film (ABF)-type polymer layer, while microvias of 10 μm were formed by excimer laser drilling. Fine pitch RDL wiring capacity lies in fine feature sizes in trenches, metal pad, and microvias. In our case, a liquid photoimageable dielectric film was used for both the fabrication of microvias and RDL trenches with a UV lithography tool.

The fabrication process has two steps for two layers of Cu dual damascene RDL. A two-step lithography process was performed to form microvias first. This is followed by Ti/Cu seed metal sputtered on top of microvias and RDL trenches. Next is the copper electroplating step to fill the microvias and RDL trenches. Lastly, to provide a planarization surface, a Cu CMP process was applied to remove the excess Cu overburden and seed metal from the dielectric surface. Up to now, one layer of fine embedded copper dual damascene RDL is completed. With Cu CMP planarization, the Cu pad surface smoothness is ensured to proceed with the second layer of fine pitch, embedded copper dual damascene RDL. There are two major challenges. The first challenge is a lithography capability of fully exposing both metal pad and microvias in the second step of lithography process while maintaining desired line and space features. The other challenge is a 300 mm wafer Cu CMP planarization process to provide a clean and uniform surface for the next RDL stack-up. Throughout the process, proper controls of wafer warpage and uniform dielectric film thickness are critical to the aforementioned two challenges. To increase the RDL routing density, a stacked via structure is demonstrated. Cu CMP planarization on the metal pad enables a strong interface of Cu via-seed metal-Cu via structure and better resistance to thermo-mechanical stress as a result of coefficient of thermal expansion (CTE) mismatch. Also, the Cu CMP planarization can eliminate the potential voids caused by a topology recess and photoresist residuals at metal pads in the conventional SAP process.

Reliability torture tests include multi-reflow (MR) 10X + temperature cycle condition C (TCC) (−65 to 150 °C) 500 cycles, and MR 3X + unbiased highly accelerated temperature and humidity stress test (HAST) (130 °C/85% RH @33.3 psia) for 96 hours. Figure 4.5 shows the SEM image of a tested sample after MR 10X + TCC 500 cycles. Results show no delamination at the Cu-dielectric film interface and no cracks in the bulk Cu microvias and metal pads.

In packaging, resistance-capacitance (RC) delay is an index for digital signal integrity in a high-speed systems. The RC delay is within 7% of the modeling result; the difference is caused by the measurement error and process-induced damages.

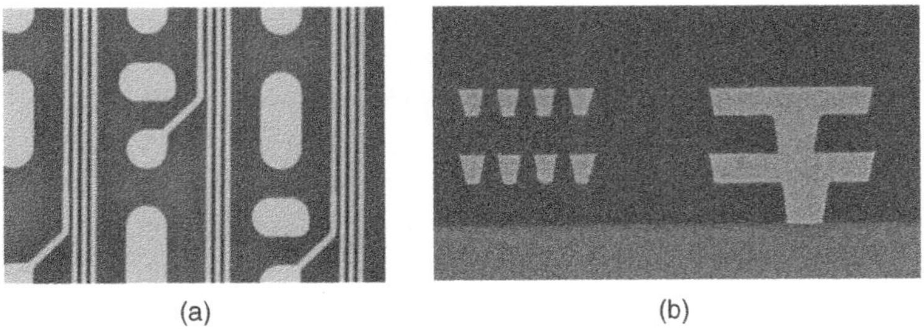

Figure 4.5 (a) Top view layout with Cu trenches, metal pad, and microvias. (b) Cross-section image of a two-layers embedded Cu dual damascene structure after MR 10X + TCC500 [6].

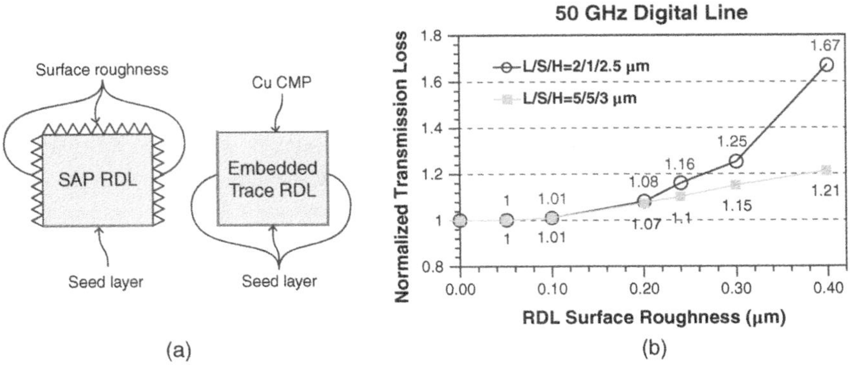

Figure 4.6 (a) Schematic cross-section view of SAP RDL (left) and embedded trace RDL (right). (b) The effect of RDL surface roughness on the normalized transmission loss for variant L/S/H @ 50 GHz for two Cu dual damascene RDL [6].

It is well known that surface roughness of RDL can be critical to the electrical performance of transmission line loss when running at high frequency, particularly for high performance FPGA and RF wireless communication in 5G and mmWave applications [13]. To facilitate the analysis, the concept of normalized transmission loss (NTL) is defined as the ratio of transmission loss of SAP RDL to the transmission loss of embedded Cu trace RDL. SAP RDL is the RDL formed by an SAP process and embedded Cu trace RDL formed by a Cu dual damascene process. Figure 4.6a shows the pictorial cross-section views of SAP RDL and embedded trace RDL. From atomic force microscope (AFM) measurement, the Cu surface roughness is about 0.5 nm for RDL sidewalls with seed metal and a few nm for RDL top surface with Cu CMP treatment.

GHz digital signal, the overall frequency band to the second harmonic frequency band (~60 GHz), and the third harmonic frequency (~90 GHz) are of importance to signal integrity. Per co-planar transmission line routing, the ground-signal-signal-ground (GSSG) routing with a differential pair design enables a better digital signal quality. The dimensions of line/space/height (or thickness) (L/S/H) of serial links need to be optimized to minimize the impact of capacitance effect on impedance matching. To understand the effects of L/S/H and surface roughness of high-speed serial link RDL on the transmission line loss at high frequency, a full EM wave simulation model was conducted. For comparison purposes, two Cu dual damascene RDL with L/S/H of 2/1/2.5 and 5/5/3 μm were selected for the analysis. Figure 4.6b shows the normalized transmission line loss of two Cu dual damascene RDL with variant surface roughness running at 50 GHz. At high frequencies, the larger the surface roughness, the stronger the electrical skin effect. From the simulation, no skin effect impact on the NTL was found, as the surface roughness is less than 0.1 μm.

A fine pitch, multi-layer embedded Cu dual damascene RDL with stacked vias on a 300 mm wafer using a single lithography dielectric film has been successfully demonstrated. Each RDL layer is composed of sub-5 μm microvias and a 2 μm/1 μm line/space escape routing using a Cu dual damascene process. The key aspects of the embedded Cu dual damascene RDL such as structure, fabrication process steps, integration challenges, test vehicle (TV) design and fabrication comparisons, reliability, R, C, and electrical transmission loss at high frequency were examined.

In the future, for logic-to-logic and logic-to-memory integration, novel ultra-high-density InFO (InFO_UHD) technology with submicron RDL is being developed to provide high interconnect density and bandwidth for logic-to-logic systems. The bandwidth density can achieve a record high 10 Tbps/mm at line width and spacing (L/S) of 0.8/0.8 μm and length of 500 μm for a logic-to-logic system using a simplified IO driver [14]. Using the technology in a logic memory system, it is found that the scaling of RDL thickness, L/S, and dielectric thickness can mitigate ring-back problems in the eye diagram of an organic substrate. Given the HBM2 specification, the bandwidth density can achieve more than 0.4 Tbps/mm from dramatically improved signal integrity [14]. The InFO_UHD structure and its process and metrology tool sets are leveraging existing InFO manufacturing infrastructure, instead of those for Cu/low-k BEOL used in semiconductor foundries. Hence, this technology enhances die-to-die communication performance for multi-die configuration and exhibits cost advantages for competitiveness in the commercial market. A plain view micrograph and the birds-eye view of the InFO_UHD structure with L/S = 0.8/0.8 μm of RDL are shown in Figure 4.7a.

For logic-to-logic interfaces, the total bandwidth density is one criterion to determine the system performance. Two important factors affecting bandwidth density are maximum data rate and line density. In I/O interface design, the maximum data rate is affected by the line width/spacing and length, while the line density

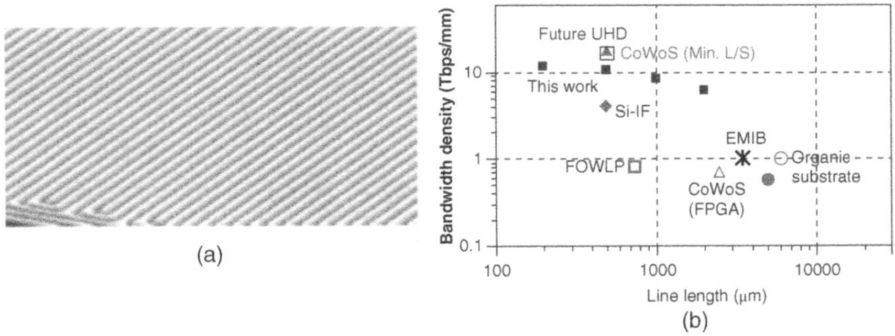

Figure 4.7 (a) RDL line-width of 0.8 μm with RDL line-spacing of 0.8 μm is demonstrated. (b) Bandwidth density of InFO_UHD versus various interconnect technologies [14].

is determined by the feature size of line width/spacing of the interconnect process. In practice, narrower lines may result in lower maximum data rates, but the bandwidth density could still be higher since the line density is much higher. As shown in Figure 4.7b, in the InFO_UHD technology, the bandwidth density can achieve a record high 10 Tbps/mm, assuming two critical routing RDL layers. However, in other technologies, such as CoWoS (FPGA), the bandwidth densities range from 1 to 4 Tbps/mm. Hence, the InFO_UHD is the technology to provide the highest bandwidth density. Furthermore, comparing InFO_UHD to CoWoS with minimum L/S using the same simplified IO driver, the bandwidth density of the InFO_UHD is lower than CoWoS with minimum L/S. Power efficiency, in terms of energy per bit, is another important metric for HPC system performance. A 0.06 pJ/bit of power efficiency was reported in a silicon interconnect fabric (Si-IF) technology [15], where 2.5x line width and 500 μm line length are used, for logic-to-logic interconnect.

In the InFO_UHD technology, the power efficiency is simulated using simplified transmitter and receiver models with pseudorandom binary sequence (PRBS) input. The power efficiency is 0.061 pJ/bit at 1.0x line width and 500 μm line length condition, based on 0.1 pF capacitance specified for both the transmitter and the receiver. Even though the power efficiency is similar in the two interconnect technologies, the line width of InFO_UHD is two fifths of that of the Si-IF technology. The InFO_UHD fine line does not consume more power. For logic-to-HBM2 memory interface, InFO_UHD interconnect of 2.5x line width for typical RDL length exhibits 0.62 pJ/bit of power efficiency, based on the memory driver model with 2 Gbps data rate. For comparison, power efficiency of a silicon interposer interconnect under typical dimensions is studied. The power efficiency is 0.83 pJ/bit, as shown in Figure 4.8. The power efficiency of the InFO_UHD technology is 25%

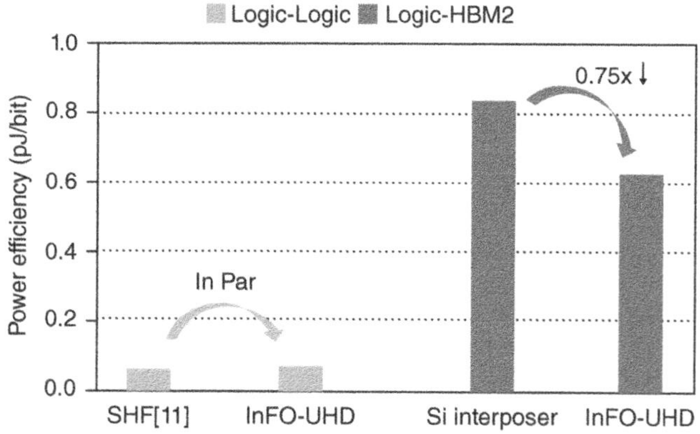

Figure 4.8 Power efficiency performance of logic-logic and logic-memory interfaces. Note: Si-IF silicon interconnect fabric [14].

better than that of the silicon interposer technology in logic-to-HBM2 memory system due to lower capacitance in the interconnect [16].

The criterion to evaluate the metallization reliability is to study chip-package-interaction between the Cu BEOL and the package. Both the wafer-level reliability tests for Cu/low-K, such as electromigration (EM), stress migration (SM), time dependent dielectric breakdown (TDDB), and breakdown voltage (V_{bd}), and component and package level reliability, namely MR, temperature cycle (TC), and thermal shock (TS), have been used for wafer-level RDL reliability and have all been confirmed that they have passed [16].

4.6 Multi-Stack System Integration (MUST) and Must-in-Must (MiM)

An advanced 3D multi-stack system integration (MUST) technology, 3D MUST-in-MUST (3D-MiM) fan out package, has been developed as next generation wafer-level fan-out package technology. 3D-MiM technology utilizes a simplified architecture that eliminates BGAs between packages for system-level performance and power and form-factor (PPA) optimization. This technology also makes use of a modularized approach for both design and integration flow to improve design flexibility and integration efficiency. Known good pre-stacked memory cube and/or logic-to-memory cubes are fabricated by leveraging the established InFO technology platform in tools, materials, design rules, and processes for cycle time and cost saving. Two 3D-MiM fan-out examples are discussed in this session. The first 3D-MiM package integrates an SoC with 16 memory chips in a $15 \times 15\ mm^2$ footprint with 0.5 mm package height (final BGA

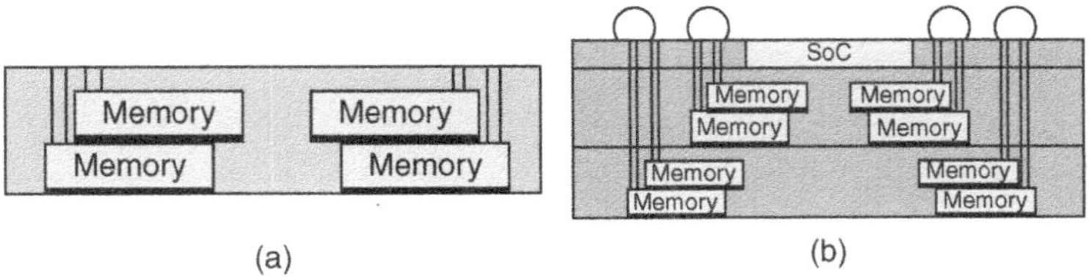

Figure 4.9 (a) A MUST memory module (building blocks). An integration building block, the modularized memory unit integrates multiple vertically stacked memory chips by InFO wafer-level system integration, with memory I/O pads at one side of the chip. (b) 3D-MiM fan-out package for edge computing [17].

included) for mobile application. The other 3D-MiM package integrates 8 SoCs with 32 memory chips in a $43 \times 28\,mm^2$ footprint to mimic a system integration of multiple logic cores and multiple memory chips for HPC applications [17].

Serving as an integration building block, the modularized memory unit integrates multiple vertically stacked memory chips by InFO WLSI, with memory I/O pads at one side of the chip. Embedded memory chips are shown in Figure 4.9a. The memory module is applicable to generic memory commodities, including LPDDR4/5, SRAM, wide I/O in mobile, DDR in NB (notebook)/PC, and servers. Such memory module has great flexibility in horizontal footprint and extensibility in vertical stacking to meet design needs. Structure-wise, the memory module can be in 2-tiers, 3-tiers, or more tiers depending on the system needs. In each tier, there could be single or multiple memory chips. Each memory module can be tested and embedded as a known good module.

The 3D-MiM package was realized by InFO technology. First, integrate 16 memory chips in the first two fan-out tiers and then cascade a SoC in the third fan-out tier, as shown in Figure 4.9b. This 3DMiM is designed as an alternative solution to the current FC-PoP and fan-out PoP, for use in mobile or computing devices requiring a thinner profile, with high memory capacity and memory bandwidth. Similarly, the second, super large, 3D-MiM fan-out package was realized by integrating 32 memory chips in the first-tier fan-out and then integrating 8 SoCs in the second-tier fan-out, as shown in Figure 4.10. This 3D-MiM fan-out is designed as a low cost alternative to the current 2.5DIC, for use in 5G/AI-driven HPC and server/data center, which require high computing performance, memory bandwidth, as well as low power consumption and latency in applications such as machine learning and AI training. On mobile application, in comparison to the FC-PoP, the 3D-MiM fan-out offers a thinner package profile (~0.5 mm z-height), higher data bandwidth (2X–4X), with lower latency (0.2X) and thermal resistance to meet demands on future 5G/AI-driven edge computing. On HPC, in comparison to 3DIC HBM, the 3D-MiM fan-out offers a lower cost alternative with new memory to memory, and SoC to memory integration architecture with

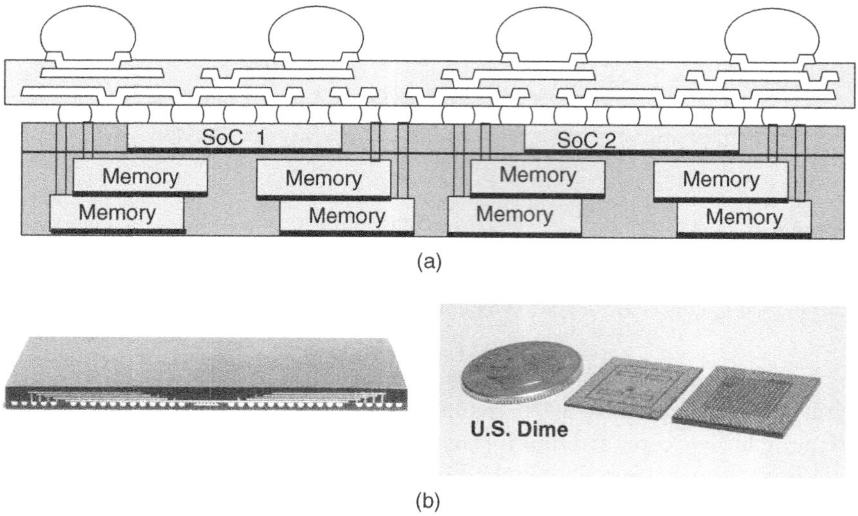

Figure 4.10 (a) 3D-MiM fan-out package for 5G/AI-driven HPC and server applications. 3D-MiM fan-out for mobile application. (b) Profile comparisons among US dime coin, 3D-MiM fan-out, and FO-PoP [17]

promising electrical and thermal performances. On manufacturing considerations, 3D-MiM fan-out technology leverages the well-established infrastructure of WLSI in capacity and materials, tools, processes, design rules for yield and competitive cost. 3D-MiM fan-out is part of the WLSI technology family in future 5G/AI-driven applications [17].

4.7 InFO on Substate (InFO-oS) and InFO Local Silicon Interconnect (InFO-L)

Placing InFO on an organic substrate (InFO on substrate or InFO-oS) combines the benefits of fan-out simplicity with traditional FCBGA for large system packages, as shown in Figure 4.11. InFO_oS offers fine pitch RDL to integrate logic and memory chips on substrate with Cu pillar bumps. HPC systems larger than a reticle size can be achieved thanks to a combination of high-density RDL plus low cost, lower end organic substrates. Reticle sizes from 1x to 1.7x are production ready and larger than 2.5x are on their way to production.

In additional to InFO-oS, when local ultra-high-density interconnection is needed, an LSI can be added into InFO, and thus InFO-L is formed. InFO-L can

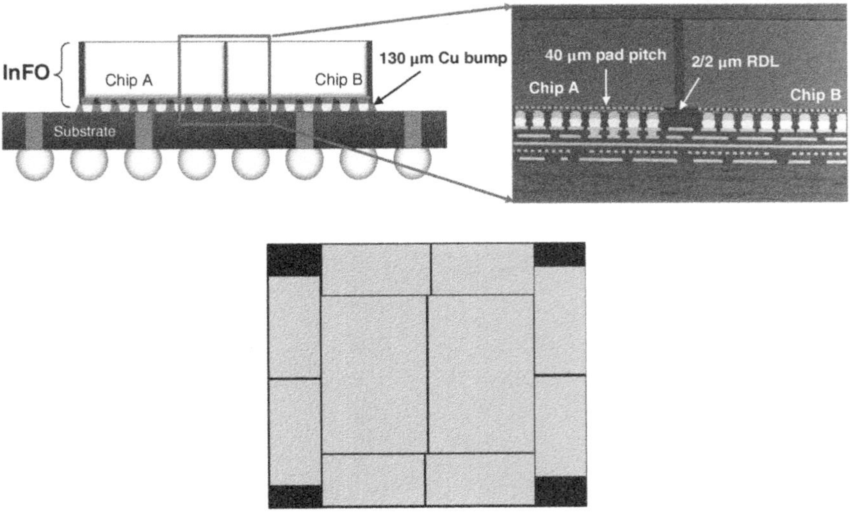

Figure 4.11 InFO-oS combined fine pitch RDL with large FCBGA substrate to enable high performance system with size larger than a reticle size.

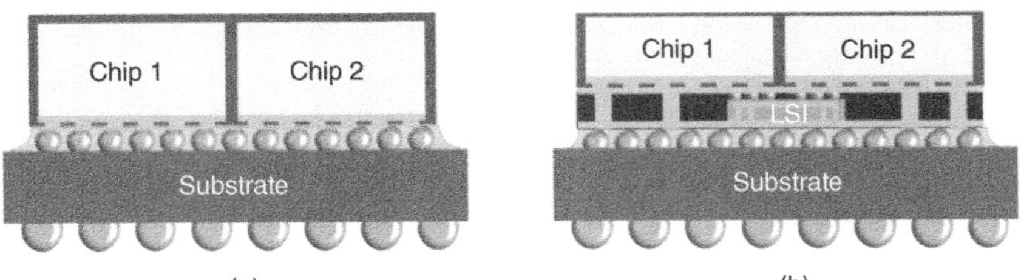

Figure 4.12 (a) InFO-oS and (b) InFO-L offer ultra large computing system integration with ultra-high density local interconnection available as an option.

be combined with InFO-oS to form ultra large HPC system with ultra-high local interconnect density, as shown in Figure 4.12.

InFO-oS is a proven technology and has evolved over the years [8, 17, 18]. With data centers data switching capacity increasing multi-fold, for example from 6.4 Tb/sec to 25.6 Tb/sec, the InFO-oS platform can be adapted to high-speed switch applications and extends the InFO-oS envelop to 1.7x reticle size. Thereby, the data center needs can be fulfilled, as proposed in Figure 4.13.

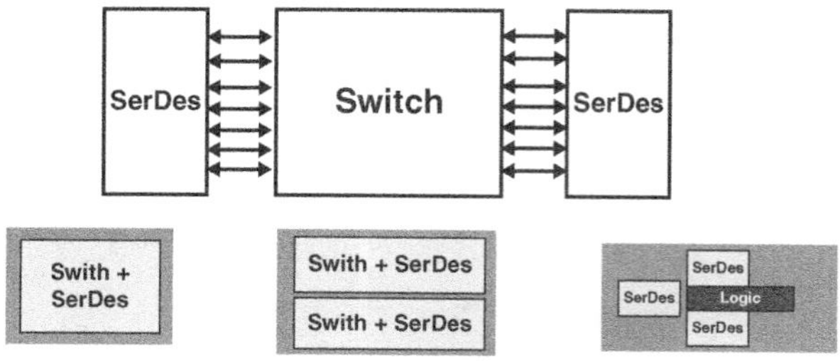

	2016	2018	2019–2021
	Flip Chip	InFO_OS 1.0	InFO_oS 2.0
Si	N16	N16	N7/N5
InFO size (reticle)	1×	1–1.5×	~1.7×
IO Speed (Gb/sec)	10–28 (25G)	28 ~ 56 (50G)	56–112 (100G)
IO number (lane)	256 (128 × 2)	512 (256 × 2)	512–1024 (512 × 2)
Data rate (Tb/sec)	3.2	6.4–12.8	25.6
Min RDL pitch (μm)	20	4	3

Figure 4.13 A proposal using advanced node silicon technology integrated with InFO-oS to fulfill the network switching needs.

4.8 InFO with Memory on Substrate (InFO-MS)

One of the biggest challenges in design and building HPC systems is to integrate high band width memories (HBM) in large capacity and with in-package proximity. This is typically done using CoWoS with multiple HBMs sitting on a silicon interposer next to the SoC. However, as InFO RDL line width and pitch improves, it is becoming feasible and attractive to build the same logic-to-memory integration using the InFO platform. InFO with memory on substrate, InFO-MS, has been developed and demonstrated in such system. Figure 4.14 compares InFO-MS and CoWoS-R for the same system. It is worthwhile to note that there are unique challenges in InFO-MS since chips, including that HBMs are embedded into the mold compound before RDL. There are stringent geometry tolerance and manufacturing challenges in dealing with third-party HBM from different suppliers to incorporate into the InFO structure. All technical challenges are manageable but not without turnaround time and cost penalties. In production planning, such penalties need to be managed and taken into consideration carefully.

			CoWoS-R	InFO_MS (Chip first)
Illustration			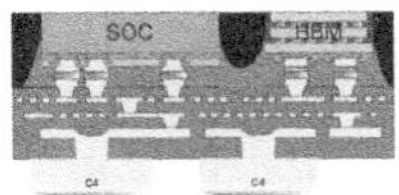	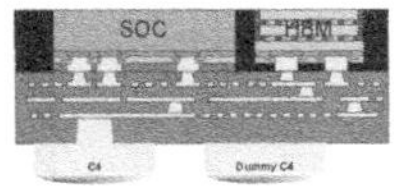
Assumed power (SoC/memory)			SoC 180 W (max.); HBM2 × 2, 16 Watts	
Performance	PI	PDN impedance	1.0 ×	0.99 ×
	SI	RC delay	1.0 ×	0.99 ×
	Thermal (SOC 180 W)		SOC 108.4 °C (spec. 115 C), Memory 88.3 °C(spec. 95 C)	
	Reliability	μBump fatigue	higher (solder IMC in μBump)	lower
		ELK de-lam	higher (@ μBump reflow)	lower
		EM	poor (1.0 ×)	better 10.0 × (pure Cu via)
		warpage	higher wafer & PKG warpage (UF)	lower wafer & PKG warpage
	Design rule scalibility		≤ 36 μm μBump pitch: solder joint quality	≤ 36 μm μBump pitch is doable
Area/HI	InFO size		larger (reserved for underfill fillet)	smaller
	HBM height limit		if HBM Hi >775 μm, needed bare Si bonded on SoC WF before μBump	
Intrinsic cost (USD/wafer)			1.0 ×	~0.8 × (under checking)

Figure 4.14 A benchmarking between InFO-MS and CoWoS-R for logic-HBM integration. The main differences is chip-last CoWoS-R requires solder micro-bumps and underfill, while InFO-MS does not have such joints.

4.9 InFO 3D Multi-Silicon (InFO-3DMS) and CoWoS-L

As mentioned before, it could be useful to combine chip-first and chip-last into one production flow. Likewise, it could also be handy to combine both RDL and multiple silicon chips, including active chips, passive chips, and local silicon bridge chips. This is basically a CoWoS-L or InFO 3DMS. A schematic of a large system integration using 3D multi-silicon (3DMS) is shown in Figure 4.15, where

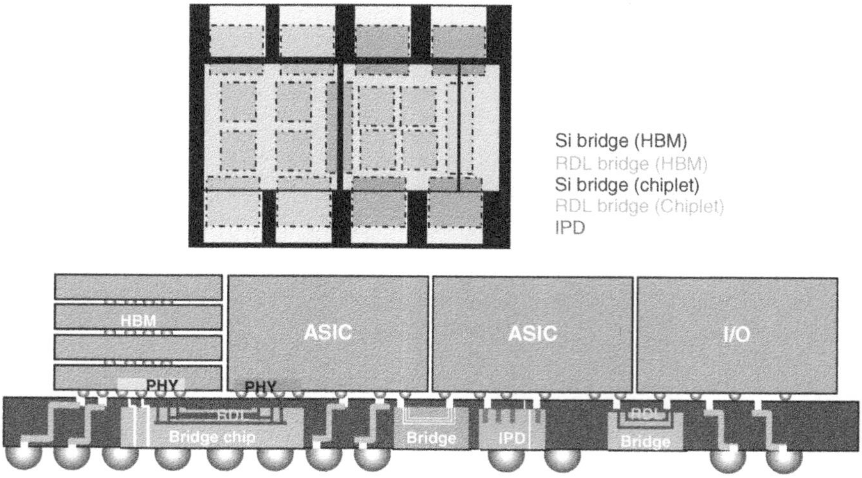

Figure 4.15 In an RDL area, 70 × 75 mm^2, 3 chiplets, 8 HBM2e, RDL bridges for HBM and high speed I/O, Si bridge for chiplet, and IPD + TSV are combined into one computing platform.

integrated passive devices (IPD), silicon bridges, and RDL bridges are made separately and then integrated using InFO. This integrated InFO now acts as an interposer. Multiple chiplets and HBM are assembled onto InFO, using a CoWoS process. Throughout the processes, known good dies, known good InFO, and known good chiplets are used with the higher value components being integrated in the final steps on tested parts, to minimize yield lost. Along the processes, system performance is optimized by carefully choosing local interconnection schemes, materials, and density.

Figure 4.16 shows a qualitative comparison among the best-known integration platforms in the market. The process flows are also being carefully thought through based on existing and proven process modules and steps.

4.10 InFO System on Wafer (InFO_SoW)

As computing systems have grown enormously and continuously in recent years, InFO technology provides the possibility of using the whole 300 mm wafer area as one system. The industry's first wafer-scale system integration package with InFO technology is called InFO_SoW. It has 15% power saving of the interconnects with length of 30 mm, thanks to lower surface roughness of wafer-level RDL. Thermal management of such high power in a compact system has been validated through a scalable proof-of-concept (POC) thermal solution. The POC thermal solution is

	CoWoS	RDL interposer	InFO_3DMS
Connection	Si BEoL	RDL	Hybrid of Si BEoL and RDL
SI chiplet IO or routing density	Good	Moderate	Good (Si BEoL)
chiplet freq	moderate	Good	Good (RDL)
HBM/ Serdes	moderate	Good	Good (RDL)
high speed external signal	moderate All SGN thru TSV	Good	Good TSV for PWR/GND. SGN thru TIV
PI DTC/IPD	interposer iCap	NA	Discrete IPD

Figure 4.16 A benchmarking among CoWoS, RDL interposer, and a mix-and-match InFO-3DMS platform.

demonstrated with the capability of dissipating 7000 W in a 2-by-5 array heater test structure, whereby the maximal temperature of the heater is kept below 90 °C. In addition, InFO_SoW structural robustness has been validated through both InFO wafer-level torture and system-level reliability tests. Despite its super large package size, a thermomechanical chip-package-interaction (CPI) study revealed that InFO_SoW has relatively low risk when compared to traditional flip-chip package with advanced Si-node [19].

InFO_SoW technology involves a wafer-scale system that integrates InFO and power and thermal modules, as shown in Figure 4.17. Connectors and power modules are solder-joined to the InFO wafer, which is then followed by the assembly of thermal module. An electrical performance comparison between a conventional system (i.e. flip-chip multi-chip module) and InFO_SoW technology through electrical simulation is presented in Figure 4.17 The line width and space (L/S) for the flip-chip substrate and InFO_SoW used in this study are 10/10 and 5/5 μm, respectively. Based on this configuration, InFO_SoW provides two times the routing density, which translates to two times higher line density and thus two

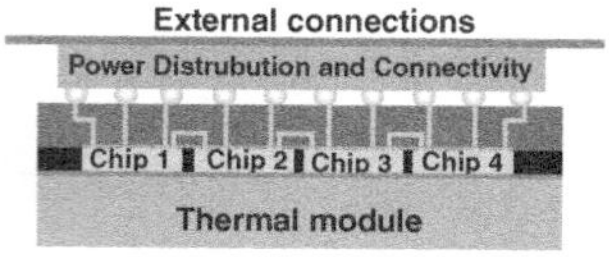

	Flip-chip MCM	InFO_SoW
Line width/space (μm)	10/10	5/5
Line density	1×	2×
Bandwidth density	1×	2×
PDN impedance	1×	0.03×

Figure 4.17 Schematic cross section of InFO_SoW structure along with a traditional flip chip MCM and performance comparison between flip-chip MCM and InFO_SoW [19].

times bandwidth density given a similar data rate. Insertion loss is a key aspect in signal integrity for high-speed differential lines, typically with more than 10 mm in length on organic substrate and even longer on PCB. It is well known that insertion loss worsens as frequency increases. This is mainly due to the skin effect and roughness effect. Longer lines will accumulate the impact of roughness. A differential pair of 100 Ω impedance is simulated using InFO RDL with a line width of 15 μm. In general, the roughness (R_q) of trace in substrate is higher than that of RDL line in InFO_SoW.

Figure 4.18 shows the line loss of substrate and InFO_SoW with respect to various lengths ranging from 5 to 30 mm at 28 GHz. It is found that InFO_SoW RDL outperformed a line loss of 0.4 dB for 20 mm length, which is equivalent to 10% power saving of the interconnects. Higher impact on longer RDL is also observed. RDL with a length of 30 mm corresponded to a line loss difference of 0.7 dB due to the roughness effects, and this translated to a power saving of 15%.

Compared to the conventional motherboard-based HPC system configuration, the InFO_SoW must dissipate much higher heat from a volumetric perspective. With the same total thermal design power (TDP), the former will distribute the compute units into sparse locations horizontally in a rack as well as vertically into different racks, while the latter condenses that onto a wafer-scale real estate. For that, a novel cooling structure based on microchannel liquid cooling solution was used. The 2-row proof of concept (POC) thermal solution is a copper-based

Figure 4.18 Line-length dependent roughness effects of line loss at 28 GHz [19].

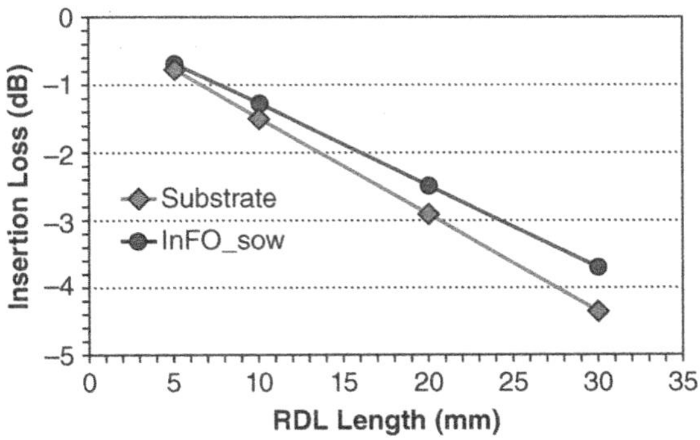

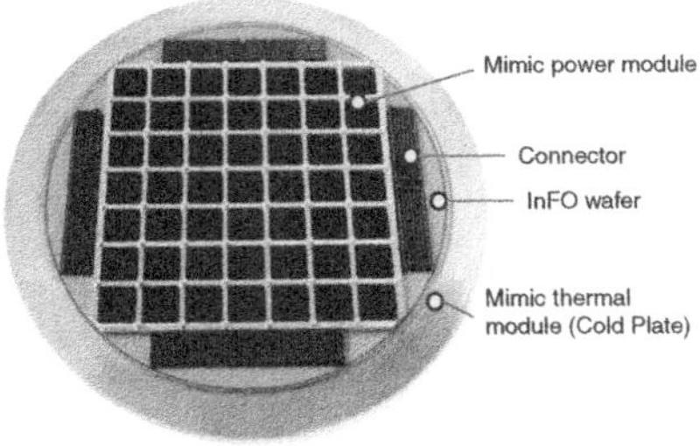

(a)

Reliability Test	Test item	Test mode	Results
Wafer-level Quick Torture	MR3X (Reflow 260 °C) + TCC 200 cycles (–65 to 150 °C)	Offline	Passed
	MR3X (Reflow 260 °C) + uHAST 96hr (130 °C/85% RH)		Passed
System-level	HTS 1000hr (125 °C)	Offline	Passed
	Vibration (Level: 0.001G)		Passed
	THB 1000hr (85 °C/85% RH)		Passed
	Temperature cycling-J 1000 cycles(0 to 100 °C)	In-situ	Passed
	Mechanical Shock (Condition C)		

(b)

Figure 4.19 (a) InFO_SoW system assembly demonstration. (b) Reliability results of InFO-SoW test wafer [19].

microchannel cold plate. For better utilization of the cooling capacity of the thermal solution, the contact resistance between cold plate and dummy heater must be minimized. A thermal grease has been spread onto the exposed surfaces of dummy heating units before the heater was attached to the POC thermal solution to reduce the thermal contact resistance. With TDP 7000 W or power density of 1.2 W/mm^2 imposed onto the 2-by-5 array heater, the maximum temperature of the dummy heater can be kept below 90 °C using water with inlet temperature of 16 °C and minimal flowrate of 4 l per minute (LPM). Figure 4.19a shows such a system built and has been demonstrated to pass engineering qualifications, as shown in Figure 4.19b.

In comparison to the conventional system (i.e. flip-chip MCM), fine pitch (L/S 5/5 μm), InFO RDL provides high density die-to-die interconnections, which leads to two times higher bandwidth density. With the absence of substrate and PCB, power distribution network (PDN) impedance of InFO_SoW is 97% lower than that of conventional systems. Furthermore, electrical characterization of the InFO_SoW demonstrated good process uniformity across the super large package.

4.11 System on Integrated Substrate (SoIS)

Larger size organic substrate, for instance, > 80 mm × 80 mm, has become necessary in recent years for HPC applications including networking. However, the larger the package size, the worse the manufacturing yield, cost, and delivery cycle time for conventional high layer count organic substrates. In the meantime, electrical performance becomes increasingly challenging. For example, higher data rate (112 Gbps) of SerDes interface will lead to worse insertion loss and crosstalk at Nyquist frequency in package interconnects. An innovative SoIS technology is proposed to mitigate the performance and cost dilemma. SoIS technology leverages wafer processes and new materials. This innovative integrated substrate presents significantly higher yield than conventional substrate solutions on a larger than 8000 mm^2 substrate size. The results show that the insertion loss is 25% lower than that of the most updated GL102 organic substrate at 28 GHz for 112 Gbps SerDes application. Furthermore, by leveraging wafer fab processes, SoIS could also provide powerful yet flexible combinations in interconnects and dielectric layers with more aggressive design rule than conventional organic substrate. Especially, for high bandwidth routing density applications, SoIS can enhance the routability by two to five times compared to conventional organic substrates, thus reducing substrate layer counts and also keeping the same impedance matching performance without adding extra cost. Figure 4.20 shows a comparison between organic substrate and SoIS [20].

Using POC TVs, the mechanical/electrical TV has passed package-level reliability tests including MSL4 and temp cycle condition G (−40 to 125 °C) after 1500 cycles (TCG), MSL4 and uHAST after 360 hours, and high temperature storage (HTS) after 1500 hours, as shown in Figure 4.21. Microstructure analysis on cracks, voids, and defects after reliability torture tests pass quality and reliability criteria.

4.12 Immersion Memory Compute (ImMC)

As mentioned in previous sections, SoIC is a front-end CoW and/or WoW chip stacking technology. Chips made by different technologies are tested KGD before SoIC stacking and integration. An SoIC integrated chip behaves like an SoC chip in terms of electrical performance and mechanical integrity and can be subsequently assembled using conventional packages or back-end WLSI platforms, such as chip on wafer on substrate (CoWoS) or InFO. Figure 4.22 shows a holistic 3D integration with front-end 3D (FE-3D) and back-end 3D (BE-3D) for "more Moore" and "more than Moore" system level scaling [21].

Structure			Organic SBT	SoIS

Design rule	Line width / space		1.0 ×	**0.3 ×**
	Via size		1.0 ×	**0.4 ×**
Electrical perfor-mance	SI	Insertion loss (dB @112Gb/s)	1.0 × (real measurement data)	**0.75 ×** (real measurement data)
		Insertion loss (dB @200Gb/s)	1.0 × (real measurement data)	**0.69 ×** (real measurement data)
		X-talk	1.0 × (simulation)	**~0.3 ×** (simulation)
	PI	Suppressed freq. range, MHz (with IPD)	1.0 × (simulation)	**4~7 ×** (simulation)
		IR Drop	1.0 × (simulation)	**0.9 ×** (simulation)
Yield projection (91mmSQ body size)			Challenge (~20~30%?)	Effective yield proven >90%
Acc. Bump strain		CL TCG	1.00 ×	**0.84 ×**
PKG warpage		@ RT	1.0 ×	**0.91 ×**

Figure 4.20 A power/performance/area/cost comparison between an 18-layer organic substrate and an SoIS with only 14 layers in the substrate [20].

Items		Pre-qual	Qual
Component scheme		InFO_oS 3×RDL with min. L/S 2/2 μm	
C4 pitch/size		130 μm/70 × 90 μm (Oval bump)	
SoIS scheme		6 × RDL+ 8× organic RDL	
SoIS SBT	Min. L/S	11/11 μm	
	SRO	70 μm	
PKG	Ring	12 × 18 × 2.5 mm	
Pre-con	MSL4a	Pass (0F/10)	Pass (0F/46)
MSL4a + TC	TCG 850X	Pass (0F/5) TCG2000 pass (0F/4)	Pass (0F/24)
MSL4a + uHAST	110C; 85%RH; 264 hrs	Pass (0F/5) uHAST 360 pass (0F/4)	Pass (0F/22)
HTS	150C; 1000 hrs	Pass (0F/5) HTS1500 pass (0F/4)	Pass (0F/23)

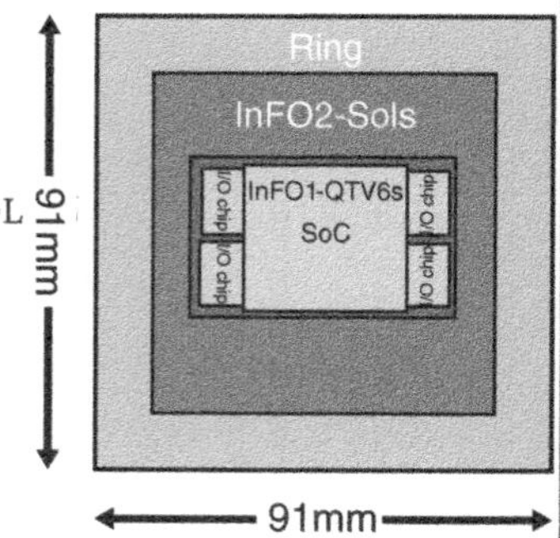

Figure 4.21 Test vehicle PoC qualification on large size SoIS has passed the qualifications [20].

A new ultra-high-density logic-to-memory stacking structure called immersion memory compute (ImMC) has been proposed based on FE-3D + BE-3D integration, as seen in Figure 4.23.

The technology provides multiple compute and memory chips to interconnect each other to gain computing power and memory bandwidth. The interconnect parasitics, bandwidth density, and power efficiency are analyzed using N7 light I/O transceiver. The ImMC compared to 3DIC with bridge and with shared die, respectively, using micro-bump and TSV, are studied. Figure 4.24 shows 10 different interconnect structures for comparison. Total interconnect parasitics,

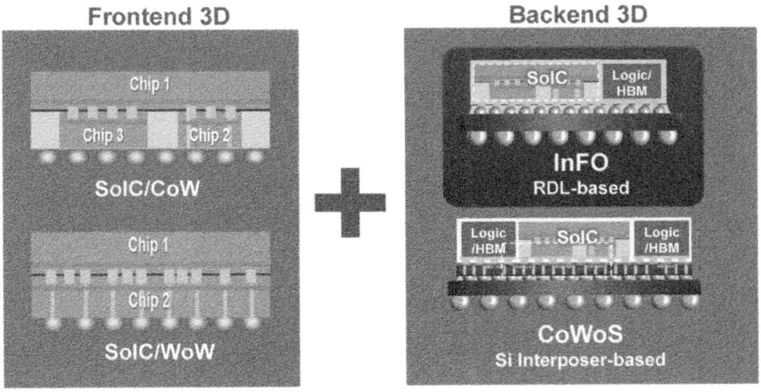

Figure 4.22 Front-end 3D SoIC integrated with back-end 3D CoWoS/InFO as a new heterogeneous integration platform for more Moore and more than Moore system scaling with best optimized system PPAC [2].

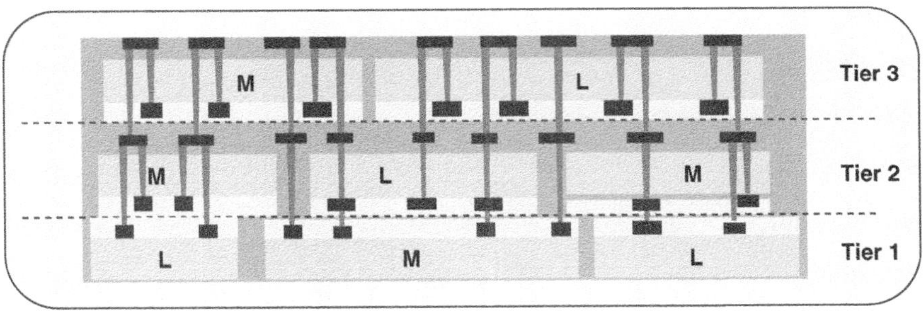

Figure 4.23 Immersion memory compute (ImMC) using FE-3D + BE-3D for ultra-high-density logic-to-memory integration. Depending on the design and system needs, the ratio of memory chips to logic chips can vary with both logic and memory chip size. Both through silicon via (TSV) and through mold via (TMV) are available and both face-to-face and face-to-back stacking are also possible [2].

system bandwidth (BW) density, and power efficiency with N7 transceiver for the processor-to-processor and processor-to-memory are analyzed.

The BW density and power efficiency results are shown in Figure 4.25. The C-II ImMC has the highest maximum data rate, 14 times that of A-I 3DIC, shown in Figure 4.25a. By multiplying the bump density, the BW density of each structure is obtained in Figure 4.25b. The C-II ImMC has 224 times higher BW density than A-I 3DIC. Moreover, the power efficiency in energy/bit for each structure at the same data rate is simulated. The ImMC has the lowest energy/bit, as shown in Figure 4.25c. Besides, the shortest interconnect simplifies the transceiver design and results in the smallest transceiver power and size. The transceiver power and

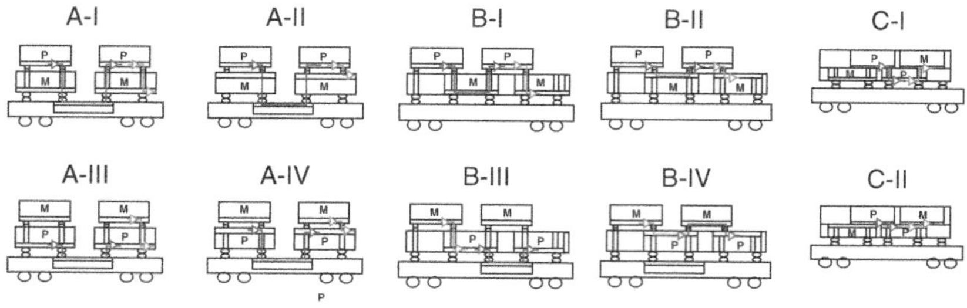

Figure 4.24 Interconnect of 10 package structures: (A-I) 3DIC with bridge (1 mm BEOL) F2B and (A-II) 3DIC with bridge F2F with processor on top, (A-III) 3DIC with bridge F2B and (A-IV) 3DIC with bridge F2F with processor at bottom, (B-I) 3DIC with shared die (0.8 mm BEOL) F2B and (B-II) 3DIC with shared die F2F with processor on top, (B-III) 3DIC with shared die F2B and (B-IV) 3DIC with shared die F2F with processor at bottom, and (C-I) ImMC F2B, (C-II) ImMC F2F [21]).

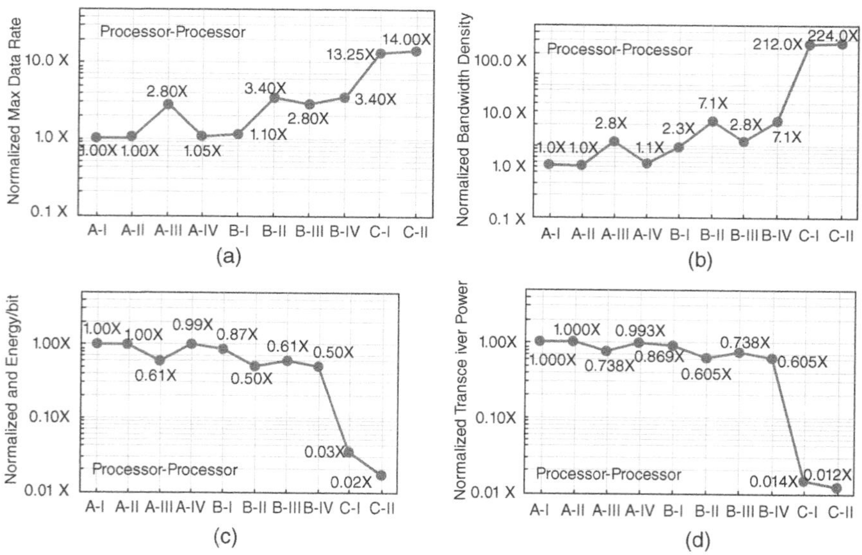

Figure 4.25 Processor-processor system (a) maximum data rate, (b) bandwidth density, (c) energy/bit, and (d) transceiver power consumption [21].

size for the ImMC are 1% of those for A-I 3DIC, as shown in Figure 4.25d. From the SoIC technology, the C-I and C-II ImMC have the lowest parasitics among the structures. The maximum data rate, BW density, power efficiency, and transceiver power for processor-memory systems are shown in Figure 4.26. The ImMCs are superior to the other structures. Finally, current-resistance (IR) drop for processors at different structures are analyzed. The structures with a face-down processor

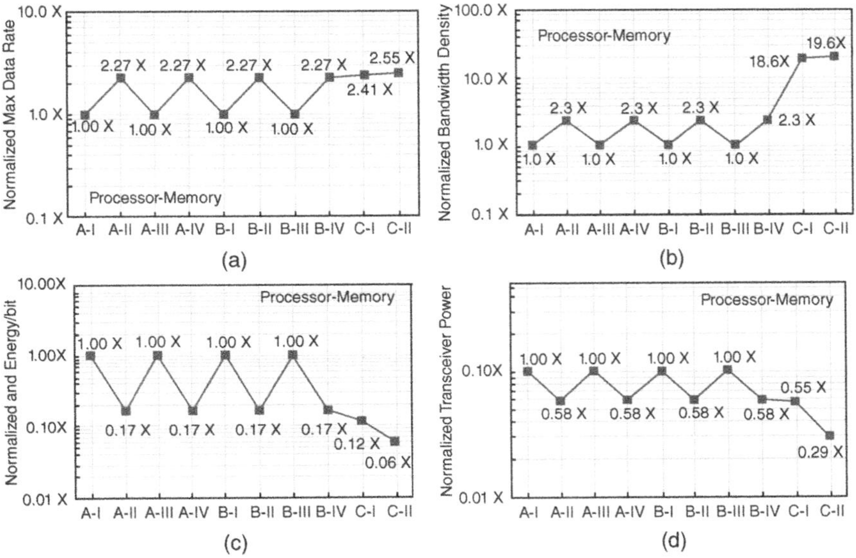

Figure 4.26 Processor-memory system (a) maximum data rate, (b) bandwidth density, (c) energy/bit and (d) transceiver power consumption.

at the bottom have lower IR drop. For face-up processor structures, the ImMC is better than other structures due to lower die thickness. To further improve the system performance, a new SoIC bond is developed. It achieves 20×, 318×, and 0.04× better than micro-bump in speed, BW density, and energy/bit, respectively.

In summary, the ImMC is 16×, 14×, and 224× better than the 3DIC with bridge in bump density, data rate, and bandwidth density, respectively. The transceiver power and size for the ImMC are only 1% of those for the 3DIC [21].

4.13 Summary

Various wafer level 3DFabric platforms are introduced. These are, by no mean, the only options available in the market. As our computing needs are evolving and changing, so does the way we integrate and package them in the best conceivable way. Figure 4.27 summarizes the possible logic-to-memory integration at different scales and possible timeline. Logic-to-memory integration is used as an example, as it is by far the highest pitch density integration in the high volume production market. As Moore's law transistor scaling becomes more challenging, logic-to-logic deep partitioning and integration may require even higher pitch density in the near future. The industry is now focusing on memory partitioning and

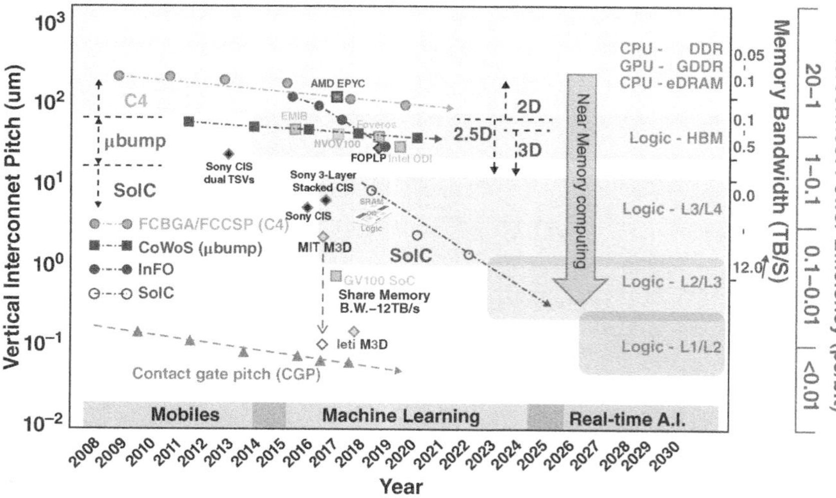

Figure 4.27 3DFabric system scaling platform. Logic-memory integration is used as an example as this is, by far, the highest pitch density system in the market. Logic-logic integration could require more pitch density in the near future.

integration for system scaling performance/energy/cost improvement. The trend may go on until logic-to-memory scaling benefits diminish. What is next for scaling becomes an important question. Fundamental computing architecture, logic IP deep partitioning, non–von Neumann computing, in-memory compute, or even quantum computing are among the visible candidates.

Revolutionary changes have been taking place in microelectronics computing systems over the last few years and are ongoing today. Transistor scaling (Moore's law 1.0) and system scaling (Moore's law 2.0) together are the technology drivers. 3DFabric WLSI enables multi-dimensional and multi-scalar scaling, whether it is for performance scaling up, transistor density scaling up, system energy scaling down, package form factor scaling down, or cost per computing scaling down. 3DFabric will be there to complement Moore's law and ensure the microelectronics computing systems will continue to improve our life and humanity for the foreseeable future.

References

1 Moore, G.E. (1965). Cramming more components onto integrated circuits. *Electronics*, McGraw Hill, Inc. 38 (8).

2 Yu, D. (2019). *IEEE IEDM 2019*, Panelist presentation, December.

3 Chen, Y.H., Yang, C.A., Kuo, C.C.et al. (2020). Ultra high density SoIC with sub-micron bond pitch. *IEEE 70th Electronic Components and Technology Conference*, pp. 576–581.

4 Liu, M. (2020). SEMICON Taiwan, keynote, Taipei, September.

5 TSMC (2021). 3DFabric. https://3dfabric.tsmc.com/chinese/dedicatedFoundry/technology/3DFabric.htm (accessed January 2021).

6 Yu, C.H., Yen, L.J., Hsieh, C.Y. et al. (2018). High performance, high density RDL for advanced packaging. *IEEE 68th Electronic Components and Technology Conference*, pp. 587–594.

7 Liu, F., Kubo, A., Nair, C. et al. (2016). Next generation panel-scale RDL with ultra small photo vias and ultra-fine embedded trenches for low cost 2.5D interposers and high density fan-out WLPs. *Proceedings of the Electronic Components and Technology Conference (ECTC)*, Las Vegas, USA, May 31–June 3, 2016, pp. 1515–1521.

8 Chen, Y.-H., Cheng, S.-L., Hu, D.-C., and Tseng, T.-J. (2015). L/S ≤ 5/5um line embedded organic substrate manufacturing for 2.1D/2.5D SiP application. *IMAPS*, November 2015.

9 Ishida, M. (2014). APX (Advanced Package X) – advanced organic technology for 2.5D interposer. *CPMT Seminar, Latest Advances in Organic Interposers*, Lake Buena, Vista, FL, USA, May 27–30, 2014.

10 Liu, F., Nair, C., Sundaram, V., and Tummala, R.R. (2015). Advances in embedded traces for 1.5um RDL on 2.5D glass interposers. *Proceedings of the Electronic Components and Technology Conference (ECTC)*, San Diego, USA, May 26–29, 2015, pp. 1736–1741.

11 Suzuki, Y., Furuya, R., Sundaram, V., and Tummala, R.R. (2015). Demonstration of 10-um microvias in thin dry-film polymer dielectrics for high-density interposers. *IEEE Transactions on Components, Packaging and Manufacturing Technology* 5 (2): 194–200.

12 Hu, D.-C., Yeh, W.-L., Chen, Y.-H., and Tain, R. (2016). 2/2um embedded fine line technology for organic interposer applications. *Proceedings of the Electronic Components and Technology Conference (ECTC)*, Las Vegas, USA, May 31–June 3, 2016, pp. 147–152.

13 Nair, C., Lu, H., Panayappan, K. et al. (2016). Effect of ultra-fine pitch RDL process variations on the electrical performance of 2.5D glass interposers up to 110 GHz. *Proceedings of the Electronic Components and Technology Conference (ECTC)*, Las Vegas, USA, May 31–June 3, 2016, pp. 2408–2413.

14 Wang, C.-T., Hsieh, J.-S., Chang, V.C.Y. et al. (2019). Signal integrity of sub-micron InFO heterogeneous integration for high performance computing applications. *IEEE 69th Electronic Components and Technology Conference (ECTC)*, pp. 688–694.

15 Jangam, S.C., Bajwa, A., Thankkappan, K.K. et al. (2018). Electrical characterization of high performance fine pitch interconnects in silicon-interconnect fabric *IEEE 68th Electronic Components and Technology Conference (ECTC)*, IEEE Press, May 2018, pp. 1283–1288, doi: https://doi.org/10.1109/ECTC.2018.00197.

16 Pu, H.-P., Kuo, H.J., Liu, C.S., and Yu, D.C.H. (2018). A novel submicron polymer re-distribution layer technology for advanced InFO packaging. *IEEE 68th Electronic Components and Technology Conference*, pp 45–51.

17 Su, A.-J., Ku, T., Tsai, C.-H. et al. (2019). 3D-MiM (MUST-in-MUST) technology for advanced system integration. *IEEE 69th Electronic Components and Technology Conference (ECTC)*, pp. 1–6.

18 Yu, D.C.H. (2014). Wafer level system integration for SiP. *IEEE IEDM*, 27.1.1–27.1.4, pp. 626–629.

19 Chun, S.-R., Kuo, T.-H., Tsai, H.-Y. et al. (2020). InFO_SoW (System-on-Wafer) for High Performance Computing. *IEEE 70th Electronic Components and Technology Conference (ECTC)*, pp. 1–6.

20 Wu, J.Y., Chen, C.H., Lee, C.H. et al. (2021). SoIS – an ultra large size integrated substrate technology platform for HPC applications. Accepted and to be published in 2021 ECTC.

21 Wang, C.T., Chang, W.L., Chen, C.Y., and Yu, D. (2020). Immersion in memory compute (ImMC) technology. *IEEE 2020 Symposia on VLSI Technology & Circuits*, TH1.5.

5

Adaptive Patterning and M-Series for High Density Integration

Benedict San Jose, Cliff Sandstrom, Jan Kellar, Craig Bishop, and Tim Olson

Deca Technologies, Tempe, AZ, USA

5.1 Technology Description

M-Series is a chips-first, face-up FO-WLP (fan-out wafer-level packaging) technology having the semiconductor device active surface and vertical sidewalls fully encapsulated within the epoxy molding compound (EMC). The device interconnect is enabled by Cu studs through the EMC as shown in the cross section [1] in Figure 5.1. M-series package encapsulation comes in two types, namely, protected fan-in (Figure 5.1a) and protected fan-out (Figure 5.1b).

The M-Series package has the EMC layer between the active region of the chip and the end electronic appliance printed circuit board (PCB), providing an intermediate stress buffer. This feature has significantly extended BLR (board level reliability) performance as compared to conventional WLP (wafer-level packaging) with only polyimide (PI) or polybenzoxazole (PBO) on the chip surface.

For both wafer-level and panel-level fan-out technologies, contending with the natural variation during manufacturing, die-shift in particular, is a critical challenge. Conventional fan-out processes have relied on very high-accuracy chip-attach equipment, at the expense of throughput and total capital cost. Rather than fight variation by using high accuracy, lower throughput equipment, adaptive patterning (AP) takes a different approach and accommodates variation through real-time design during manufacturing (DDM) [2].

First a high-speed optical scanner measures the actual position of each die (die measurement), next the AP software system generates a unique per-unit lithography pattern adapted to its individual measurement (GDSII file creation), and finally the specific per-unit patterns are implemented with a single-pass, maskless lithography system. Integration techniques such as adaptive alignment for single-die and adaptive routing for multi-die are used to enable the highest density design rules.

Embedded and Fan-Out Wafer and Panel Level Packaging Technologies for Advanced Application Spaces: High Performance Compute and System-in-Package, First Edition. Edited by Beth Keser and Steffen Kröhnert.

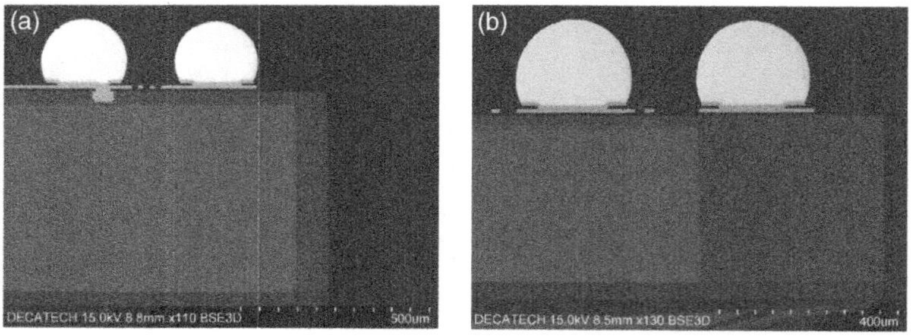

Figure 5.1 (a) Cross-section of M-Series protected fan-in. (b) Cross-section of M-Series protected fan-out.

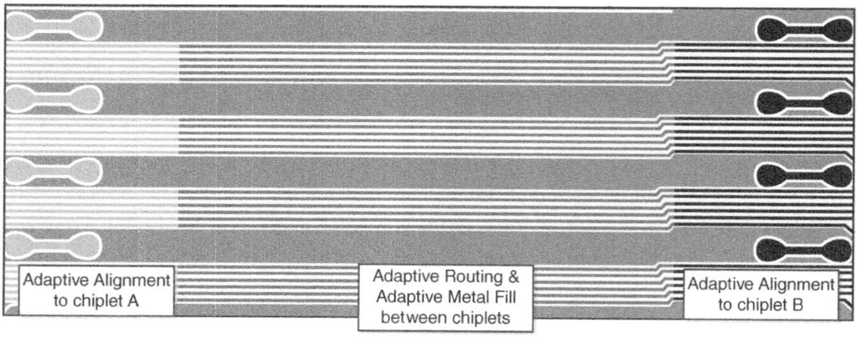

Figure 5.2 Adaptive patterning used for die-to-die connections between two chiplets with dynamically generated dummy metal fill around the connections to minimize topography for multiple RDLs.

The evolution of AP to support high density and advanced applications has also driven development of new techniques, including adaptive metal fills (Figure 5.2) to dynamically regenerate dummy metal fill in adaptive regions to optimize the topography in multi-redistribution layer (RDL) products, delivering real-time design for manufacturing (DFM). A detailed description of AP and how it supports high-density applications follows in Section 5.6.

In combination with AP, the design rules for M-Series provide the largest via contacts on the smallest chip interconnects. This will be further described in Section 5.5.

Scaling of M-Series for high-density integration will also be described later. The Gen2 M-Series consists of 2 μm lines and spaces using laser direct imaging (LDI) with up to four layers of Cu RDL allowing die-to-die interconnect of > 200 wires/mm/layer. Deca's Gen2.5 roadmap at 1.5 μm lines and spaces will target die-to-die interconnect of 250 wires/mm/layer. This number of interconnects is equivalent to that of Intel's embedded multi-die interconnect bridge (EMIB) structure with the advantage of not using a complicated bridge chip structure. A method to build a 20 μm full array bond pad pitch device that is nearly half the size of competitive technologies will also be discussed. The 20 μm bond pad pitch is made possible with AP. AP also allows scaling to ever finer bond pad pitches.

M-Series can also integrate chiplets and passive devices in a 3D package-on-package (3DPoP) structure. M-Series enabling 3DPoP and scaling to high density will be discussed in detail in Section 5.7.

5.2 Applications and Markets

The key commercial applications being designed into fan-out include 5G, artificial intelligence (AI), high-performance compute (HPC), and radar. 5G is being rolled out with higher data rate transfers and wider bandwidths. Fan-out technology improves the signal loss from system integration, so better signal performance is the key reason it will be adapted for 5G applications.

Another key benefit is providing a solution to front-end scaling limitations. Large die are undesirable in the front end due to yield and cost. For these reasons, the industry is trending away from large monolithic system-on-chip (SoC) designs toward partitioning into multiple smaller chiplets and integrating them using fan-out technologies. Additionally, high-end HPC and AI applications require interconnects larger than a single silicon interposer reticle, which necessitates complicated techniques such as reticle stitching, making a maskless (reticle-less) fan-out technology very desirable.

Advanced driver assistance systems (ADAS) is also a key technology to enable a fully autonomous car. 77 GHz automotive radar offers a key advantage because of range and multiple depth detection. Fan-out packaging is established in 77 GHz radar applications over competing technologies because of routing of the RF signals in RDL, enabling low loss wiring with better RF performance.

In summarizing the market, front-end chip and system scaling will continue to require new creative breakthroughs in chiplets and heterogeneous integration, which continue to drive strong interest in advanced fan-out technologies such as M-Series with AP. System-in-package (SiP) modules are a natural extension of this technology. This is the reason fan-out is gaining momentum. More information about the applications and markets of FO-WLP are described in Chapter 1.

5.3 Basic Package Construction

M-Series is fundamentally an embedded-chip fully molded fan-out technology, which provides complete molded protection to the active semiconductor device

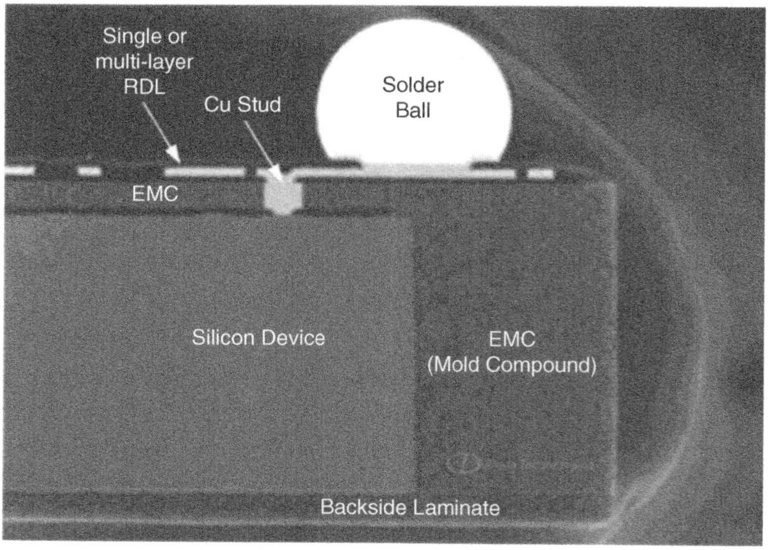

Figure 5.3 Basic package construction of M-Series.

while utilizing the massively parallel processing capability of wafer- or panel-level processing to create the interconnect layers. The basic structure with a single layer of Cu RDL is shown in Figure 5.3.

M-Series provides several important reliability and manufacturability benefits compared to traditional fan-in (WLP) and conventional fan-out [3, 4]. The primary reliability benefit includes lowering of e-CBI (chip-board interaction) stresses applied to the device [5] due to the presence of the molded stress buffer layer between the active region of the semiconductor device and the PCB. In Figure 5.4 we show the M-Series molded stress buffer layer enabling superior BLR performance exceeding a 200% improvement as compared with traditional fan-in WLP technologies. Conventional fan-out technologies such as WLP are expected to have higher e-CBI stresses since they utilize only spin-on polymers as the stress buffer between the active device and the PCB.

Another benefit of the M-Series structure over chip-first face-down process is the elimination of the EMC layer from the physical vapor deposition (PVD) process during the first level interconnect to the die aluminum bond pads. Since connection of the Cu stud to Al bond pad is performed on native silicon wafer, this allows a wider process window ensuring low contact resistance (R_c). In addition, M-Series also embeds sawn device edges with the EMC layer to prevent metallic buildup from laser grooving and dicing processes to protrude through the PI dielectric layer over the surface of the Si-EMC interface. This feature has the benefit of preventing

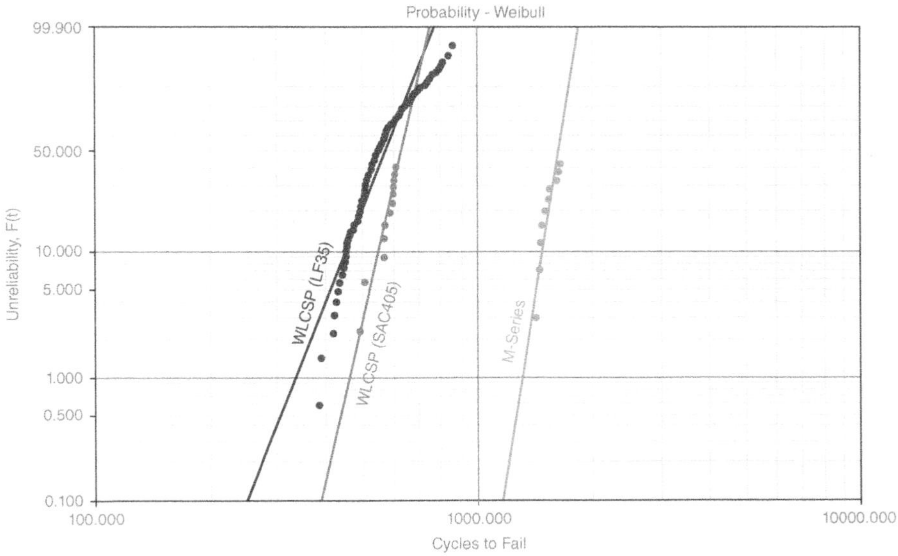

Figure 5.4 BLR Comparison of M-Series versus WLP.

electrical shorting failures at the device edge observed in conventional fan-out structures such as WLP.

A key advantage of M-Series versus conventional fan-out is its planarized structure above the encapsulated active die. This planarized structure eliminates the discontinuity from the active device surface to the EMC surface preventing non-planarity of the first dielectric layer. When two semiconductor devices are placed in close proximity to each other, the non-planar topography from the silicon surface to the EMC surface in a conventional fan-out can be significant, as shown in Figure 5.5.

In the case of conventional fan-out such as embedded wafer-level ball grid array (eWLB), the 5.5 μm of dielectric non-planarity in Figure 5.5 will severely limit the ability of scaling to fine lines and spaces (2 μm and below) because of the requirement of a tightly controlled DOF (depth of focus) in the lithography process. In contrast, the advantage of the planarized nature of M-Series, as shown in Figure 5.6, allows for straightforward scaling to 2 μm and below since planarization is built into the process flow.

A crucial advantage of M-Series over other chip-first face-up FO-WLP technologies such as integrated fan-out (InFO) is AP enabling a significantly larger via contact area for the same bond pitch. Figure 5.7 compares the InFO and M-Series stack-up. The InFO structure requires a large Cu capture pad to provide an interconnect between the RDL and the Cu stud [6]. The via size is limited by die shift

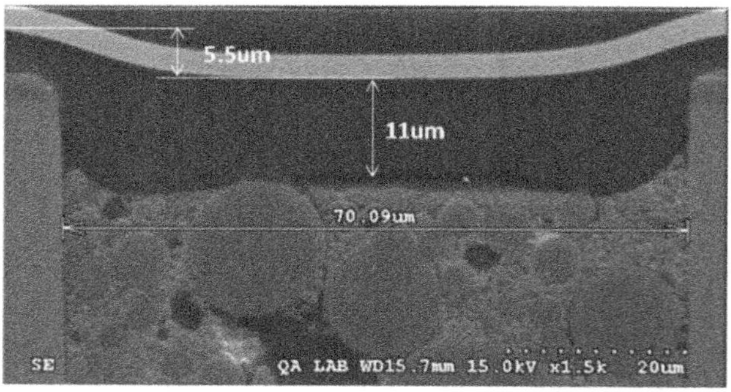

Figure 5.5 Non-planarity of silicon to EMC surface in conventional fan-out. (Source: courtesy of ASE.)

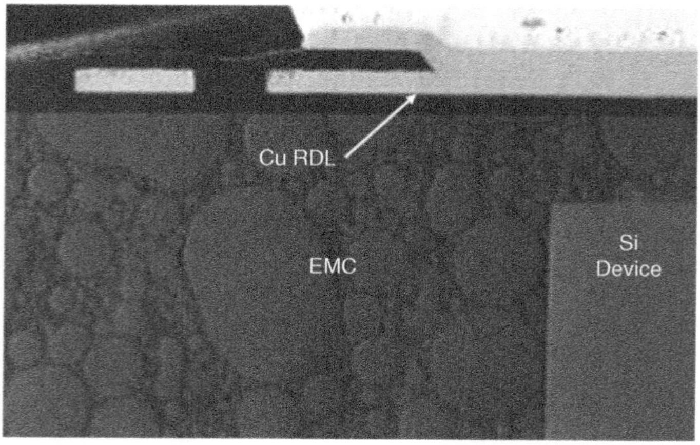

Figure 5.6 Planar surface over silicon device to EMC interface on M-Series.

brought about by die placement variations and die shift during molding. In the case of the M-Series structure, AP enables a large via size connection by accounting for die shift. This has an added advantage of not requiring an extra Cu layer as a capture pad. The combination of M-Series and AP allows for a >300% increase in via contact area for the same bond pitch with excellent electrical performance and yield.

In addition to enabling a large via contact area for the same bond pitch, AP is also essential in scaling of M-Series for high-density integration. The Gen2 M-Series starts with 2 μm lines and spaces using LDI with up to four layers of Cu RDL. In Section 5.7, we highlight 1.5 μm lines and spaces with a die-to-die spacing of 50 μm approaching die-to-die interconnect of 250 wires/mm/layer. As previously

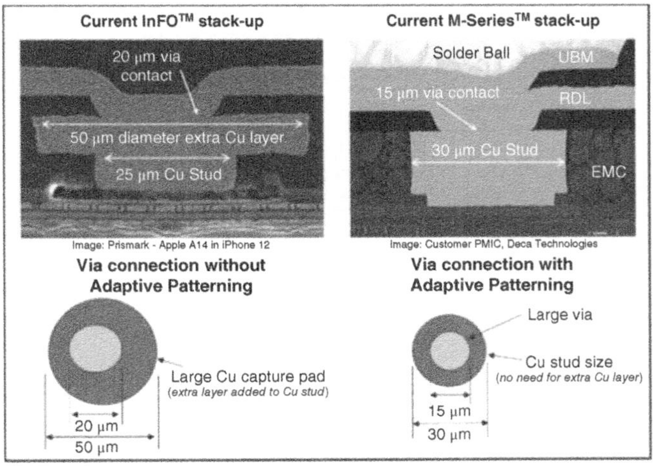

Figure 5.7 M-Series and AP enabling a significantly larger via contact area given the same bond pad pitch and scaling to fine bond pitch.

mentioned, this number of interconnects is equivalent to that of EMIB structure with the advantage of not using complicated bridge chip structures.

5.4 Manufacturing Process Flow and BOM

As described earlier, M-Series is a chips-first, face-up FO-WLP technology having the semiconductor device active surface and vertical sidewalls fully encapsulated within the EMC. A high-level flow for Deca's M-Series has been described previously [7], but here is a more detailed description (Figure 5.8). The basic flow can be broken down into four key modules: Cu stud, panelization, fan-out fab processing, and test and finish (backend2, or BE2).

The purpose of the Cu stud is to make electrical contact between the incoming Al pads, which is typically done on 200 or 300 mm wafers supplied by front-end fabs through the mold to the plated RDL. Depending on the pitch of the incoming Al pads, an added dielectric layer may be necessary. The Cu stud formation is in the first column of Figure 5.8 and entails a standard seed deposition followed by a photo process using a thick dry film or liquid resist. After imaging the resist on the seed layer, the wafer is plated. Typical features sizes in this operation are 60 μm round studs, but the use of oblong, octagon, or ground planes have also been demonstrated. After plate, the resist is stripped followed by etch of the seed layer and then by a wafer clean step, which removes any residue. Then a quality check for defects and Cu stud thickness uniformity is done. Not shown in this integration

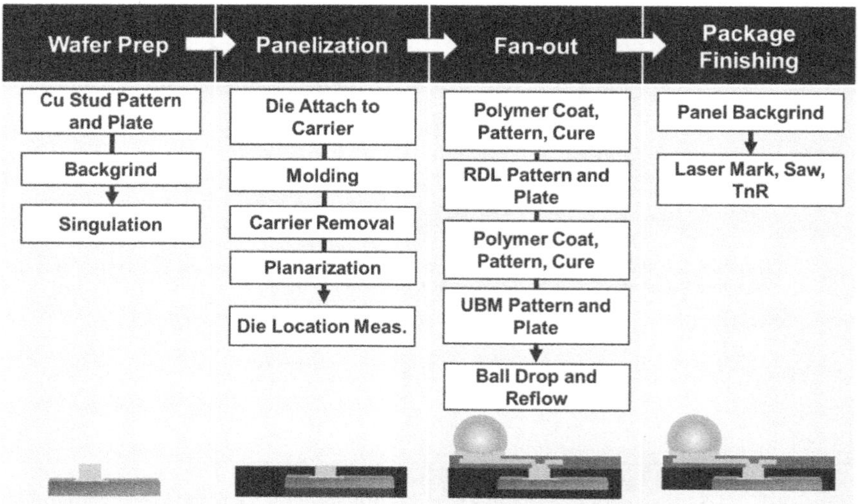

Figure 5.8 M-Series process flow for a fully molded FO-WLP.

is the option to include a fan-in RDL step before the Cu stud process if additional routing is needed.

Part of Deca's Cu stud module is backgrind and wafer saw. Typical incoming wafers are on the order of 780 μm thick and backgrinding is done to thin the wafer for better saw performance in terms of chipping and operational throughput. This is followed by the wafer saw process itself and depending on the front-end silicon node may include laser groove (not shown in Figure 5.8). The final step in the Cu stud module is chip expansion where the saw tape is stretched to increase the spacing between adjacent die to prevent handling issues during the next operation, chip attach, in the panelization module.

The panelization module is shown in the second column of Figure 5.8. The input to this module is an expanded saw film frame with mounted die. The output of this module is a 300 mm or other fully molded panel (see Section 5.7) with the Cu studs revealed for future contact between the Al pads and the RDL interconnect. The first operation before chip attach is UV treatment of the film frame to allow the die to be easily removed. The input-to-chip attach is the film frame, and the output is a carrier with die attach tape and die. The goal of chip attach is to take the die from the film frame and carefully and accurately place it on the carrier for later molding. The heart of the panelization process is chip attach as it drives large capital purchases and enables complex SiP (system in package) with multi-die placement opportunities. In addition to die, Deca also places electrical test structures into the molded panel, described in Section 5.10.

After the die are placed on the carrier, Deca uses a compression mold process with granular material for better material properties than could be achieved with liquid mold compound. The mold is thicker than the Cu studs in what the industry refers to as overmold. This is done by design to adjust for variations in the Cu stud and backgrind processes, as discussed earlier. After molding, the carrier is removed through both a heated and mechanical process called debond. This process separates the carrier and the die attach tape from the newly formed molded panel. The next operation is backside laminate (BSL) followed by a post-mold cure. The BSL is an epoxy base film laminated on the back of the panel as a temporary film for warpage control. The warpage of the panel is a key metric in building a fan-out device, and many factors are involved such as mold material properties and silicon content in the mold. Silicon content or ratio of silicon to the mold in terms of volume is a large factor in determining warpage. Film stresses related to later processing such as seed and dielectrics need to be balanced to minimize warpage. By applying a BSL, a balance of these stresses can be achieved. After the BSL, the panel goes through a cure process to cure both the laminate and the molded material.

As mentioned above, the panel is over-molded, meaning the Cu studs are not revealed. The next step is front planarization, which reveals the Cu stud for later connection to the RDL. After this operation, a light Cu clean is done to clean the surface of the studs for die measurement and clean smeared Cu due to the front planarization operation. This is commonly known as a smear etch. The key to Deca's M-Series technology are the last two operations called die measurement and GDSII file creation, which were briefly described in Section 5.1.

The fan-out module in the third column of Figure 5.8 is a standard WLP process, but on molded panels and involves creating vias and RDL for connection between the solder balls down through the Cu stud to the Al pads on the die. The number of RDL layers varies by application but typically is one to two. The first step is a via formation using a PI or other suitable dielectric materials followed by a cure step. A light plasma is typically done to remove any residue in the via post develop pre-cure. To form the via, Deca uses a unique exposure technology in which a GDSII file is created for each panel. After cure, a seed layer is applied in much the same way as the Cu stud module, patterned with either a dry film or liquid resist, plated, photoresist strip, and seed etch to form either the RDL or underbump metal (UBM) layers. A clean and quality check for defects typically follows each RDL etch for multi-layer packages. One RDL and one UBM layer is shown in Figure 5.8. The panel then follows a standard ball drop, reflow, flux clean, and another quality checks for defects and ball height and width before the electrical test.

The panel, currently running in high-volume production on 300 mm round format at multiple factories now ready for electrical test (see Section 5.10). Known good units are identified during the electrical test, and then the panel goes through a backgrind process. This process removes the BSL film applied in the panelization module and thins the package to the requirement desired by the customer. After backgrind, the panel is mounted on a film frame in the same manner as the wafer, as discussed in the Cu stud module section. The panel is marked either on the backside of the exposed silicon die or a black laminate film is added for more traditional laser marking. The panel is then singulated and packaged in tape and reel for shipment.

The key BOM (bill of materials) for the FO-WLP is the granular EMC. The EMC properties for M-Series were selected to minimize panel warpage during manufacturing and have excellent molding performance. Key properties for the EMC having an impact on panel warpage are the coefficient of thermal expansion (CTE), Young's modulus, and silica filler content. In addition, the size distribution of the silica fillers will greatly influence mold flow and mold package filling.

5.5 Design Features and System Integration Capability

The standard M-Series cross-sectional stack-ups are shown in Figure 5.9. The M4 stack-up contains a total of four patterned layers (polymer 1, Cu RDL, polymer 2, and UBM). The M6 structure contains an additional RDL and polymer layer for a total of six patterned layers.

Several different via stack-up options are possible within the M-Series fan-out buildup structure. In many designs a staggered via structure is employed, as showed in Figure 5.10. In this case, the vias connecting the Cu RDL and UBM layers are offset with respect to each other. It is possible to position the UBM and BGA ball directly over a Cu stud. Figure 5.11 shows this embodiment. Via in pad structures wherein the via connecting the UBM capture pad to the first RDL layer (in the case of the M6 stack-up) or the Cu stud (in the case of the M4 stack-up) are also possible, as shown in Figure 5.12. Finally, fully stacked vias with the UBM and solder ball positioned directly over and directly connected to one or more Cu studs through both RDL layers is possible, as shown in Figure 5.13.

M-Series provides mechanical isolation between the solder ball and Si die by providing a rigid molded epoxy layer between the die and fan-out buildup structure. As a result, there is no design rule restriction with respect to placement of BGA balls in the "die shadow" region of the package. BGA pads can be placed directly over the die edge, as shown in Figure 5.14.

Since the M-Series structure uses a plated Cu interconnect without solder to connect the die bond pads to the fan-out routing layer, there is a lot of flexibility

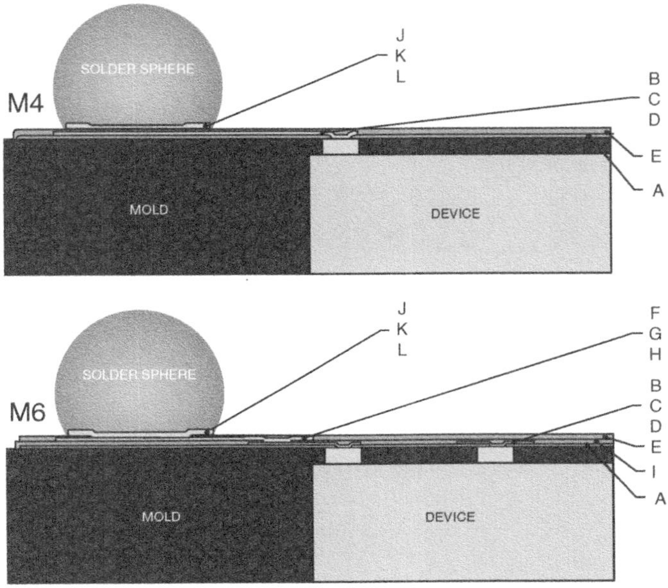

A	Fan-out Polymer 1	B	RDL1 Barrier Ti	C	RDL1 Cu Seed
D	RDL 1 Plated Cu	E	Fan-out Polymer 2	F	RDL 2 Barrier Ti
G	RDL 2 Cu Seed	H	RDL 2 Plated Cu	I	Fan-out Polymer 3
J	UBM Barrier Ti	K	UBM Cu Seed	L	Plated Cu UBM

Figure 5.9 Standard M-Series cross-sectional stack-ups.

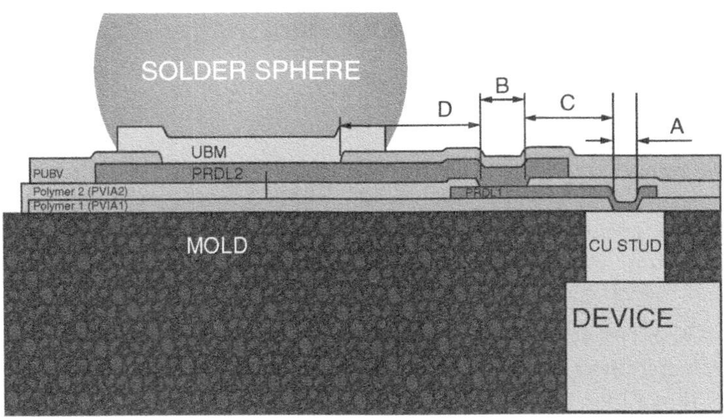

Figure 5.10 M-Series buildup with staggered vias.

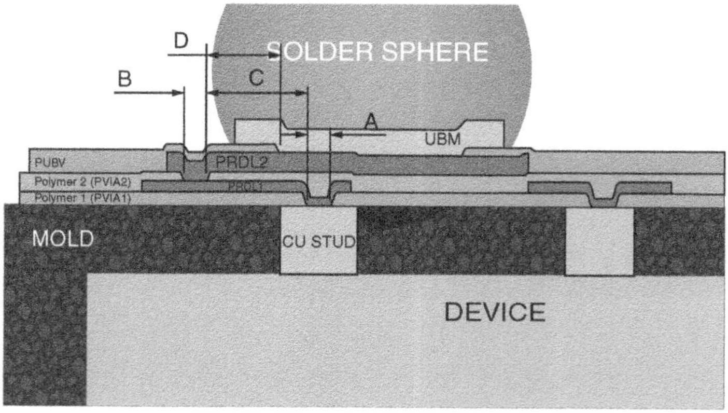

Figure 5.11 M-Series buildup with UBM and BGA ball directly over a Cu stud.

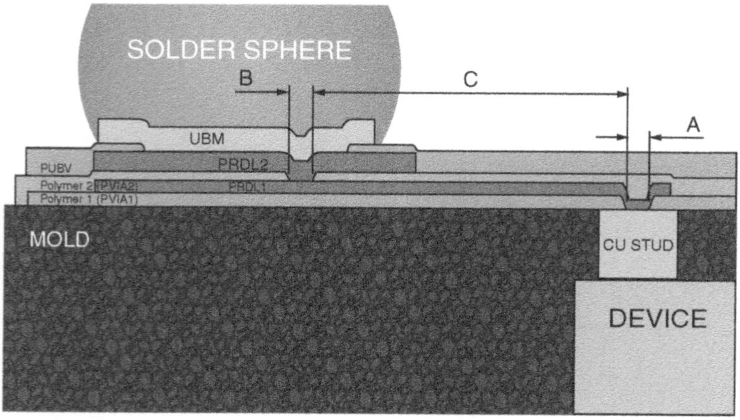

Figure 5.12 M-Series build-up with UBM via in RDL2 via.

on designing the Cu stud layer. Different interconnect sizes and shapes can be used within the same design, as shown in the die layout drawing in Figure 5.15. 3D package designs are also possible using the M-Series structure. Common 3D structures include peripheral PoP (package-on-package), as shown in Figure 5.16.

A summary of M-Series critical design rules, roadmap design rules, and design features are given in Table 5.1. Gen1 are current rules in HVM today while Gen2 are high-density integration rules, which will be available in 2022. These rules are discussed more in Section 5.7. An interim of 5 μm line and space will be ready in 2021.

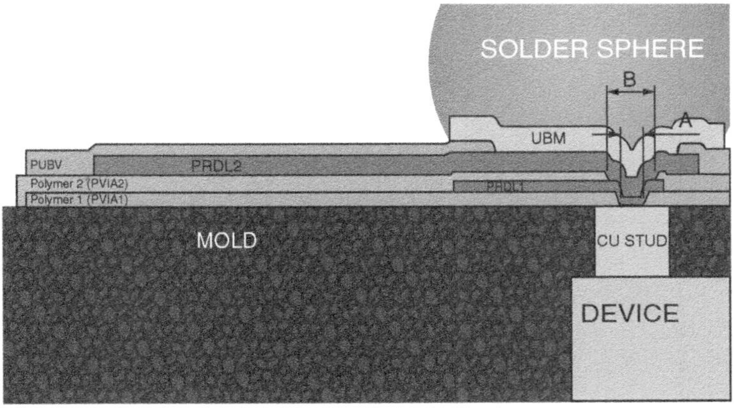

Figure 5.13 M-Series build-up with fully stacked vias.

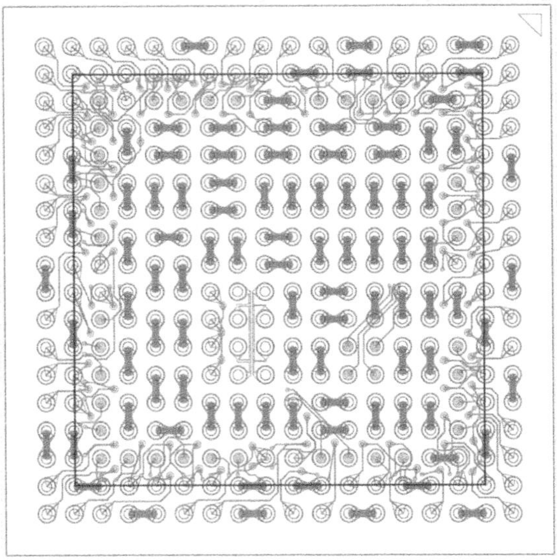

Figure 5.14 M-Series BGA view showing BGA balls over die shadow, shown as black line.

5.6 Adaptive Patterning

Manufacturing embedded die structures, such as the M-Series, with reliable yield and robust quality requires overcoming the critical impact of natural variation from multiple sources. The first source of random variation is chip-attach, where singulated die from one or more source wafers are picked and placed onto a carrier coated with an adhesive film. The resulting position of each die has a random X, Y, and rotational offset from its designed position due to mechanical and

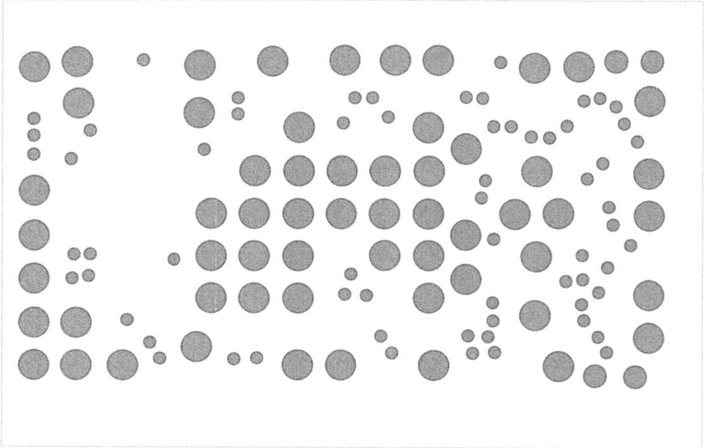

Figure 5.15 M-Series Cu studs view showing different interconnect sizes and shapes.

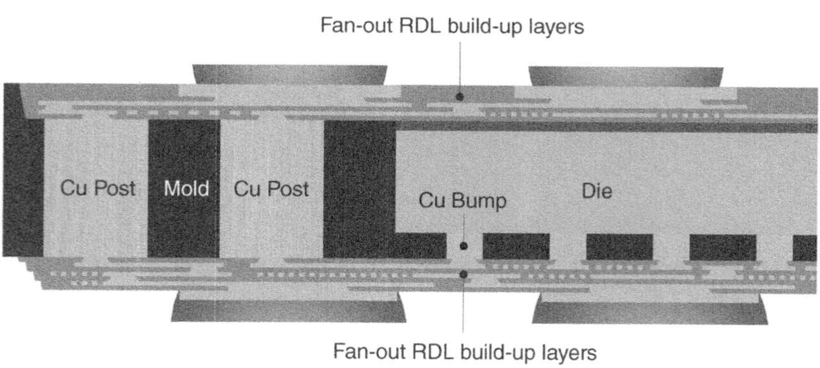

Figure 5.16 M-Series 3DPoP schematic.

alignment tolerances of the equipment. The next step, where the placed die are compression-molded into a panel format, also contributes significant variation. In practice, the effect from molding varies considerably depending on the layout of silicon in the molded panel and can account for up to a fifth of the total die displacement (Figure 5.17). After planarization to reveal the copper stud contacts to the die, the cumulative die-shift from both chip-attach and molding is visible (Figure 5.18).

Conventional approaches offer limited options for accommodating the misalignment of contacts from their design positions. Often, such methods require adding additional area (also called enclosure) to the die contact and using a limited-size via to ensure connection. However, any extra space in design rules

Table 5.1 M-Series design rules.

M-Series design feature roadmap		
Design feature	**Gen1**	**Gen2**
Cu stud layer		
Maximum stud thickness (μm)	25	100
Minimum stud thickness (μm)	20	10
Minimum stud diameter (μm)	35	12
Minimum stud spacing (μm)	20	8
Minimum stud pitch (μm)	60	20
First fan-out polymer via layer		
Minimum via diameter (μm)	20	6
Minimum Cu stud enclosure (μm)	7.5	3
Minimum PI thickness, first layer (μm)	7.5	5
Maximum PI thickness (μm)	12	15
Fan-out RDL layers		
Minimum trace width (μm)	10	2
Minimum trace space (μm)	10	2
Maximum RDL thickness (μm)	9	12
Minimum Via 1 enclosure (μm)	15	3
RDL layer count	2	4
Final fan-out polymer via layer (under UBM)		
Maximum PI thickness (μm)	12	12
Via 2, RDL enclosure (μm)	7.5	3
UBM layer		
Maximum UBM thickness (μm)	9	9
Minimum BGA pitch (μm)	350	100
Package outline		
Minimum die thickness (200 mm) (μm)	170	60
Max fan-out ratio	3.5	3.5
Minimum die-to-package edge (μm)	75	60
Max die edge length (mm)	6	40
Max package edge length (mm)	8	80
Total package thickness (mm)	0.45	0.17
Land side caps		✓

(continued)

Table 5.1 (Continued)

M-Series design feature roadmap		
Design feature	**Gen1**	**Gen2**
3D features		
Minimum Cu post pitch (μm)	NA	180
Minimum Cu post diameter (μm)	NA	100
Minimum Cu post spacing (μm)	NA	80
Back-side routing layer count	NA	4
Back-side routing minimum feature size (μm)	NA	2
Embedded passives	NA	✓
Die count	NA	12+

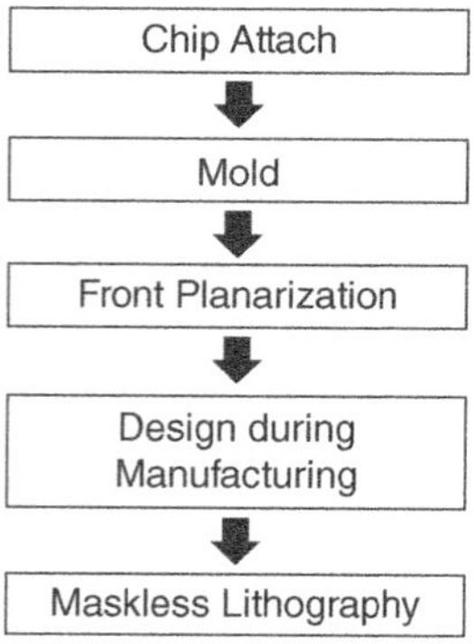

Figure 5.17 High-level adaptive patterning process flow.

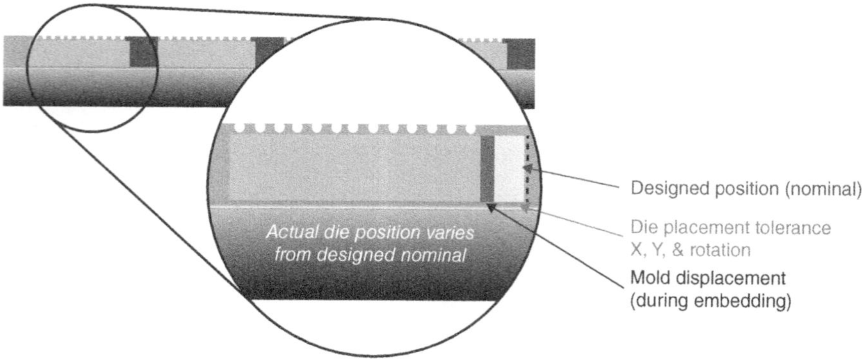

Figure 5.18 Illustration of combined die-shift resulting from chip-attach tolerances and displacing during compression molding.

limits the minimum achievable bond-pad pitch, and any reduction of the via's electrical contact area affects electrical performance.

Rather than fight the natural variation inherent to the process, Deca's AP approach generates a unique pattern for each unit in real time that dynamically accounts for the die-shift. First, a high-speed optical scanner is used to measure the final position of each die in the panel and determine its X, Y, and rotational (θ) offset from its designed position. Next, a DDM software system processes the measurement data to generate per-unit layouts using design rules and constraints setup ahead of time within an electronic design automation (EDA) system. Finally, a maskless lithography tool is used to expose the unique patterns in photosensitive materials. Since the software system generates the layout for the entire panel at once, analogous to a panel-sized reticle, stitching methods are not required for large unit sizes.

The generation of per-unit adapted layouts by the DDM software is controlled by a specialized design data package that includes a partitioned version of the original artwork, design rules to enforce in real time, and pre-verified techniques for adjusting to die-shift. Two methods, previously published in detail [8], have been used to great success on a wide variety of applications. In Figure 5.19, the first technique,

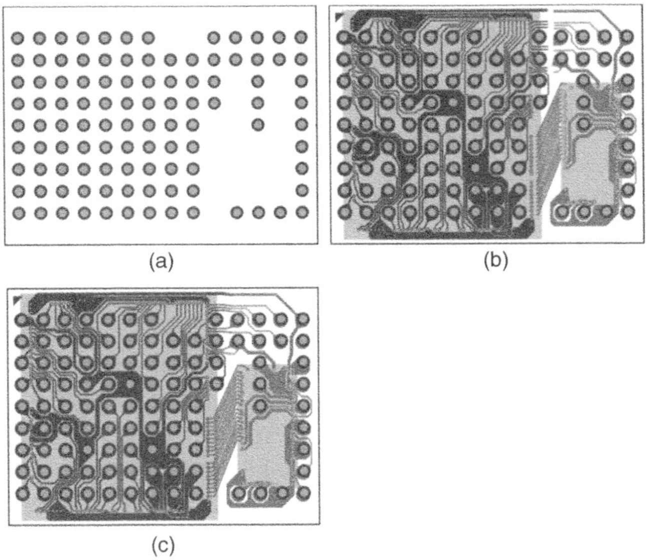

Figure 5.19 First-generation M-Series application with AP for a two-die device; in (a) the UBM and under-bump via openings are fixed relative to the package outline to maintain a constant BGA footprint; in (b) adaptive alignment translates and rotates the via and RDL pattern for each die to match its measured position; in (c), adaptive routing dynamically adjusts traces between the die to account for the combined displacement of both die.

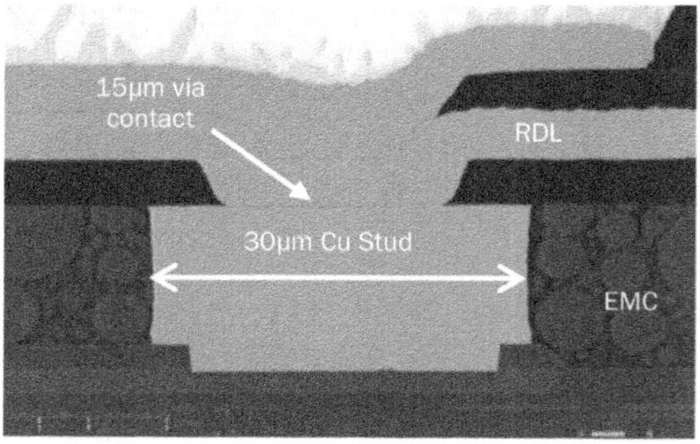

Figure 5.20 First-generation M-Series cross-section highlighting the via-to-copper stud interface without extra enclosure for die misalignment.

adaptive alignment, is used to translate and rotate the via and RDL to match the measured location of each die. Since the vias are precisely aligned to the die contacts, the pad-stack does not require extra enclosure for die-shift and only the lithographic tolerances are considered, as shown in Figure 5.20. The second technique, adaptive routing, is applied to dynamically regenerate the die-to-die connections, accounting for the combined displacement of both die. Critical design rules and constraints, as well as setup ahead of time in an EDA system, are maintained in real time.

These same techniques have also been extended to second-generation M-Series designs with greatly increased complexity. For example, the chiplet-based test vehicle design in Figure 5.21 requires more than 5600 die-to-die with connections RDL to connect nine chiplets using 2 μm lines and spaces. Adaptive alignment is employed to separately align nine RDL and via layer regions to each die, and then adaptive routing is again used to dynamically adjust the connections between the chiplets, accounting for the combined die-shift.

New designs also tend to integrate far larger die sizes, 25–35 mm on the largest sides in this case. Due to the rotational component of the die-shift, contacts near the corner of the die will have a displacement from nominal designed position that is much larger than the X and Y values alone, plotted in Figure 5.22. Without AP, rotation makes achieving fine bond-pad pitch unpractical. For example, a 35 mm die with contacts at 55 μm pitch to mirror the JEDEC standard HBM footprint [9] can expect a combined displacement to 24 μm, even for a tightly controlled shift distribution. Maintaining full contact of the via would require 48 μm of extra enclosure (24 μm per side), leaving only 7 μm for the via and lithographic tolerances. At 20 μm pitch, increased enclosure cannot accommodate a 24 μm displacement. With adaptive alignment, the RDL and via pattern have matching translation and rotation applied per die, so only the measurement and lithographic tolerances must be considered, making full via size achievable.

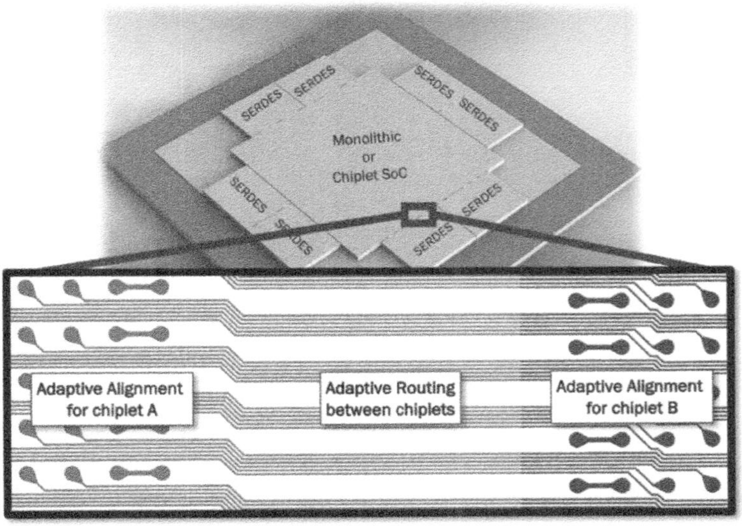

Figure 5.21 Second-generation fan-out on substrate test vehicle design with nine chiplets and more than 5600 2 μm die-to-die connections per RDL that are dynamically adapted in real time.

Figure 5.22 The displacement of die contacts from the nominally designed position increases with die size because of the rotational component, shown computed at 3σ for several example post-mold distributions (includes chip-attach plus mold variation).

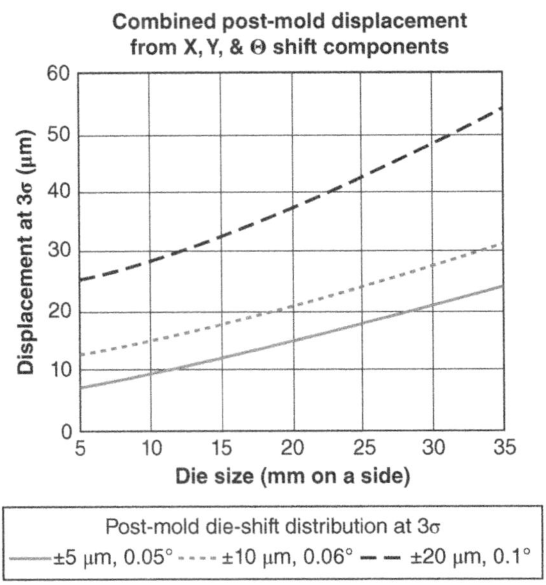

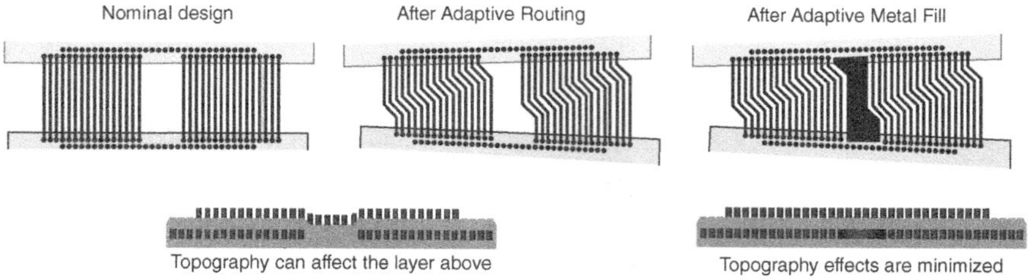

Figure 5.23 For multi-RDL designs, the metal fill for adaptive routing must be recalculated in real time, in addition to the traces between die. Without metal fill, topography can cause variation or defects in metal layers built above. With dynamically generated fill, the impact of topography is minimized.

Beyond accommodating die-shift, DFM concerns have become important when scaling AP to 2 μm and finer geometries with multiple RDLs. Nonuniform metal density on RDL layers can cause topography effects, such as CD variation or RDL metal defects, during the buildup of subsequent layers. Conventionally, these topographic effects are minimized during design by fully filling unused space with additional pattern. Since adaptive routing dynamically adjusts design geometry to account for die-shift, this operation must also be performed in real time. A recent extension to AP adds the capability to recalculate metal fill areas on the fly around other adaptive features. During the design stage, fill zones are specified, but the final geometry is not fully computed. During production, the DDM software first applies alignment and adaptive routing and then calculates fill geometry around the adapted features. Results from initial demonstrations of this capability are displayed in Figure 5.23. Beyond this initial work for real-time DFM, development is underway to apply a similar approach to signal and power planes that cross between die, meaning they also need to be computed in real time.

5.7 Manufacturing Format and Scalability

As FO-WLP has gained familiarity throughout the electronics industry and semiconductor supply chain, rapid growth in demand for the technology continues. To create a large panel fan-out standard for future capacity, Deca established the 600 mm × 600 mm square panel format in 2010 with first equipment and sample realization in 2013.

The 600 mm square format was created with considerations for both short-term implementation and long-term optimized productivity. The ability to segment the 600 mm format into four 300 mm square subpanels for use with conventional 300 mm round wafer probe test equipment was a driving factor, as shown in Figure 5.24. Extending process capability for key lithography, metal deposition,

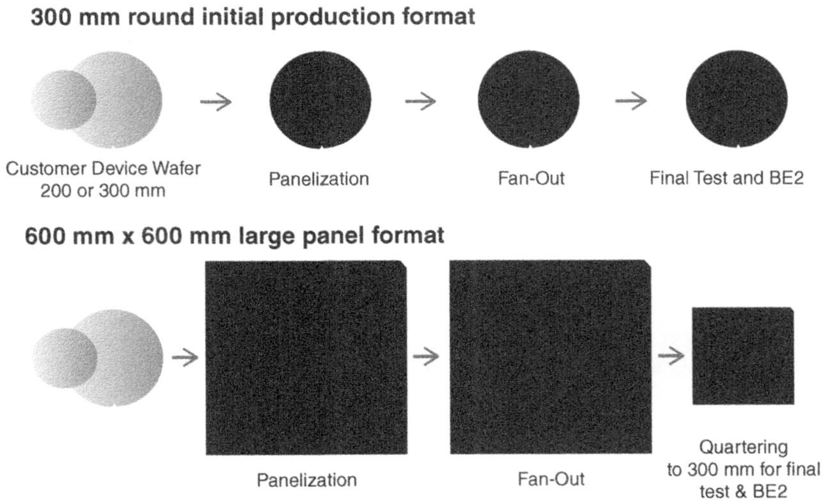

Figure 5.24 M-Series manufacturing formats.

and other processes to complete a 600 mm line provided a challenging yet achievable target for equipment suppliers. The key driver in going to a large panel format is cost. Building large packages on a 300 mm round format is inefficient since the packages are square or rectangular. Given the same package area, a square panel format will have a higher utilization of molded area compared to a round format, which will translate to lower cost.

Another key concern is capital and material cost of a 600 mm or other large panel formats (Figure 5.25). In the top left corner of Figure 5.25, the x-axis represents the area in k mm^2 of the different formats used in the industry. The y-axis represents the relative cost, referenced to 300 mm round. The 600 × 600 mm has a 50% cost reduction over a round 300 mm format. Material cost is driven by two factors: coating and lamination. A square format is ideal for a slot or slit die coater (vs. a spin on with a round 300 mm wafer) as the amount of material is conserved on the square panel. A 600 mm with over five times the area will use two times the dielectric material such as PI. PI or other dielectrics are generally the largest material cost component for a fan-out package. Another cost advantage is lamination. Films are supplied in rolls, and with a round wafer or panel, waste occurs. Capital is also conserved as seen in front-end fabs, which has been the basis for scaling to larger formats, that is, going from 150 to 200 to 300 mm wafers. In very general terms, the 600 mm tools are slightly more expensive and have about half the throughput over a 300 mm line but produce five times the area for packages. The downside to a larger format is the need to be supported by volume and the lower efficiency for new product development. Large panels also run the risk of yield in much the same way as front-end fabs suffer from similar scaling.

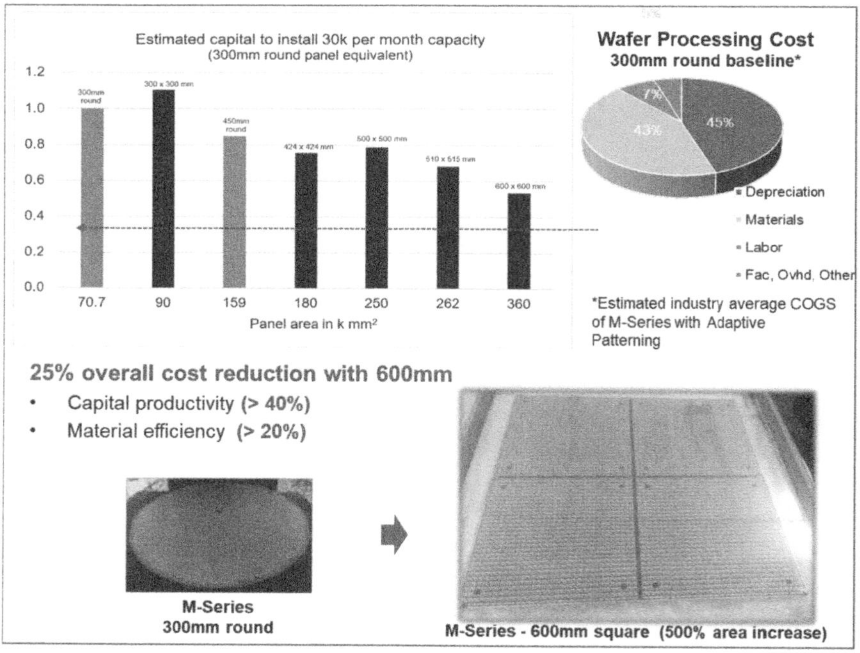

Figure 5.25 Cost reduction and capital cost of M-Series 600 mm × 600 mm format.

Besides format, another key technology that enables scaling to high-density integration is the use of LDI with AP for exposing fine lines. The main advantages of LDI over other exposure methods such as mask projectors or steppers is a high DOF that allows scaling of lines to 2 μm and below with ability to have thicker Cu RDL traces. In Figure 5.26, the size of a focus-exposure process window for an 9 μm RDL line was measured. The DOF is its ability to maintain a desired amount of image quality, typically +/−10% of the CD target. Larger DOF is desired for panel processing due to warpage. Figure 5.26 shows LDI at 9 μm line and space (with a DOF of ±19 μm vs. high NA stepper with a DOF of ±2.1 μm). A balance of numerical aperture (NA), imaging capability, and DOF is needed to create the right processing conditions. Front-end tools have high NA lenses with low DOF that prevent adequate exposure of thicker films with enough image contrast to meet resolution requirements such as linewidth and sidewall angle. Aligners also struggle with thicker or higher aspect ratio imaging due to focus offset limitations.

Deca's planar M-Series fan-out surface is ideal for building highly integrated SoC structures. Scaling to 2 μm lines and spaces and multiple RDL provides powerful new possibilities for IC designers. Chiplets can now be fabricated using the optimum wafer fab technology node providing the best performance with the most desirable commercial terms. Deca's 1st generation process is currently limited

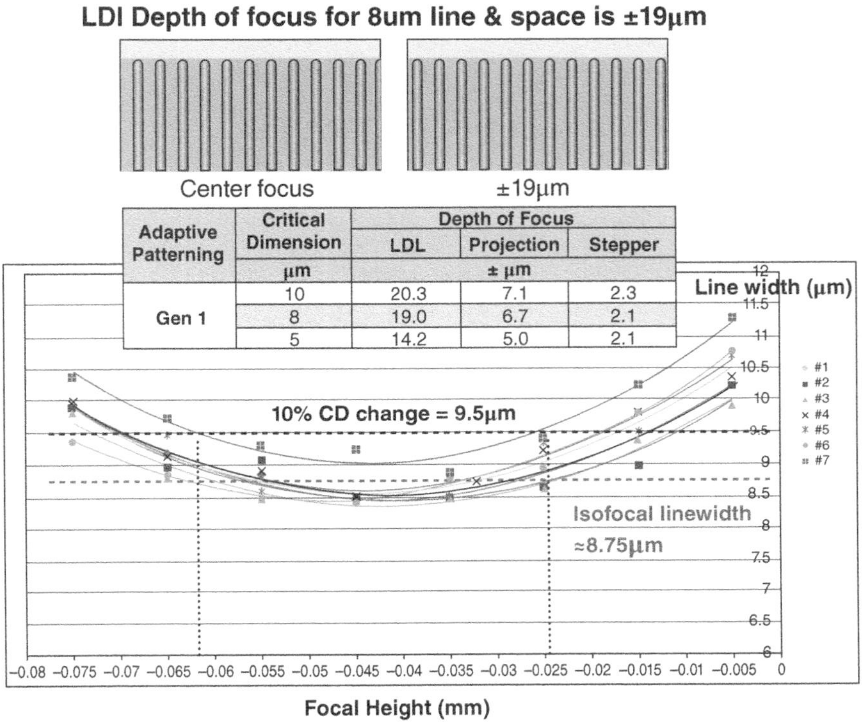

Adaptive Patterning	Critical Dimension	Depth of Focus		
		LDL	Projection	Stepper
	μm	± μm		
Gen 1	10	20.3	7.1	2.3
	8	19.0	6.7	2.1
	5	14.2	5.0	2.1

Figure 5.26 Laser direct imaging (LDI) shows an advantage of high depth of focus (DOF) compared to mask projection and stepper systems.

to 10 μm RDL lines and spaces with extension worked being done to achieve 5 μm in 2021.

The M-Series Gen2 process will target 2 μm lines and spaces with future roadmap activities to achieve 1.5 μm (Gen 2.5) down to 0.8 μm (Gen 3.0) using LDI with up to four layers of Cu RDL. Figure 5.27 describes a wiring density graph showing M-Series Gen2 at 2 μm line and space with a shoreline spacing of 150 μm approaching die-to-die interconnect of 200 wires/mm/layer. Deca's Gen 2.5 roadmap at 1.5 μm line and space will target die-to-die interconnect of 250 wires/mm/layer where high-density routing can be achieved on a simple molded chips-first, chips-up structure without the need for complicated bridge chips in substrates.

The 2 μm lines and spaces using LDI will use either a negative tone photoresist, which utilizes acrylate groups that cross-link upon exposure or a positive tone photoresist, as shown in Figure 5.28, developed in an industry standard 2.38% TMAH solution. The photoresist is coated (spin or slot die) to a thickness of around 7 μm, but thicker films are under development.

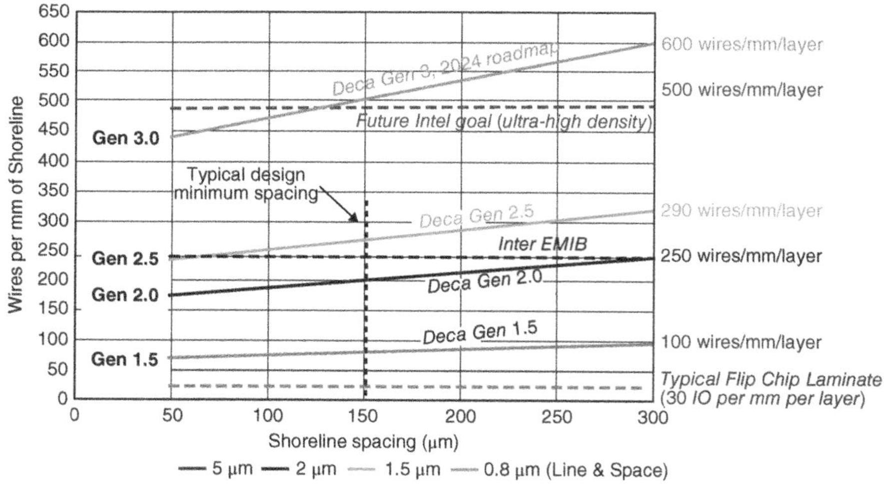

Figure 5.27 Gen2 M-Series with AP enables high density of 200 wires/mm/layer without the need for bridge chips.

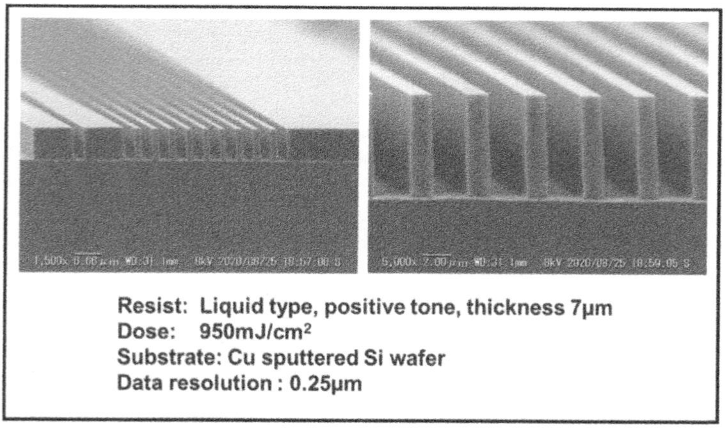

Figure 5.28 Development of liquid photoresist achieving 2 μm lines and spaces.

In Figure 5.29, we describe a device interface density with 20 μm pitch full array bond pads made possible with AP. Comparing Gen1 M-Series to Cu pillar flip chip on substrate, Gen 1 M-Series with a die pad pitch of 45 μm delivers a density of 492 I/O per mm^2 compared 100 μm die pad pitch with 105 I/O per mm^2 for flip chip. The roadmap of EMIB is at 36 μm bond pad pitch with 806 I/O per mm^2 while Gen2 M-Series with 20 μm pitch full array has a density of 2518 I/O per mm^2. Furthermore, M-Series third Gen allows scaling to ever finer bond pad pitches.

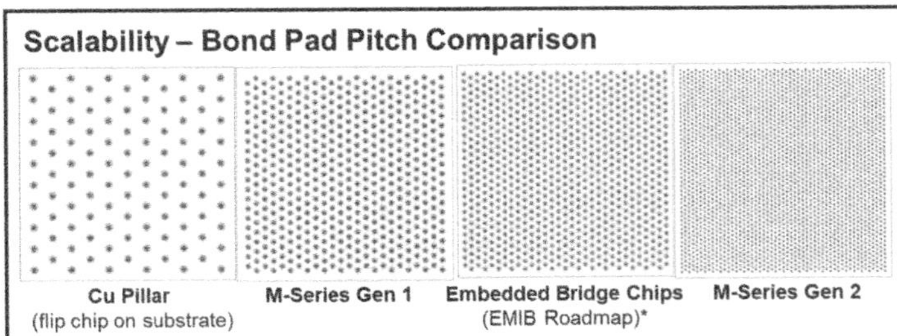

	Flip Chip Cu stud	M-Series Gen 1	Embedded Bridge Chips	M-Series Gen 2
Die pad Pitch	100 μm	45 μm	36 μm	20 μm
IO per mm²	105	492	806	2518

Figure 5.29 Scaling of Gen1 and Gen2 M-Series compared to flip chip on substrate and EMIB.

5.8 Package Performance

The die in M-Series is fully encapsulated on all sides with EMC and a BSL film. This structure provides excellent optical isolation for photo sensitive dies. In Figure 5.30, we show optical transmissivity data from UV to infrared light of the encapsulation materials, namely PI, EMC, and the BSL film. In the basic M-Series package, EMC at 25 μm thickness provides optical isolation for the active die surface at 0.9–2.5% from UV, 2.8–5% from visible range, and 5–8% transmission at 700–1200 nm infrared range. The BSL at 40 μm thickness on the other hand provides 0.01% transmission from UV to infrared from the die backside surface. The EMC and BSL shows 10x and 8000x, respectively, reduction in optical transmissivity in contrast to PI, which is used to encapsulate the die surface in WLP.

Emergent 5G millimeter wave (mmWave) systems will require major advances in package design rules and process development to integrate and connect transceiver, passive elements, and antenna array in a miniaturized antenna package [10]. The two key requirements to accomplish next generation 5G mmWave systems are low electrical loss or dissipation factor (Df) and high-density interconnects between integrated components. In contrast to silicon interposers with high-density interconnections using through silicon vias (TSV), FO-WLP is emerging as the next generation 5G format because of lower cost, better electrical performance in the 5G mmWave spectra, and bump-less interconnection using Cu pillars [11–13].

The application of M-Series for 5G mmWave packaging will require dielectric materials with low dielectric constant (Dk), low Df, as well as high resolution for photoimageable dielectric such as PI. Table 5.2 shows the Dk and Df of the BSL, PI, and EMC used in M-Series at high-frequency conditions.

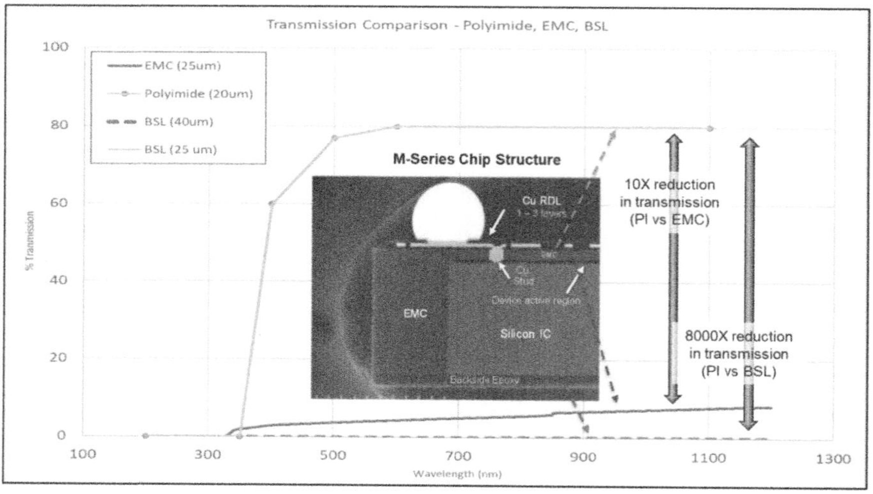

Figure 5.30 Optical transmissivity spectra from UV, visible, to infrared of the M-Series package using epoxy mold compound (EMC), backside laminate (BSL), and polyimide (PI).

Table 5.2 Dielectric constant (Dk) and dissipation factor (Df) values of M-Series materials showing compatibility for 5G mm Wave applications.

Material	Dielectric constant (Dk)	Dissipation factor (Df)	Method (resonance frequency)
PI	3.28 (10 GHz)	3.5×10^{-3} (10 GHz)	Ring resonator model (0–110 GHz)
	3.0 (105 GHz)	4.0×10^{-3} (105 GHz)	
BSL	3.9 (10 GHz)	3.4×10^{-2} (10 GHz)	Split post dielectric resonator (10 GHz)
EMC	4.3 (1 MHz)	1.3×10^{-2} (10 GHz)	Capacitance method (1 MHz^{-1} GHz)
	3.85 (1 GHz)	1.45×10^{-2} (1 GHz)	
	3.8 (1 GHz)	1.75×10^{-2} (1 GHz)	Strip line resonator method (1 GHz–15 GHz)
	3.75 (15 GHz)	1.1×10^{-2} (15 GHz)	
	3.8 (70 GHz)	–	Fabry-Perot resonator method (70 GHz)

5.9 Robustness and Reliability Data

M-Series has been fully qualified according to JEDEC and original equipment manufacturer (OEM) reliability standards for mobile phone applications. As of March 2018, M-Series packages up to 8 mm × 8 mm body size and fan-out ratio as low as 1.05 have been qualified through MSL1 component level and BLR requirements. Table 5.3 shows results for package sizes ranging from 5 mm to 8 mm per side across four different wafer foundries and three device technology nodes. Since M-Series was designed to meet MSL level 1 preconditioning requirements, more extensive package level testing was completed through larger sample sizes as compared to MSL level 3.

Table 5.4 shows the BLR performance of M-Series PMIC devices describing the number of TCT cycles until first failure and 5% failure for different die sizes tested under standard JEDEC test conditions.

Figure 5.31 shows that front side EMC thickness from 10 μm to 40 μm has no measurable effect on BLR performance of M-Series. The 40 μm curve is the left curve while the 10 and 25 μm curve are on top of each other on the right. This shows that even a small volume of EMC over the active die has a big advantage in improving BLR performance.

Table 5.3 M-Series Reliability Results.

M-Series Package-Level Reliability Results							
Device	**Si Node**	**Fab**	**Test type**	**MSL 1**		**MSL 3**	
				TCG	**uHAST**	**TCG**	**uHAST**
A	150 nm	1	electrical	0/216	0/220	0/68	0/66
			CSAM	0/216	0/220	0/68	0/66
		2	electrical	0/240	0/240	0/80	0/80
			CSAM	0/240	0/240	0/80	0/80
B	40 nm	3	electrical	0/400	0/400	0/240	0/240
			CSAM	0/400	0/400	0/240	0/240
C	14 nm	4	electrical	0/240	0/240	0/80	0/80
			CSAM	0/240	0/240	0/80	0/80
Test conditions		MSL 1 and MSL 3 – JEDEC/IPC joint industry standard J-STD-020A TCG = 1000 cycles air to air temperature cycling from −40 to +125 °C uHAST = 96 hours at 130 °C in 85% R.H., 33.3 psi Electrical = open, short or leakage failures CSAM = visible delamination as detected with 20 μm resolution					

Table 5.4 BLR performance of M-Series PMIC—number of TCT cycles until first failure and 5% failed.

Device	Die area (mm^2)	# of BGAs	First failed (TCT cycles)	5% failed (TCT cycles)
Device A	18	126	1112	1149
Device B	20	144	1015	1143
Device C	23	161	876	953
Device D	25	177	787	832

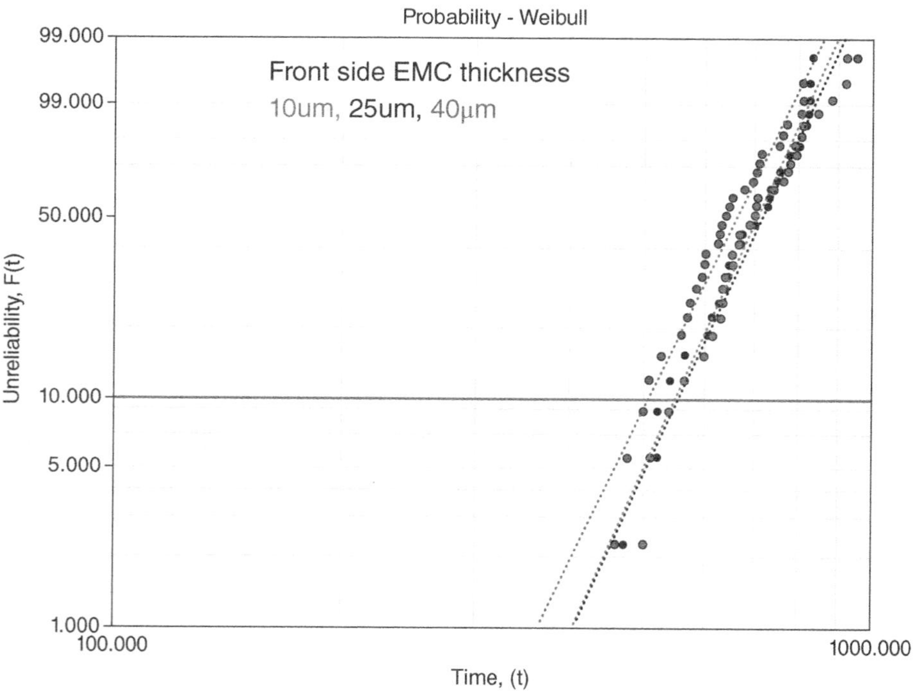

Figure 5.31 Weibull probability plot showing BLR impact of front side EMC thickness.

The significant improvement in BLR temperature cycle capability of M-Series versus conventional WLP and, by similarity, conventional fan-out constructed only with spin-on dielectric stress buffer layers is shown in Figure 5.4.

5.10 Electrical Test Considerations

The application of test to fan-out typically involves taking known good die from the wafer fab foundries and embedding within the mold and testing in panel format before singulation and tape and reel. Initially probers struggled to handle panels with > 1 mm of warpage, but the industry has evolved to be able to manage up to 5 mm of warpage with the latest generation of probers. Testing can also be done after singulation in a frame prober. Singulated testing has an advantage of testing for defects that occur in the dicing process that would otherwise escape if doing a full panel probe.

As mentioned earlier, the 600 mm format was chosen in part due to test. Segmenting a 600 mm panel into four equal 300 mm square subpanels allows the reuse of current probers with small modifications to handle the square format. Another test consideration is testing during the manufacturing process. As mentioned earlier in Section 5.4, after the Cu studs are built, an electrical test is done to measure contact resistance to the Al pad. This is repeated at RDL and UBM allowing quick feedback on potential processing issues. As packages become more complex with 3D integration using interposers, bridge chips, customer die, or multi-die, testing within the manufacturing process is of greater importance.

5.11 Summary

This chapter has described in detail the features of M-Series as a chip-first face-up FO-WLP technology that provides several important reliability and manufacturing benefits compared to traditional WLP and conventional fan-out packaging technologies. These features of Gen1 M-Series deliver superior BLR performance, while Gen2 M-Series will enable scaling to 2 μm line and space features for chiplets and heterogenous integration. AP as a real-time DDM tool enabling single- and multi-die integration makes possible the highest density design rules and support high-density applications was explained. The development of new AP techniques such as adaptive metal fills to optimize topography in multi RDL products was also described. M-Series enabling a 3DPoP structure as well as scaling to high-density interconnect was discussed in detail. This chapter also highlights Deca's next-generation M-Series roadmap that enables die-to-die interconnect of 250 wires/mm/layer where high-density routing can be achieved without the need for complicated bridge chips in substrates. These key technologies of M-Series and AP will be essential as the industry move toward chiplets and high-density interconnect.

References

1 Rogers, B., Scanlan, C., Olson, T. (2013). Implementation of a fully molded fan-out packaging technology. *IWLPC Proceedings*, San Jose, USA (November 2013).

2 Bishop, C., Olson, T., Scanlan, C. (2016). Adaptive patterning design methodologies. *2016 IEEE 66th Electronic Components and Technology Conference (ECTC)*, Las Vegas, USA (May 2016).

3 Brunnbauer, M., Fugut, E., Beer, G. et al. (2006). Embedded wafer level ball grid array (eWLB). *Electronics Packaging Technology Conference 8th Proceedings*, Singapore (December 2006).

4 Keser, B., Amrine, C., Fay, O. et al. (2007). The redistributed chip package: a breakthrough for advanced packaging. *2007 Electronic Components and Technology Conference*, Nevada, USA (May 2007).

5 Zhao, W., Nakamoto, M., Dhandapani, K. et al. (2017). "Electrical chip-board interaction (e-CBI) of wafer level packaging technology." *IMAPS Advancing Microelectronics* (October 2017).

6 Prismark Partners LLC (2021). Apple A14: Customer teardown analysis. pp. 1–50.

7 Olson, T. and Scanlan, C. (2018). M-series fan-out with adaptive patterning. In: *Advances in Embedded and fan-out Wafer Level Packaging Technologies* (eds. B. Keser and S. Kroehnert), 117–140. New Jersey: Wiley.

8 C. Bishop, B. Rogers, C. Scanlan et al. (2016). Adaptive patterning design methodologies. *2016 IEEE 66th Electronic Components and Technology Conference (ECTC)*, Las Vegas, NV, USA (May 2016).

9 Ramalingam, S. (2016) HBM package integration: technology trends, challenges and applications. *2016 IEEE Hot Chips 28 Symposium (HCS)*, Cupertino, CA, USA (August 2016).

10 Thai, T., Dalmia, S., Hagn, J. et al. (2019). Novel multicore PCB and substrate solutions for ultra broadband dual polarized antennas for 5g millimeter wave covering 28ghz 39ghz range. *2019 IEEE 69th Electronic Components and Technology Conference (ECTC)*, Las Vegas, NV, USA (May 2019).

11 Wang, C., Tang, T., Lin, C. et al., (2018). InFO AiP technology for high performance and compact 5G millimeter wave system integration. *2018 IEEE 68th Electronic Components and Technology Conference (ECTC)*, San Diego, CA, USA (May 2018).

12 Lim, J., Pandey, V., Oo A. K. et al. (2017). Fan-out wafer level eWLB technology as an advanced system-in-package solution. *2017 14th International Wafer-Level Packaging Conference (IWLPC)*, San Jose, CA, USA (October 2017).

13 Kim, J., Choi, I., Park, J. et al. (2018). Fan-out panel level package with fine pitch pattern. *2018 IEEE 68th Electronic Components and Technology Conference (ECTC)*, San Diego, CA, USA (May 2018).

6

Panel-Level Packaging for Heterogenous Integration

M. Töpper, T. Braun, M. Billaud, and L. Stobbe

Fraunhofer IZM, Berlin, Germany

6.1 Introduction

Heterogeneous integration is one of the most promising ways to bridge the gap between emerging microelectronics and its derived applications, and both are pushing new packaging technologies [1–4]. New technology architectures are needed to integrate the progress made in nanoelectronics, wireless technologies, and photonic component technologies into electronic systems.

A strong link exists between the development of components and system-oriented integration technologies reflected by the SiP (system-in-package) approach. SiP is concentrated on a nonspecific approach based on a widespread generation of packages with more than 1000 package families with subgroups that can be found on the market, some highly specialized to niche markets, others generic to serve multiple industrial areas [5]. The definition of SiP in IEEE EPS' "Heterogeneous Integration Roadmap 2019" refers to a package (such as small-outline [SO], quad flat package [QFP], ball grid array [BGA], chip size package [CSP], or land grid array [LGA]) that has multiple die (Si, SiGe, SiC, III/Vs like GaAs or GaN) plus optional passives integrated together.

The increasingly central role of electronics in ensuring system reliability and functionality will further push the reliability of electronics to a higher degree in the future. Reliability standards will see failure rates expressed as failures per billion parts rather than per million parts, as is currently the case.

The concepts of "More Moore" as well as "More than Moore" are aiming at single component solutions such as SoC (system-on-chip), but increasingly new applications are demanding product-specific integration technologies with different functions. Thus, the future of electronic system integration will see a combination of "More Moore" and "More than Moore" components, combined in one

Embedded and Fan-Out Wafer and Panel Level Packaging Technologies for Advanced Application Spaces: High Performance Compute and System-in-Package, First Edition. Edited by Beth Keser and Steffen Kröhnert.

package, beyond the SiP concept. Therefore, heterogeneous integration is the next challenge exceeding the boundaries of current SiPs. Such a heterogeneous integration approach provides lower cost and risk compared to a SoC including a shorter time to market and a higher degree of flexibility. One of the biggest advantages of heterogeneous integration technologies is the possibility to integrate components based on vastly different technologies and materials such as power devices (GaN, SiC or others), photonic or RF components (using InP, GaAs or others), energy harvesting, MEMS, or smart displays. There is no limitation to pure CMOS, which is a must for SoC.

Fan-out wafer-level packaging (FO-WLP) or panel-level packaging (FO-PLP) is now the optimal result of the merge between single-chip and multi-chip packages [6, 7]. The evolution of single-chip packages (SCPs) started from small metal boxes and developed into dual inline packaging (DIP) for through-hole assembly and ultimately surface mount technology (SMT) packages such as the QFP to the BGA. BGA packages use a rigid or flexible interposer for the redistribution from the peripheral pads to the area array (Figure 6.1).

The interconnection to the chip can be done by wire bonding (WB) or flip chip (FC). A further miniaturization to a maximum of 1.2x the chip size resulted in the concept of the chip-scale package (CSP). DIP and QFP represent packages with peripheral I/Os while BGAs and CSPs are area array packages, which are assembled with FC interconnects. FC is a face-down assembly technique originally

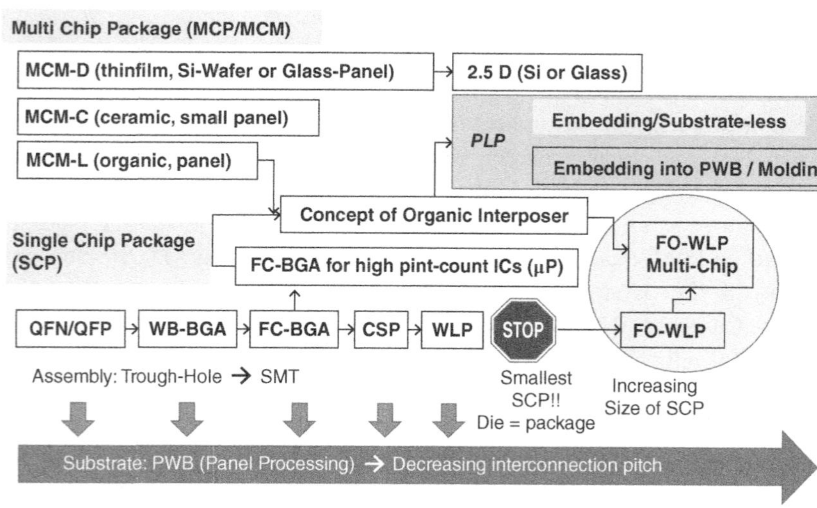

Figure 6.1 Evolution of single-chip packages (SCP) and multi-chip packages/modules (MCP/MCM) resulting into packages based on embedding technologies (FO-WLP/panel-level packaging).

developed by IBM as C4 (controlled collapse chip connection). Low cost and miniaturization with increased functionality have driven the packaging industry toward adopting the concepts of wafer-level packaging (WLP). The concept of WLP is to complete most of the packaging steps at wafer level. WLPs are true CSPs using only a fan-in routing, which is also a limitation. The minimum package size has been achieved with WLP because the package size is equal to the die size.

Substrate-based packages (e.g., FC-CSP, FC-BGA) can be laminate-based such as printed circuit boards/printed wiring boards (PCB/PWB) or a multi-layer ceramic (MLC) such as low-temperature co-fired ceramics (LTCC). These substrates combine several packaged ICs and passive components to build the final microelectronic system for the users. The types of the substrates that have been used for the classifying MCMs are:

- MCM-D: D for deposited, i.e., thin film processes usually on Si-substrates, but also glass may be used;
- MCM-C: C for ceramics, i.e., using ceramic (pure Al_2O_3 or AlN) but also multi-layer LTCCs or HTCCs (high-temperature co-fired ceramics) as the substrates;
- MCM-L: L for laminated substrates, i.e., substrates usually called PWB or PCB.

A major difference for these types has been the temperature budget for the multilayer process on top of the substrates. There is nearly no limitation for the process on ceramic, but the laminates are normally restricted to process temperatures below 200 °C. The MCM-D is in a range of the BEOL (back-end of line) for CMOS technology.

6.2 Fan-Out Panel-Level Packaging

The migration of WLP to FO-WLP to extend the routing area opened the idea for embedding active and passive components into molding compound [6]. FO-WLP has the potential for significant package miniaturization in both package volume and thickness. The main advantages of FO-WLP and FO-PLP are the substrate-less package, lower thermal resistance, higher performance due to shorter interconnects together with direct integrated circuit (IC) connection by thin film metallization instead of wire bonds or FC bumps, and lower parasitic effects [7, 8]. Especially the inductance of FO-WLP is much lower compared to FC-BGA packages. In addition, the redistribution layer (RDL) can also provide embedded passives (R, L, C) as well as antenna structures using a multi-layer structure. Therefore, FO-WLP are used for multi-chip packages and can be viewed as a synergy between optimal chip package and the MCM concepts. Hence, technology is well suited for heterogeneous integration.

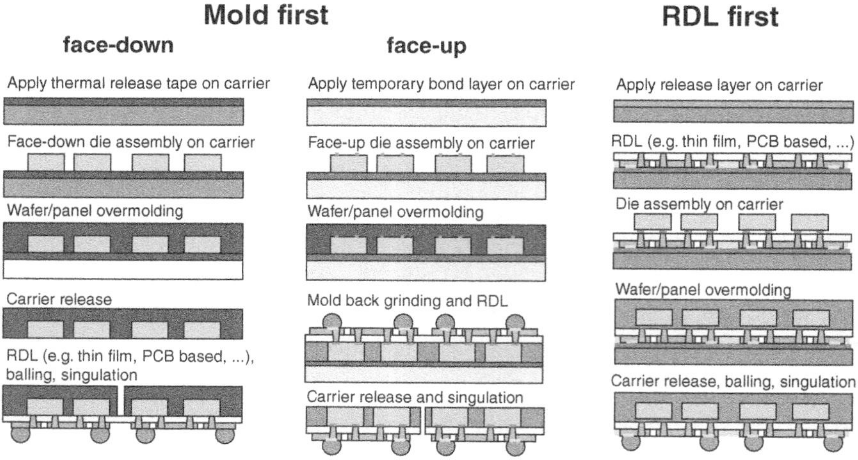

Figure 6.2 Comparison between mold first and RDL first.

There are two main approaches for embedded die technologies: FO-WLP or FO-PLP, where die are embedded into polymer encapsulants or 3D vertical integration in organic substrates, where die are embedded into PCBs. For FO-WLP, two basic process flows are encountered the mold first or the RDL first approach. Process flows of both options are depicted in Figure 6.2.

Mold first face-down starts with die assembly on an intermediate carrier followed by overmolding and debonding of the molded wafer/panel from the carrier [9]. The RDL, typically based on thin film technology, is finally applied on the reconfigured molded wafer/panel. The face-up approach also starts with die assembly on a carrier with a temporary adhesive layer. Here die have a copper (Cu) bump and are placed face-up on the carrier. After overmolding, backgrinding opens access to the Cu-bumps of the die again. The redistribution is applied, and finally the wafer is released from the carrier and diced for package singulation.

The RDL-first process is comparable to an advanced FC assembly. Here the RDL is applied first on an intermediate carrier, and the bumped die are assembled by chip-to-wafer bonding on the RDL. Afterward the assembly is underfilled and overmolded, and the molded wafer including RDL must be released from the carrier. Mold first face-down has the shortest interconnect with a direct plated via. This may lead to best RF performance at higher frequencies due to lowest loss especially when chip-to-chip connection is considered. Mold first face-up needs a Cu pillar, and RDL first requires a soldered interconnect. Additionally, the last two options have an additional polymer/underfill layer between die and RDL. Therefore, the mold first face-down approach is the most promising solution for RF and mm-wave applications. Besides the shortest interconnection, this technology has

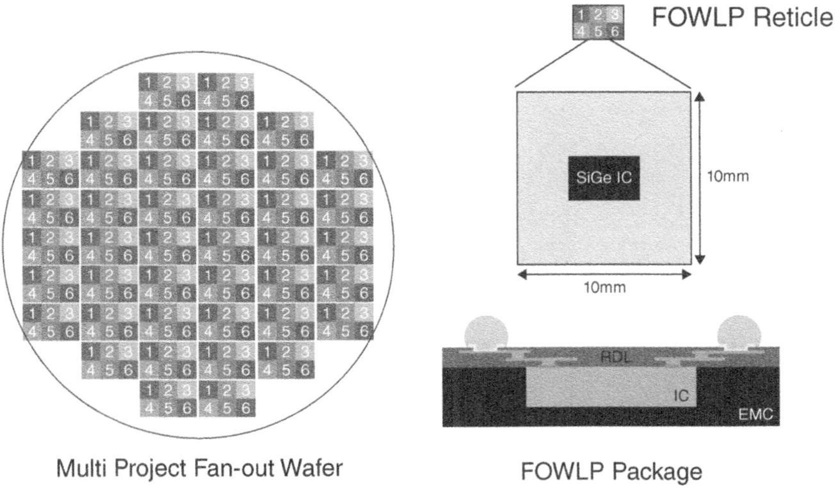

Figure 6.3 Principle of the MPW concept for FO-WLP.

the highest flexibility and highest potential for heterogeneous integration. Die and passive components can be integrated without any additional preparation such as bumping as well as components from different suppliers or even from different compounds as, for example, Si, SiGe, or GaN can be combined in one package. This has been demonstrated using multi-project wafer (MPW) processing, which is an established approach in semiconductor manufacturing for fast and low-cost prototyping [10]. This idea has been demonstrated with FO-WLP. Here die from different sources or different technologies with varying thicknesses and sizes can be handled and packaged with one integration technology. In Figure 6.3, the concept of MPW for fan-out is shown. The wafer is distributed in a reticle layout, and each of these reticles consist of six different designs as an example. Especially for RF applications, this might offer a path to down-select the optimal components and design to perfect product performance.

For the technology demonstration, five functional die and one test die for package RF characterization and reliability testing have been selected. The different ICs have been fabricated in a 250/130 nm high-performance SiGe BiCMOS technology [11]. The MPW FO-WLP have a large variety of ICs with different functionalities including a 60 GHz low-noise amplifier (LNA), a 120 GHz transceiver, a 50 GHz IF-IF (intermediate frequency) converter, a DC-15 GHz variable gain amplifier (VGA) and a phase shifter working at 10 GHz. Reticle and package design are depicted in Figure 6.4.

A dielectric layer with a curing temperature below 250 °C was selected due to the low dielectric constant and loss of the material. Low cure dielectrics are important in FO-WLP because the reconfigured wafer is based on epoxy mold compounds

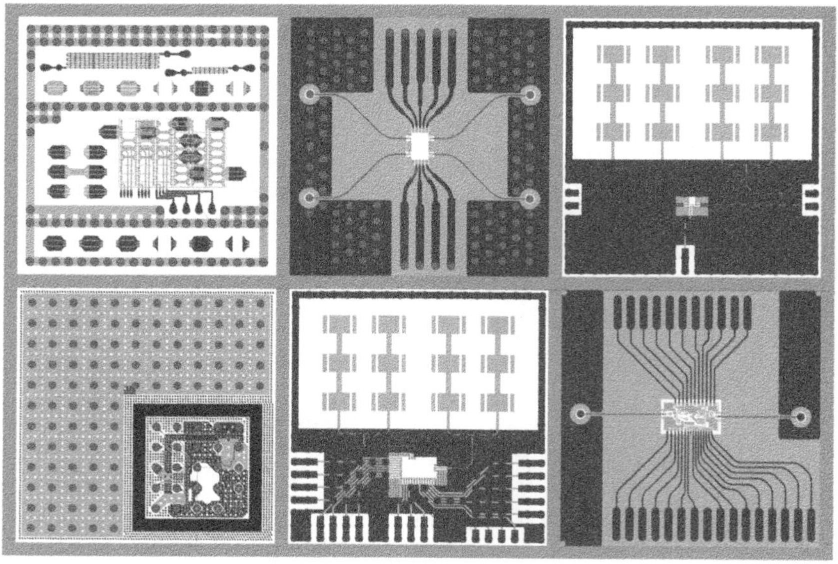

Figure 6.4 Multi-project FO-WLP reticle layout with six different package types.

Figure 6.5 Multi-project FO-WLP packages with RDL before bumping (left) and after solder attach (right).

(EMC) with glass transition (T_g) temperatures less than 200 °C. The RDL consists of three dielectric layers and two metal layers plus an under-bump metallization (UBM). The different package types with RDL are shown in Figure 6.5. In summary, the electrical tests have proven this packaging concept for heterogeneous integration with a focus on RF applications [12].

6.3 Economic Efficiency Analysis of PLP

In response to the emerging trends in packaging technologies, detailed cost analysis models are required to assess the economic and environmental feasibility of different technology options, business scenarios, design options, and process configurations [13]. As established models do not provide a sufficient level of granularity both in representing realistic industrial processes and detailing the cost allocation, new multi-level cost models are necessary for comparative cost assessments. Indeed, the advanced and new packaging manufacturing processes such as FO-PLP require a more accurate cost prediction. High data granularity is needed in addition to the detailed cost allocation in order to allow differentiated comparisons of, for instance, different technology options (e.g., chip-first or chip last, lithography technology, and panel design). It has to be adaptive to different products and their package design (e.g., number of die, package dimensions, and number of RDLs) and to a specific business case (e.g., production location, production volume, and production time frame). The cost model methodology should be based on a bottom-up approach. The model essentially comprises the sequence of the individual process steps and assigns the appropriate equipment to each step. Each piece of equipment includes multiple parameters such as equipment invest, floor space requirement including clean room class, processing duration per substrate (machine and operator), and power consumption. The model also has to include material type and quantities and costs per substrate, as well as the possible reusability. The scaling must be done to a specific product scenario with a chosen volume, production or production time frame, production location and conditions, productivity/maintenance time, and labor costs. In addition, facilities costs, related to the chosen location, must be considered (electricity, clean room capital expenses and operating expenses, and rent). The cost model developed by Fraunhofer IZM is shown in Figure 6.6.

In an example, we will focus on the mold-first face-down approach. The packaging process involves three RDLs with photosensitive dielectric, Ni-Au UBM,

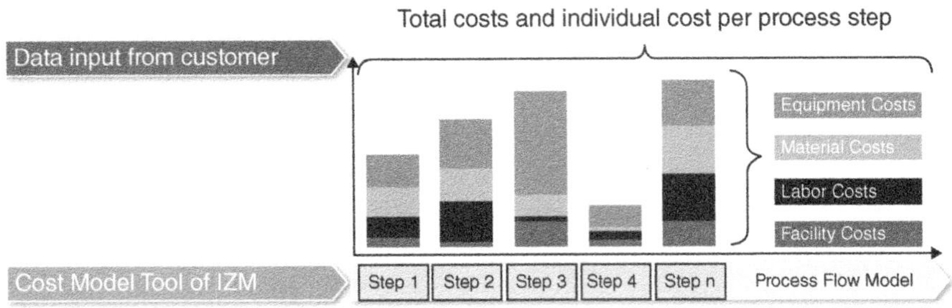

Figure 6.6 Specific cost model for PLP.

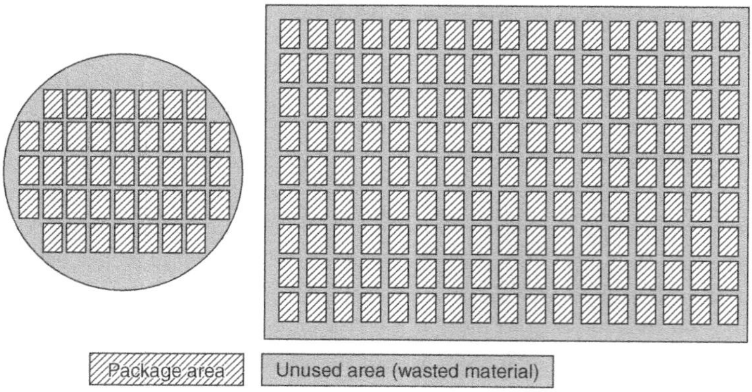

Figure 6.7 Comparison between fan-out technology on wafer vs. on panel.

and SnAgCu balls. The process flow starts with the component placement on the carrier and ends with package singulation. It is structured in four main modules (with their related challenges): assembly (die shift compensation, speed, and accuracy), molding (warpage, thickness, die aspect ratio, and die count on panel), RDL (e.g., L/S reduction), and UBM/ball attach. Test is not included in this model. The bottom-up approach of this cost model allows adjustment of relevant technical parameters related to every process step in order to evaluate its impact on total costs. Process quality measured in yield data do not exist at this level of development and cannot be properly estimated from the technology demonstration. The cost model analysis points out the beneficial effects of a panel geometry over a wafer geometry. It is already obvious on a simple drawing that PLP is more efficient in processing of materials compared to wafers (Figure 6.7), especially when the package size is large. Detailed cost analysis has been calculated for different substrate sizes and geometries vs. different package sizes (Figure 6.8).

The relative cost per substrate area is increasing steeply with decreasing package size due to the cost of the assembly process [14]. This is much more drastic for modules having three chips compared to single chip modules. Worst case is the 300 mm approach with 1 small die. This relative cost of this case is much improved for a panel size of 610 mm × 457 mm. This increase can be lowered by implementation of assembly equipment with higher UPH (units per hour). In conclusion, the relative cost is always lower for rectangular large panels compared to round wafers keeping the number of chips constant. In addition, increasing the size of the panel is further decreasing the relative cost per reconfigured panel area. Therefore, the main advantage of panel over wafer is its high flexibility for a variety of package sizes. A direct key performance indicator (KPI) is the higher area utilization (AU) of the substrate for most of the package configurations. AU describes the proportion of the substrate area that is covered by packages. Maximizing AU has two

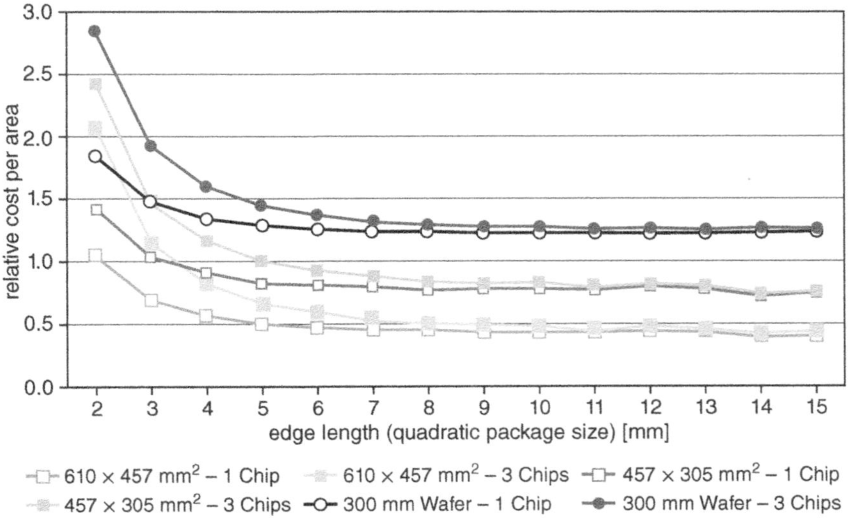

Figure 6.8 Relative cost per area for different reconfigured wafer or panel geometries and different number of die per package: comparison between panels (610 × 457 mm^2 or 457 × 305 mm^2) and wafer (300 mm).

major advantages: first, the time advantage by having more packages per reconfigured wafer or panel (the number of reconfigured panels or wafers processed to reach a defined production volume is reduced) and second, the cost and environmental advantage by reducing the amount of materials needed per package, since many materials are applied to the whole substrate and are subsequently wasted on non-package area.

Figure 6.9 shows the share of package aspect ratio combinations that allow a certain AU threshold on 300 mm wafers and on 457 mm × 610 mm panels. For this calculation, the package edge length and width vary between 1 and 50 mm (with a step of 1 mm), and the aspect ratio is limited to three. PLP enables a high AU (≥ 90%) for 70% of package size combinations. On the contrary, the 300 mm wafer leads to an AU between 85 and 88% but only for small packages. For packages larger than 11 × 11 mm^2, the AU remains under 85%. The panel achieves a much higher AU for a wider range of package size combinations and especially for large packages. Furthermore, rectangular panels outperform squared panels for the AU of rectangular packages because those can be placed either horizontally or vertically to maximize AU. Therefore, rectangular panels provide greater flexibility in package design as more edge length combinations allow for high AU. In addition, unutilized substrate area results in a materials overconsumption and waste generation at each process step (see Figure 6.10).

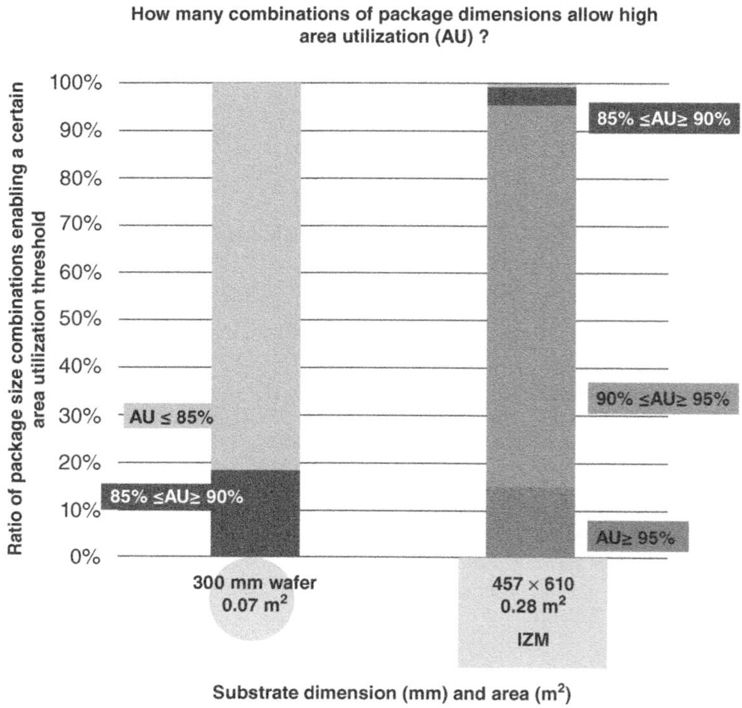

Figure 6.9 Comparison of the ratio of package size dimension combinations allowing a certain AU rate on 300 mm wafer compared to PLP by providing higher package design flexibility.

To illustrate the order of magnitude of this material loss, Figure 6.10 compares the amount of EMC loss for the production of 50 million packages on a 457 mm × 610 mm panel and on a 300 mm wafer. For this production volume, around 76 000 panels (457 mm × 610 mm) or 350 000 wafers (300 mm) with 20 mm × 20 mm packages are required. The AU is 94.7 and 82.1%, requiring 402 and 1725 kg of EMC on unused area, respectively on panel and wafer. In total, the consumption of EMC can be reduced by 14% by producing on panel instead of wafer. The EMC thickness can be reduced but is always limited by the chip thickness and handling issues. Furthermore, EMC is one of the most expensive materials in the FO-PLP process. A higher utilization of the substrate area, thanks to panel, is a simple and cost-effective solution to significantly reduce material wastes and thus production costs, as well as environmental impact.

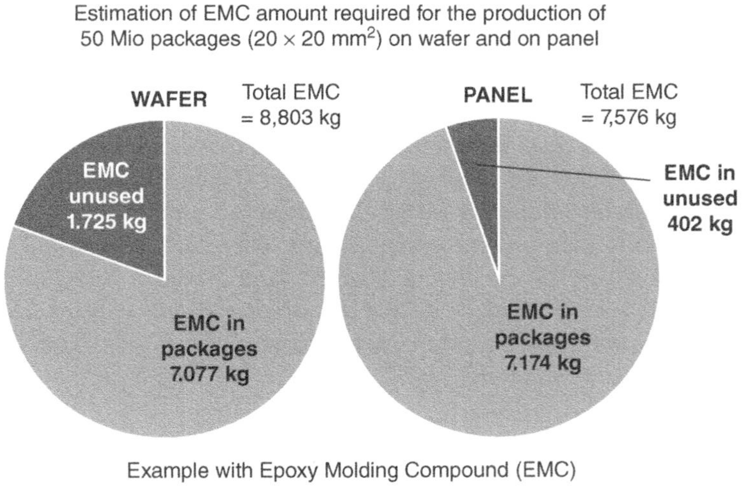

Figure 6.10 Unutilized reconfigured wafer or panel area results in a materials overconsumption and waste generation for wafer and panel processing.

6.4 Summary

There had been a steady development in the miniaturization of SCPs until the concept of WLP, the smallest possible package, had been fully commercialized. Bare die without a package had been an issue for MCMs in the 1990's due to problems in testability and assembly. Therefore WLP was the perfect compromise of package miniaturization and a well-protected die surface with large solder bumps for a high-throughput assembly process with the testability to confirm goodness. This package type is currently number one for mobile devices combining smallest form factor with high electrical functionality. The incredible progress of the CMOS technology (recent semiconductor front-end process roadmaps indicate technology nodes down to 2 nm and with integrated device manufacturers [IDM] and foundries are already producing in 7 nm) in conjunction with the demands of heterogeneous integration requires fan-out solutions. Thus, the different fan-out package types offer a cost-effective solution for multi-chip integration combining high-density thin film routing being currently used for high-performance Si-interposers with embedding chips into organic substrates or EMC.

There is a clear cost advantage of panel-level processing for large modules but the wafer-level approach might be a solution for high-performance modules of small size and with the potential of integrating chiplets due to the more aggressive wafer-level design rule capabilities to interconnect the chiplets. Further research has to be done for the packaging of chiplets [15]. The definition of chiplets can be summarized as a physical piece of silicon that encapsulates an intellectual property (IP) subsystem being designed to be integrated with other chiplets through package-level integration, typically through advanced package integration and through the use of standardized interfaces. Therefore, chiplet technology is a different way of integrating multiple electrical functions in a package or system. The chiplet approach is a fast and less expensive way to assemble various types of third-party IPs, such as I/O drivers, memory ICs, and processor cores, in a package. A chiplet is not a package type, it is part of a packaging architecture. The question remains which package type will be the most effective with respect to performance and cost in the future.

References

1 Reichl, H., Aschenbrenner, R., Töpper, M. et al. (2009). Heterogeneous integration – building the foundation for innovative products. In: *More than Moore* (eds. G.Q. Zhang and A.J. van Roosmalen), 279–303. Springer Verlag.

2 Thomas, T., Becker, K.-F., Braun, T. et al. (2009). State-of-the-art of 3D SiP technology. *Proceedings of the International Microelectronics and Packaging Society (IMAPS)* Chapter Poland (21–24 September 2009) Pszczyna, Poland.

3 Töpper, M. and Tönnies, D. (2017). Microelectronic packaging. In: *Semiconductor Manufacturing Handbook*, 2e (ed. M.H. Geng), 173–204. McGraw-Hill.

4 Garrou, P. and Turlik, I. (1997). *Multichip Module Technology Handbook*. McGraw-Hill Professional.

5 IEEE-EPS Heterogeneous Integration Roadmap (2019). https://eps.ieee.org/technology/heterogeneous-integration-roadmap/2019-edition.html

6 Töpper, M., Ostmann, A., Braun, T. et al. (2019). History of embedded and fan-out packaging technology. In: *Advances in Embedded and Fan-Out Wafer Level Packaging Technologies* (eds. B. Keser and S. Kroehnert), 1–38. Wiley.

7 Meyer, T., Ofner, G., Bradl, S. et al. (2008). Embedded wafer level ball grid array (eWLB). *Proceedings of Electronics Packaging Conference (EPTC)*, Grand Copthorne Waterfront Hotel, Singapore (9–12 December 2008). IEEE.

8 Keser, B., Amrine, C., Duong, T. et al. (2007). The redistributed chip package: a breakthrough for advanced packaging. *Proceedings of Electronic Components and Technology Conference (ECTC)*, Reno, Nevada, USA (29 May – 1 June 2007). IEEE.

9 Braun, T., Voges, S., Töpper, M. et al. (2015). Material and process trends moving from FOWLP to FOPLP. *Proceedings of Electronics Packaging Conference (EPTC)*, Marina Mandarin, Singapore (2–4 December 2015). IEEE.

10 Braun, T., Raatz, S., Töpper, M. et al. (2017). Development of a multi-project fan-out wafer level packaging platform. *Proceedings of Electronic Components and Technology Conference (ECTC)*, Lake Buena Vista, Florida, USA (30 May – 2 June 2017). IEEE.

11 Heinemann, B., Barth, R., Bolze, D. et al. (2010). SiGe HBT technology with fT/fmax of 300GHz/500GHz and 2.0 ps CML gate delay. *Proceedings of the International Electron Devices Meeting*, San Francisco, CA, USA (6–8 December 2010). IEEE.

12 Kaynak, M., Wietstruck, M., Göritz, A. et al. (2017). 0.13-μm SiGe BiCMOS technology with More-than-Moore modules. *Proceedings of the IEEE Bipolar/BiCMOS Circuits and Technology Meeting (BCTM)*, Miami, FL, USA (19–21 October 2017). IEEE.

13 Billaud, M., Stobbe, L., Braun, T. et al. (2019). Process flow and cost modelling for fan out panel level packaging. *Proceedings of European Microelectronics and Packaging Conference (EMPC)*, Pisa, Italy (16–19 June 2019). IMAPS.

14 Braun, T., Becker, K.-F., Töpper, M. et al. (2019). Panel level packaging – from idea to industrialization. *Proceedings of the IEEE CPMT Symposium Japan (ICSJ)* Kyoto, Japan (18–20 November 2019).

15 Töpper, M., Braun, T., and Aschenbrenner, R. (2020). Electronic packaging for future electronic systems. *Journal Chip Scale Review* 24 (4): 10–14.

7

Next Generation Chip Embedding Technology for High Efficiency Power Modules and Power SiPs

Vikas Gupta, Kay Essig, C.T. Chiu, and Mark Gerber

ASE Group, Dallas, TX, USA

7.1 Technology Description

Cost, performance, and package size are some of the key drivers required in the next generation of package interconnect and package structure evolution (see Figure 7.1). Embedding active die into substrates was mainly driven by package miniaturization for communication handheld devices [1]. However, in the case of power modules, miniaturization is not the only driver that enhances the need for embedded die substrate (EDS) packages. Embedding has shown to have additional beneficial effects on electrical performance and thermal dissipation, especially for mid-power modules (from a few hundred watts to a few kilowatts) [2, 3]. Enhanced reliability, electrical performance, and thermal dissipation are key benefits for EDS power modules [4].

As the switching frequencies increase from kilohertz to megahertz there is an increased focus on parasitics. This is increasingly critical for wide band-gap (WBG) gallium arsenide (GaAs), gallium nitride (GaN), and silicon carbide (SiC) power device implementation. WBG devices switch one to two orders of magnitude faster than silicon devices. To realize full dynamic performance of a WBG device, device and circuit parasitic inductance, resistance, and capacitance must be significantly reduced. Figure 7.2 shows the most critical parasitic elements on the interconnection path in the power train and the driver section of a buck converter [1]. The schematic in Figure 7.3 outlines how package parasitics can contribute to power losses by heating. Embedded packaging provides three distinct advantages:

1. Better interconnections: These reduce parasitics through implementation of shorter low resistance interconnects, shorter distance between active and passive components, and fewer capacitance areas, thereby reduced power loss due

Embedded and Fan-Out Wafer and Panel Level Packaging Technologies for Advanced Application Spaces: High Performance Compute and System-in-Package, First Edition. Edited by Beth Keser and Steffen Kröhnert.

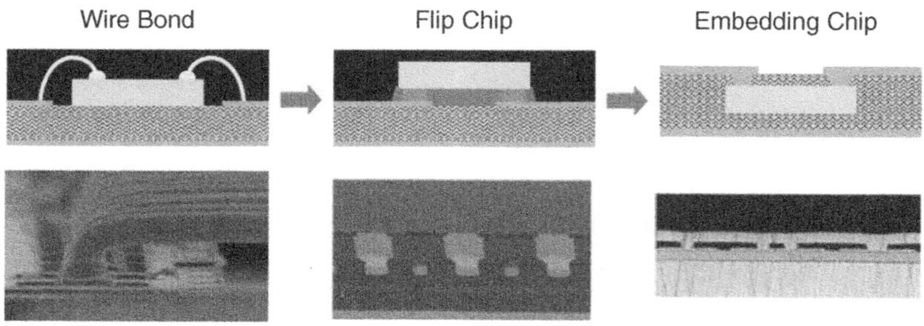

Figure 7.1 Package interconnect evolution.

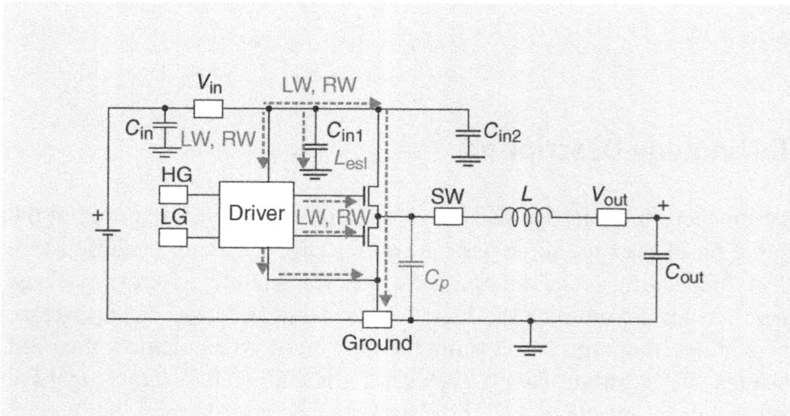

Figure 7.2 Parasitic interconnect elements in the power train and the gate driver of a buck converter [1].

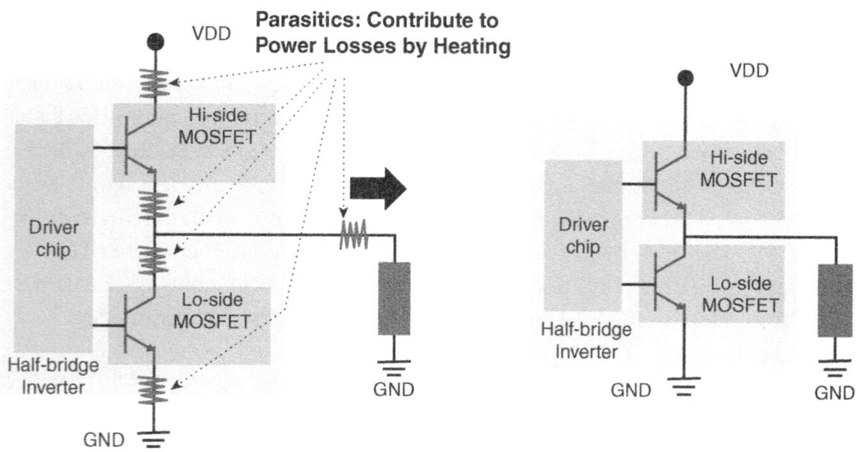

Figure 7.3 Half-bridge inverter challenges—benefits of embedding.

to heating and turn on resistance ($R_{DS(ON)}$). The short interconnection introduces low inductance, which also benefits high switching frequency operation on GaN and SiC.

2. Improved thermal dissipation: Reduced parasitics result in higher power efficiency and the embedded. Structure design has a better thermal dissipation path.
3. Smaller package: Board space savings and closer integration to the load limits losses on board outside of the package.

These benefits are especially found in EDS where the die mounted on a lead frame are embedded in organic laminate material. The connection from the printed circuit board (PCB) pad to the gate and source pads on the MOSFET side are done by copper (Cu) filled micro via, directly plated onto the Cu finished die pads. The EDS configuration eliminates wire bonds or intermetallic compound (IMC) layers with higher resistance and inductance found in traditional power packages.

A very small number of suppliers have EDS manufacturing expertise and capacity as EDS requires capabilities that span package assembly and substrate fabrication that are very rarely found all together at an outsourced assembly and test services (OSAT) or substrate supplier. While substrate suppliers lack preassembly processes, such as wafer handling, bumping, grinding, dicing, and die placement, OSATs have no expertise in substrate-related process steps. The embedded components typically require Cu terminals to withstand laser drilling and to be connected by a Cu plating process. This becomes even more challenging for thin wafers that also require back side metallization (BSM). The BSM process is completed at a bumping facility. The overall flow results in a complicated supply chain, which is not preferred by potential customers because of supply chain management, complexity of issue resolution ownership, and cost. Ideally the customers prefer a "one stop shop" experience where the wafers are shipped from the foundry to the embedding facility, which is capable of performing all the required process steps, from bumping to preassembly to die attach, followed by the embedding in substrate, substrate assembly to singulation, and final test. Current dedicated manufacturing lines are capable to perform the embedding and substrate finishing processes [4, 5].

Assembly yields is another major aspect in all chip-embedding technologies because all the embedded die packages contain one or more known good die (KGD). Hence each failing substrate unit is combined with the loss of at least one or more KGD. The substrate yield is directly linked to the design rules, especially in the field of power module devices, which use thick Cu in the redistribution layers (RDL). The combination of advanced design rules and RDL with thicker Cu could result in poor yields in high-volume manufacturing (HVM). Low yields in combination with a high number of KGDs per unit can make an embedded package solution unaffordable. Therefore, it is recommended to use conservative design rules for the substrate RDL when possible [3].

Based on the advanced embedded active system integration (a-EASI) production experience, KGD is not an issue for power nodes. The controller die and MOSFET die are typically in mature wafer node process. The KGD yield is relatively high compared to logic devices. In addition, power modules have low I/O count, and there is no high-density requirement. Thus, the thick RDL process is not a bottleneck either.

The evolution of power modules trends to smaller, lighter weight package and higher switching frequencies. Thus, the package parasitics, especially the inductance and resistivity, become a critical factor to fully realize the electrical performance of the power module. The increasing current density needs improved thermal dissipation to correlate with the enhanced reliability requirements for industrial and automotive standards. EDS is rather efficient in heat dissipation when the die is directly bonded to a large metal pad or even better to a lead frame. If the design can be satisfied by a two-layer EDS, then the thermal dissipation is increased, plus there can be significant electrical performance advantages. The embedded power substrate can also easily be expanded to multi-die or -component substrates to provide complete functional integration. In addition, additional components may be surface mounted to further enhance the power module or SiP.

In this chapter, ASE Group's trademarked embedded substrate process and package termed a-EASI™, which uses the die-first process sequence, is detailed.

7.2 Basic Package Construction

The embedded power module a-EASI was introduced in high-volume manufacturing in 2014 with a half bridge configuration—a low side and a high side MOSFET together with a driver die placed on a lead frame while embedded into organic laminate material. This structure shown in Figures 7.4 and 7.5 is called a-EASI P1 (first power module) structure.

The MOSFET drain pads were conductively connected by TLPB (transient liquid phase bonding), which forms a thin layer of about 5 μm of Cu_3Sn/Cu_6Sn_5 IMC. Driver die pads and MOSFETs gate and source pads were directly connected through microvias while the drain on the die backside is connected to the lead frame through conductive die-attach, and from there the connection to the package drain pads was realized with microvias as well. All the connections in this package consist of single or multiple Cu-filled microvia. In this interconnect scheme the low parasitic connection matrix has a lower resistance and inductance and thus improved parasitics, better reliability as compared to Al and Cu wire

Figure 7.4 Schematic drawing of the a-EASI P1 structure.

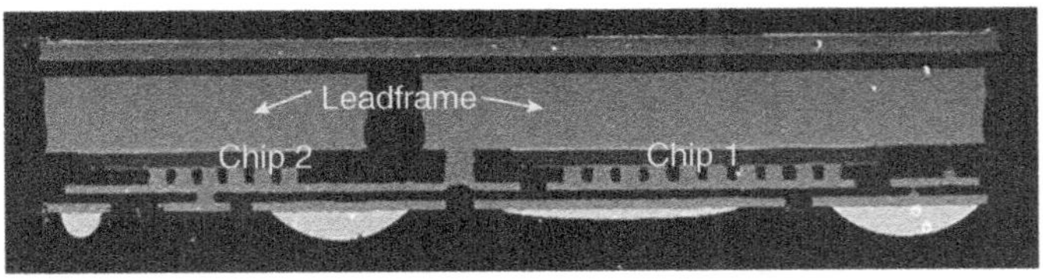

Figure 7.5 X-Section picture of an a-EASI P1 package.

bonds and solder drain connections. As is evident from the cross-section in Figure 7.5, the basic structure characteristics provide several inherent advantages:

1. Lead frame base: good thermal dissipation and native electromagnetic interference (EMI) shielding. A full thickness lead frame provides an order of magnitude thicker Cu as compared to conventional substrate trace, which results in significantly improved current-carrying and thermal dissipation capability, specifically advantageous for power devices.
2. High conductivity die-attach: direct connect of die to thick lead frame with or without back side metal using a high conductivity die-attach enables high efficiency thermal dissipation.
3. Copper via interconnect: low $R_{DS(ON)}$, low inductance, and high current-carrying capability.
4. High T_g prepreg: greater than 2.5 kV breakdown voltage, suitable for high-voltage applications.

The power modules in the P1 structure showed very good electrical performance and came in with yields well above 99%. The TLPB die attach, preformation of the prepregs, and inspection steps were further optimized in the P2 structure.

The next generation embedded power module P2 structure holds the MOSFETs and driver die in cavities of the lead frame. The drains of the MOSFETs with a vertical current flow are conductively connected to the bottom of the lead frame cavity mainly by hybrid Ag sintering (epoxy glue with high Ag content) technology; likewise, Ag sintering technology can be used [6]. A standard prepreg is used in a standard lamination process to embed the die on a lead frame into an organic substrate. As the die are attached into a cavity, the risk of die crack in the lamination process step is minimized, and the connecting micro via will all have approximately the same depth, therefore, simplifying the laser drill process.

In the P2 structure of a-EASI embedded power module with one RDL, it is possible to design and connect the package to the PCB either with the lead frame toward the bottom of the package resulting in QFN-style (quadflat, no lead) or with the lead frame to the package top and with solder balls attached to the RDL resulting in a BGA-style (ball grid array), as shown in Figures 7.6–7.8.

In both package styles (QFN or BGA), the die have good shielding, and it is possible to have passive components attached to the package top layer. Thus, the use of the real estate occupied by the power module is optimized. This package type offers the lowest thermal resistance, and hot spots cannot be expected. On the package top side, a heat sink can be directly attached using a thermal conductive adhesive, and former active cooled devices can be turned into passive cooled devices.

For the need of more routing capability on smaller package outline the a-EASI P3 structure was developed. In the P3 structure the MOSFETs are assembled on opposite sides of the lead frame (Figures 7.9–7.11), before the embedding process

Figure 7.6 Schematic drawing of a-EASI P2 structure in QFN and BGA package style.

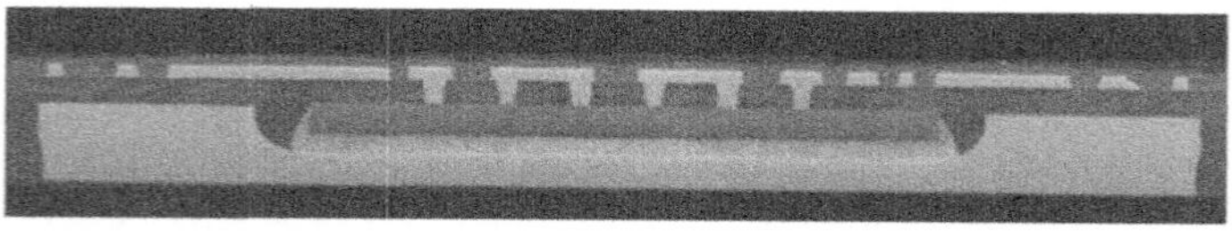

Figure 7.7 X-section picture of an a-EASI P2 structure in a QFN package style.

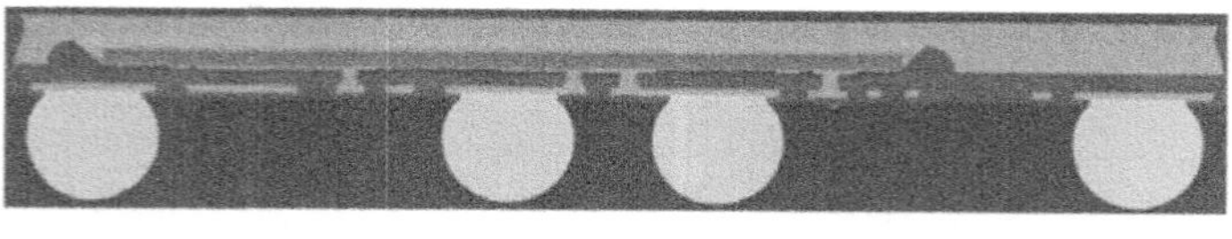

Figure 7.8 X-section picture of an a-EASI P2 structure in a BGA package style.

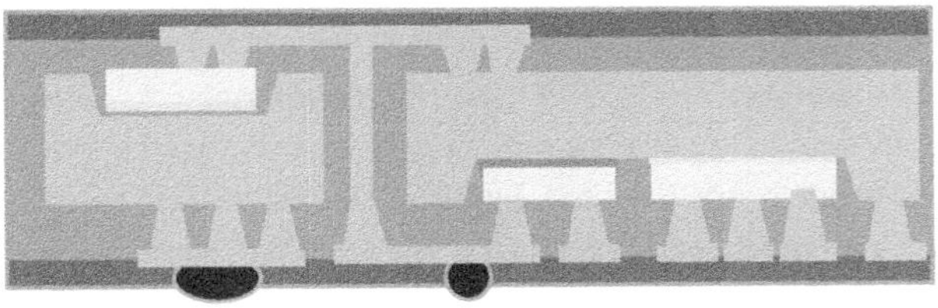

Figure 7.9 An a-EASI P3 structure with MOSFETs in cavities on opposite sides of the lead frame (side-by-side driver die).

Figure 7.10 An a-EASI P3 structure with MOSFETs in cavities on opposite sides of the lead frame and stacked driver die.

Figure 7.11 X-section picture of P3 stacked structure detail.

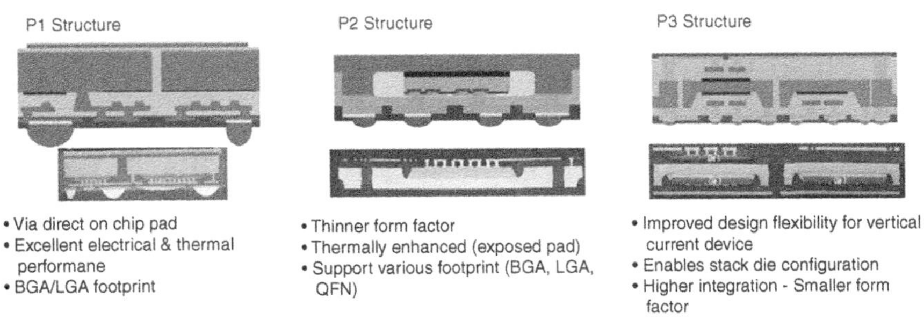

Figure 7.12 An a-EASI package portfolio.

is performed. This structure results in a power module with two RDLs on each side of the module.

With the two RDL layers the power module offers an increased area for routing and additional possibility to assemble passives and/or active die on package top by similar or even smaller outline of the package. Figure 7.12 summarizes the key attributes of the three a-EASI configurations.

7.3 Applications and Markets (HPC, SiP)

The a-EASI technology provides four distinct advantages:

1. Improved electrical and thermal performance;
2. Miniaturization;
3. Design flexibility; and
4. Improved reliability and mechanical stability.

All three a-EASI configurations—P1, P2, and P3—are fully qualified. As shown in Figure 7.13, the P1 configuration has been in high-volume manufacturing since 2013. More than a hundred million units have been shipped for the P1 configuration by the end of 2019. P2 and P3 are in low-volume manufacturing since mid-2019 and mid-2020, respectively. The foundation of a-EASI in a thick Cu lead-frame makes it very appropriate and advantageous for power applications including but not limited to multi-chip power modules, power discrete devices, and voltage regulators. The application space spans automotive, industrial, and consumer.

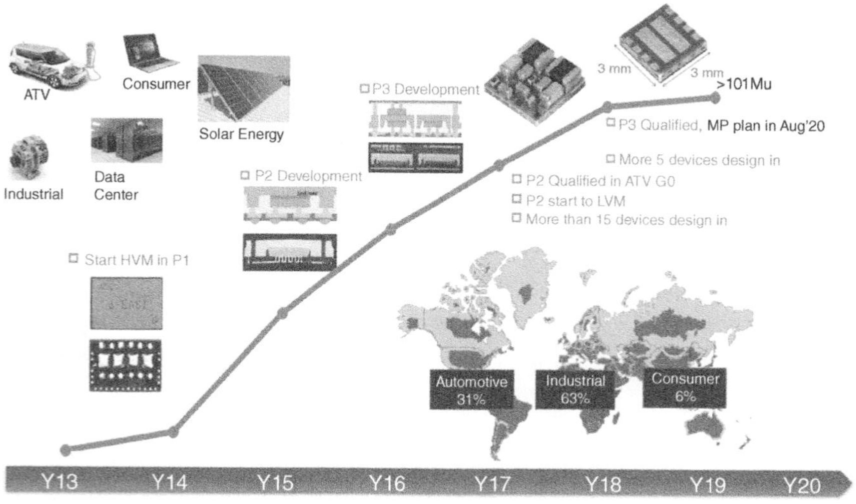

Figure 7.13 a-EASI package key milestones.

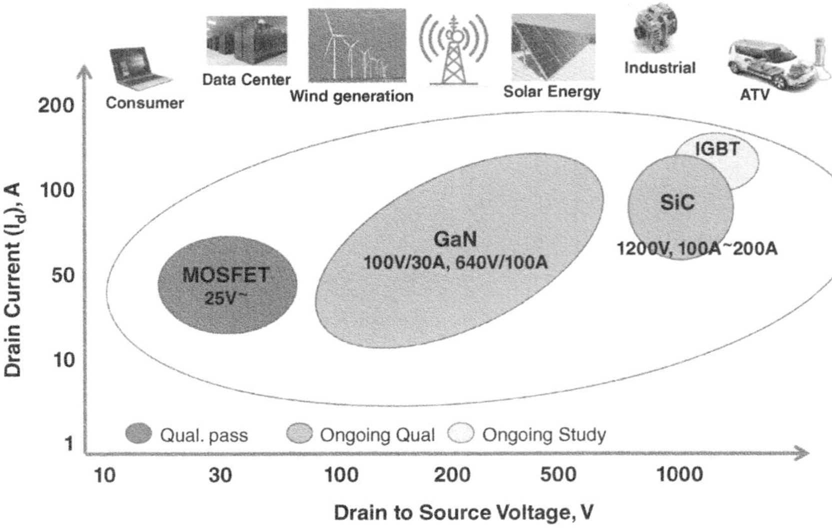

Figure 7.14 a-EASI qualified power range.

The qualified power range for a-EASI technologies is captured in Figure 7.14. Both silicon and wide band-gap devices are implemented on this platform. The low parasitic interconnect and high breakdown voltage dielectric used in a-EASI technology inherently lends to packaging requirement for improved performance of wide band-gap devices—for SiC devices and for GaN devices, which operate at higher frequencies for RF power applications in 5G and server applications, it enables greater input voltage and drives higher power outputs in electric/hybrid vehicles and solar inverters.

7.4 Manufacturing Process Flow and BOM

Figure 7.15 outlines the process flow for a-EASI P1 package configuration. The schematics on the right side of the figure show details of each process step. The process starts with a pretreated full thickness lead frame; thinned singulated die is attached to the lead frame using hybrid sintering or a silver sintering process. Die-attach is followed by lamination and a laser-drilling process. Next microvias to the die pads and Cu traces are formed using the plating and patterning process. Finally, the package pad process steps of solder mask (S/M), surface finish (such as electroless nickel immersion gold or ENIG), and solder printing are completed before singulation.

Figure 7.16 highlights the details of the microvias from die pad to Cu trace, which is the fundamental block to enable low $R_{DS(ON)}$, low inductance, and high

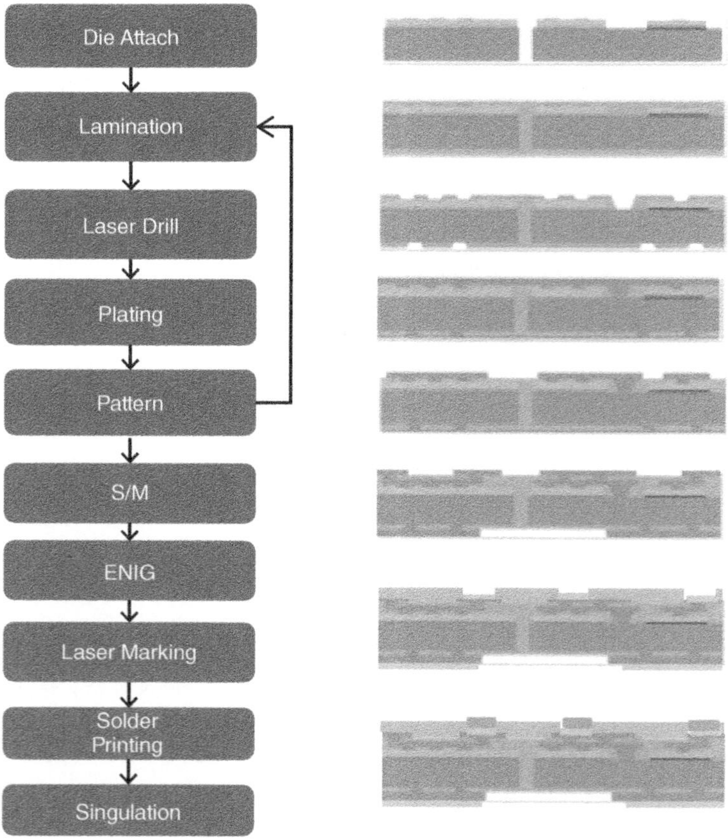

Figure 7.15 Baseline process flow for a-EASI P1 configuration.

current-carrying capability of the a-EASI package. The microvia has a Cu-to-Cu interface, which provides high-reliability interconnection under high current densities. The via is surrounded by prepreg material, which has > 2.5 kV dielectric breakdown strength for prepreg thickness of > 40 μm (Figure 7.17).

The P2 configuration has several advantages. Depending on the choice of the die attach, the thermal performance can be very effective. Hybrid silver sintering pastes, for example, can be sintered at temperatures below 200 °C and have a thin bond line and therefore have very low thermal and electrical resistance. The pressure-less die attach has a higher throughput, lower warpage, and a higher placement accuracy. The die shift measurement, typical for TLPB can be omitted as well as the preformation of the prepreg before lamination, since the die already sits in a cavity. The laser drill process is to one depth and therefore, one diameter and one parameter setting. As illustrated in Figure 7.18, the process again starts

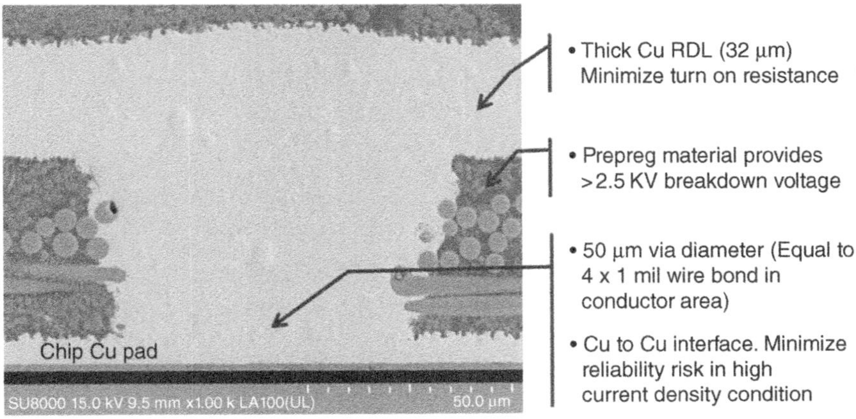

Figure 7.16 a-EASI–to-die interconnection details.

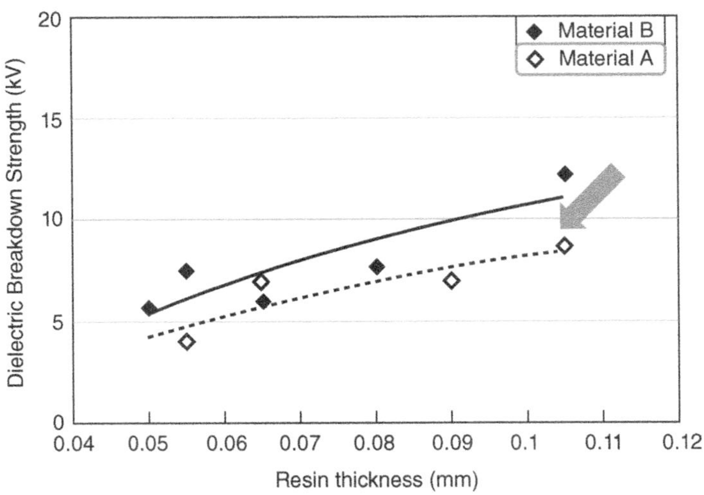

Figure 7.17 Prepreg material breakdown voltage.

with a cavity etched lead frame. The die are attached with the respective die bond material. Prepreg and Cu foil are laminated over the substructure, and laser vias are drilled to the die pads and plated. If full panel plating is used instead of pattern plating, then the circuit pattern is formed by subtractive etch, and the substrate is completed as usual with solder mask and surface finish. Representative in-process and final package examples are shown in Figure 7.19.

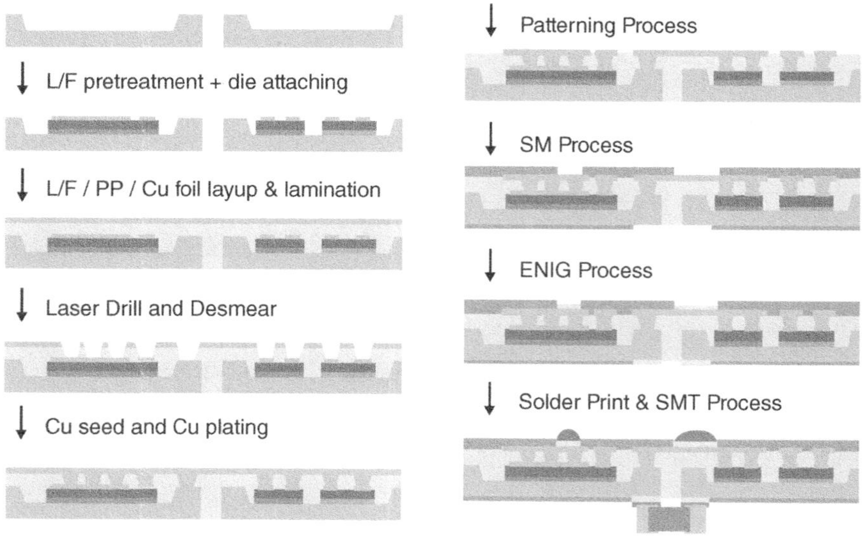

Figure 7.18 Generic process flow for a two-layer embedded die substrate according to P2 structure a-EASI technology.

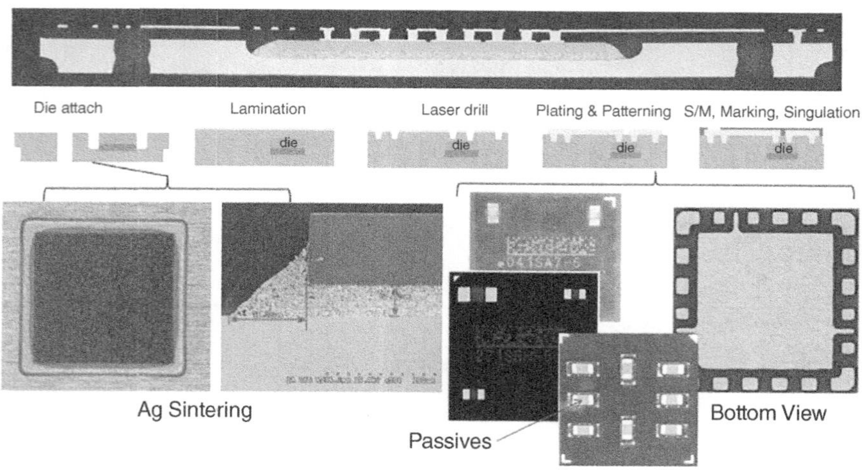

Figure 7.19 a-EASI P2 cross section and in-process example images.

7.5 Design Features

As discussed in previous sections, all three package configurations (P1, P2, and P3) are in production. All three configurations are available in QFN, land grid array

Table 7.1 a-EASI technology platform envelope (current production).

Package	Die (mm^2)	Package (mm^2)	Pin count	Min pin pitch (mm)
P1	7.5	30	~40	0.5
P2	30	172	~30	0.5
P3	5.4	24	~20	0.5

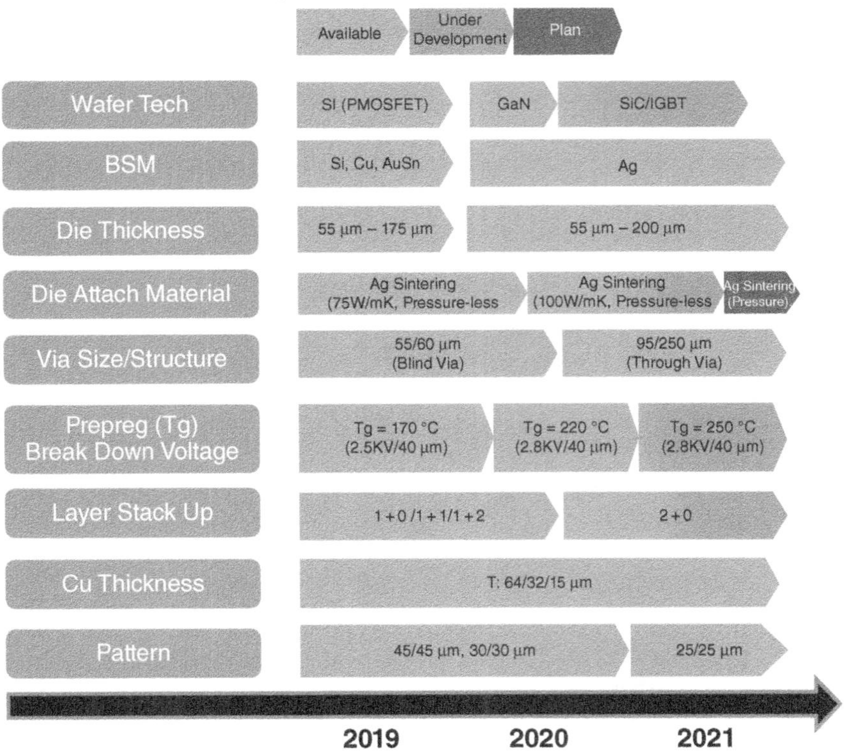

Figure 7.20 a-EASI technology attributes.

(LGA) and BGA pinout. Table 7.1 summarizes the current production envelope for a-EASI technology.

Figure 7.20 highlights the key technology attributes for a-EASI technology. The platform spans Si, GaN, and SiC devices but can be used for non-power digital devices as well; insulated-gate bipolar transistor (IGBT) integration is in development. Die with and without back side metal are supported. Different via sizes are possible on and off the die. Unlike the traditional substrate process, which requires buildup layer fabrication on both sides of the core, a-EASI allows asymmetric stack-up. Layer stack-up of $1+0/1+1/1+2$ and $2+0$ are in production. For power applications, thicker Cu pad/trace (32 μm) is the preferred choice, while 15 μm Cu thickness is the option for higher-density designs.

7.6 System Integration Capability

Figure 7.21 shows two examples of a-EASI power modules. As illustrated in Section 7.2, the embedding scheme provides capabilities of single die, side-by-side, and die stacking configuration. The technology utilizes different size microvias on and off the die as vertical interconnects and thru-vias as low parasitic interconnects. Furthermore, fanout using RDL layers provides ability to integrate passives and/or active die on top of the package (refer to power management and passive integration example in Figure 7.21). This results in a

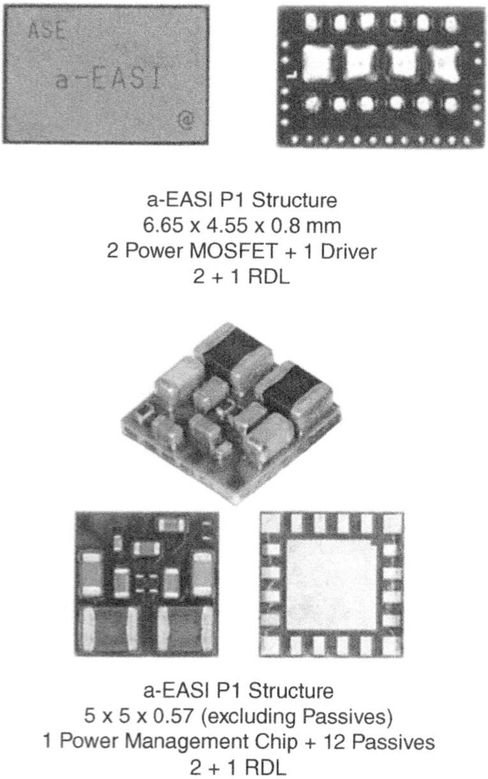

Figure 7.21 a-EASI power module examples.

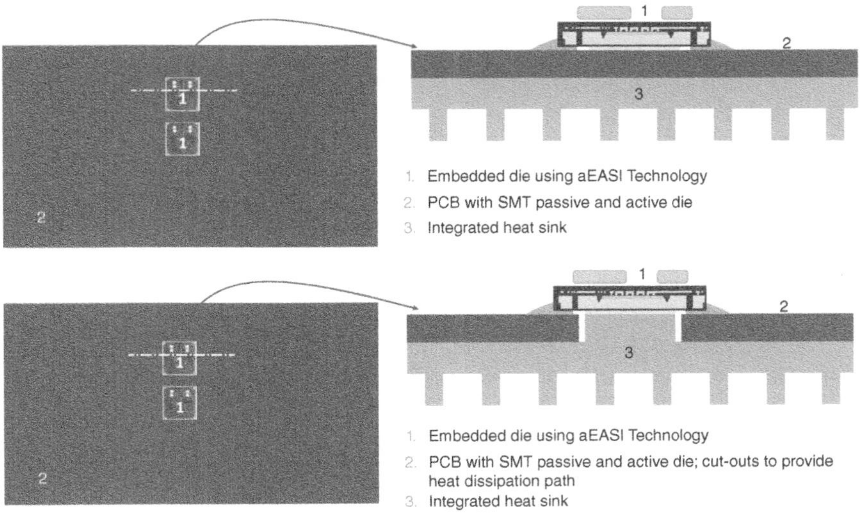

Figure 7.22 a-EASI integration using PCBA for performance and cost.

very flexible heterogeneous integration for a broad set of applications including power modules and power system-in-packages (SiPs).

Maximizing power density (W/cm^3) while minimizing parasitics are driving designs to embedded technologies. While embedding die in a system level PCB is a designer's holy grail, it comes with its own set of manufacturing challenges, yield issues, and cost implications. From that aspect, embedding of key components using a-EASI coupled with system level integration using traditional SMT and PCB assembly (PCBA) process (Figure 7.22) provide good performance and cost balance.

7.7 Package Performance

The electrical performance and thermal dissipation capabilities of such embedded die power modules have been modeled for a-EASI P1, P2, and P3 structures and compared with today's state of the art power-QFN (PQFN) power modules using Cu wire and Cu clip interconnects, shown in Figure 7.23.

The results in Figures 7.24 and 7.25 show that the parasitics (resistance [mΩ] and inductance [nH]) for the embedded die packages are more than one order of magnitude lower. In the a-EASI packages the MOSFET drain pads are full area connected via hybrid Ag sintering, while the gate and source pads are directly connected via Cu-filled micro vias where the Cu is directly plated onto the Cu die

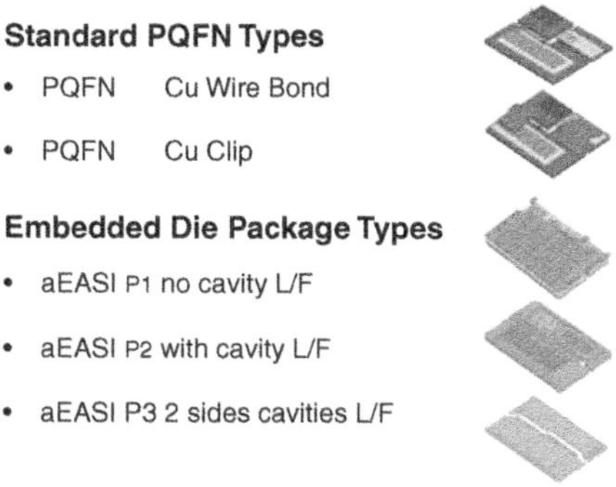

Figure 7.23 Structures of the compared packages in simulation analysis. Cu wire and Cu clip PQFN packages and embedded die power package in a-EASI P1, P2, and P3 structures.

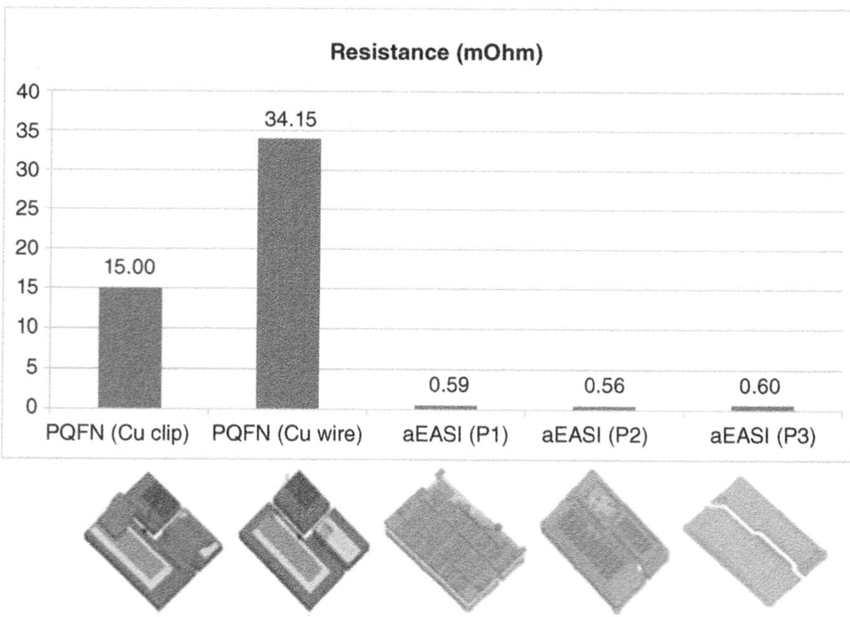

Figure 7.24 Resistance analysis of embedded die power package in a-EASI P2 and P3 structures compared to Cu wire and Cu clip PQFN packages.

pads. No thick inter metallic compound (IMC) layers or any wires are found in the a-EASI package.

Reduced parasitics were translated into power efficiency over output current (A) as shown in Figure 7.26. It can be shown that the power efficiency at 50 A is about 10–15% higher than for Cu clip PQFNs and almost 25% higher than for the PQFNs

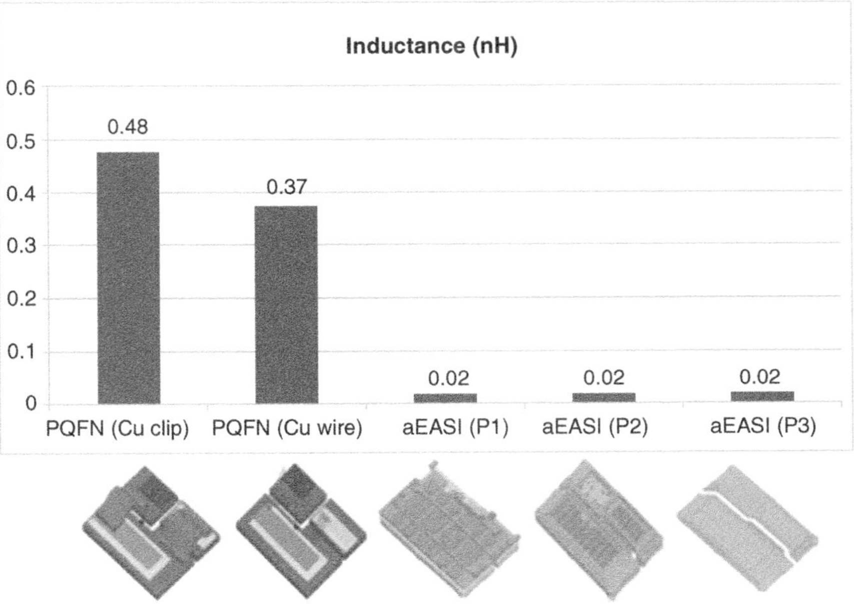

Figure 7.25 Inductance analysis of embedded die power package in a-EASI P2 and P3 structures compared to Cu wire and Cu clip PQFN packages.

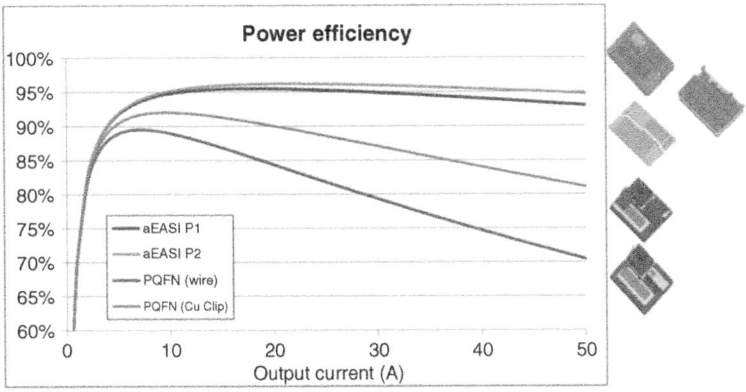

Figure 7.26 Simulated power efficiency comparison PQFN (Cu clip, Cu wire) vs. a-EASI (P1, P2, P3).

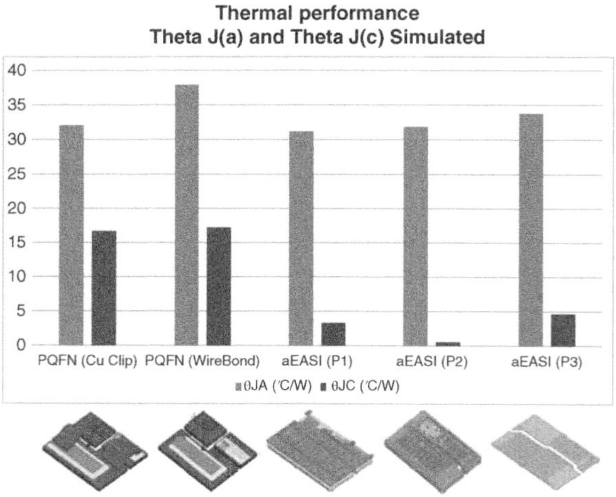

Figure 7.27 Thermal dissipation performance. Comparison of the PQFN-type packages with the a-EASI P2 and P3 structures.

using Cu wire. For the same power output of a PQFN device, the power input into an aEASI power module can be ~15% lower. The power dissipation is ~15% lower, so there is less heat that needs to be dissipated from the package that has a lower thermal resistance. This results in a lower effort to cool the system, which results in even higher energy savings.

Figure 7.27 compares the thermal dissipation of a-EASI package variants with traditional wire bond, clip PQFN packages. The simulated thermal dissipation toward the PCB (Theta-JA) is quite comparable with the PQFN (Cu Clip) package, while the thermal resistance toward the top of the package (Theta-JC) shows an order of magnitude lower values for the aEASI package structures compared with the PQFNs. With the lead frame on top of the dies in the aEASI packages the heat is spread and dissipated a lot quicker and easier than through the mold compound in the PQFNs.

The basic characteristics of the a-EASI structure, that is, direct exposed pad and shorted connection path, enable significantly better electrical and thermal performance as compared to the traditional power QFN packages. The number of vias is optimized for electrical and thermal performance and cost using simulations.

7.8 Robustness and Reliability Data

The a-EASI P2 structure styles show very promising reliability results. As a representative example, component level reliability (CLR) and board level reliability (BLR) data on a BGA pin-out test vehicle (TV) is summarized below. Package configuration is shown in Figure 7.28, die size of 2 mm × 2 mm, package size of

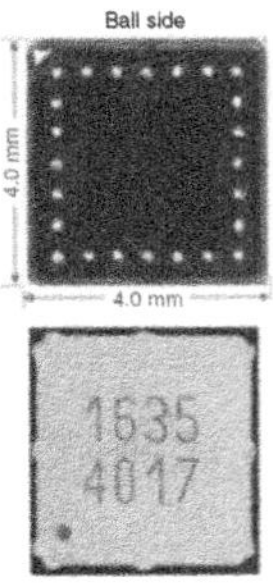

Key information:

- ✓ Size: 4.0 × 4.0 +/– 0.1 mm, TH = 0.35 +/– 0.040 mm
- ✓ Ball count: 24 (pitch: 0.5 mm)
- ✓ Blind via: 0.085 mm
- ✓ Line/Space (L/S): 60/60
- ✓ Cavity lead frame

Figure 7.28 a-EASI P2 package – BGA example.

Table 7.2 Component level reliability data for a-EASI P2 BGA package.

Test description	Abbreviation	Condition	Readout	Result
Preconditioning AEC-Q100	PC	MSL3 MSL Moisture soaking 3 30 °C/ 60% RH	Bake 125 °C/ 24 hours Moisture soak 168 hours 3x reflow (265 °C+[0/−5])	Pass
HAST AEC-Q100	HAST	130 °C/ 85% 33,3 PSIA	100 hours	Pass
			200 hours	Pass
Temperature cycling AEC-Q100 GRADE O	TCT	-65–150 °C	500 cycles	Pass
			1000 cycles	Pass
			1500 cycles	Pass
			1700 cycles	Pass
			2000 cycles	Pass
			3000 cycles	Pass
HTST AEC-Q100 GRADE O	HTST	150 °C	500 hours	Pass
			1000 hours	Pass
			1500 hours	Pass
			2000 hours	Pass

4 mm × 4 mm with 0.5 mm BGA pitch. Table 7.2 summarizes the CLR data: the TV passed up to 3000 thermocycles thermal-cycle test (TCT) (−65 to 150 °C) and 2000 hours high temperature storage (HTST) at 150 °C. Tests were completed without failure in open short electrical test and scanning acoustic tomography (SAT) test. The package met the AEC-Q100 Grade 0 reliability requirements. To further

Table 7.3 Board level reliability data for a-EASI P2 BGA package.

Test description	Condition	Read out	Sample size	Result
Drop test (JESD22-B111)	−1500 G, 0.5 ms	• maximum >30 drops • until 80% of all devices have failed.	60	Pass, >300 drops
Cyclic bend test (JESD22-B111)	Deflection: 2 mm Frequency: 1 Hz	• maximum >200 000 cycles • until at least 60% of all units have failed	36	Pass, 200K cycle
TCT (JESD22-A104)	−40–125 °C 50 min	• Above 20% resistance ohm of 5 consecutive reading scans • until 63.2% of all devices have failed	30	PASS, 1st fail sample occurred in TCT 1060

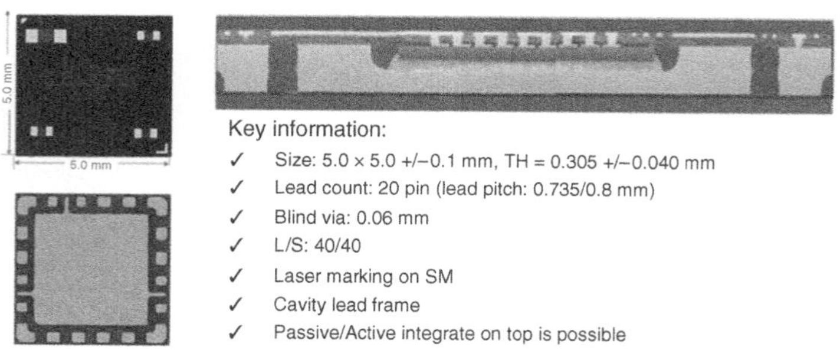

Figure 7.29 a-EASI P2 package – QFN example.

demonstrate P2 package robustness, TCT (−65 °C to 150 °C) was conducted on the same package post MSL1 and MSL2; the package passed 2000 cycles. BLR data results are shown in Table 7.3; drop, bend, and temp cycle requirements were successfully met and exceeded.

Another TV in QFN style with die size 2 mm × 2 mm and package size of 5 mm × 5 mm passed TCT 2500 and 1000 hours of HTST. Package configuration and reliability data are compiled in Figure 7.29, Tables 7.4, and 7.5 respectively.

As a final example in Figure 7.30, P3 package with 3 die (2 die stacked, third die in a side-by-side configuration to the stacked die). AEC-Q100 Grade 0 CLR data is consolidated in Table 7.6.

Table 7.4 Component level reliability data for a-EASI P2 QFN package.

Test description		Condition	Readout	Result
Precondition AEC-Q100	MSL3	30 °C/ 60% RH	TCT:SX Bake: 125 °C/ 24 hours Moisture soak: 168 hours Reflow: 265 + 0/−5 °CX3 cycle	Pass
HAST AEC-Q100	HAST	130 °C/ 85% 33,3 PSIA	96 hours	Pass
			192 hrs.	Pass
Temperature cycling AEC-Q100 GRADE O	TCT	−65 °C ~ 150 °C	500 cycles	Pass
			1000 cycles	Pass
			1500 cycles	Pass
			2000 cycles	Pass
HTST AEC-Q100 GRADE O	HTST	150 °C	500 hours	Pass
			1000 hours	Pass
			1500 hours	Pass
			2000 hours	Pass

Table 7.5 Board level reliability data for a-EASI P2 QFN package.

Test description	Condition	Read out	Sample size	Result
Drop test (JESD22-B111)	−1500 G, 0.5 ms	• maximum >30 drops • until 80% of all devices have failed.	60	Pass, >1000 drops
Cyclic bend test (JESD22-B111)	Deflection: 2 mm Frequency: 1 Hz	• maximum >200 000 cycles • until at least 60% of all units have failed	36	Pass, 200 K cycle
TCT (JESD22-A104)	−40–125 °C 50 min	• Above 20% resistance ohm of 5 consecutive reading scans • until 63.2% of all devices have failed	30	Pass, 1st fail sample occurred in TCT 2140 X

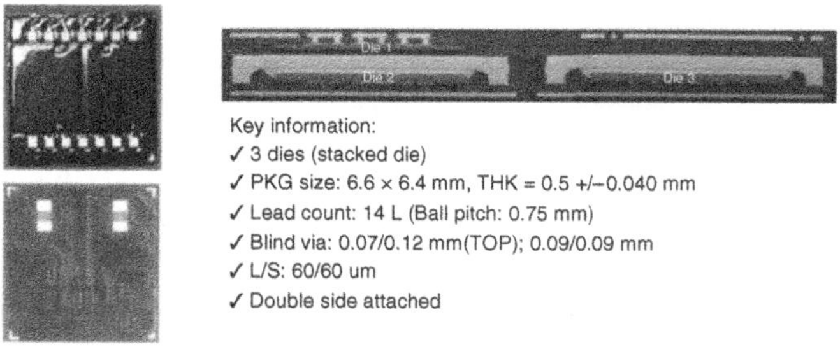

Figure 7.30 a-EASI P3 package – 3 die example.

Table 7.6 Component level reliability data for a-EASI P3 package (3 die).

Test description		Condition	Readout	Result
Precondition **AEC-Q100**	MSL3	30 °C/60% RH	TCT:SX Bake: 125 °C/ 24 hours Moisture soak: 168 hours Reflow: 265 + 0/−5 °CX3 cycle	Pass
Hast **AEC-Q100**	HAST	130 °C/85% 33,3 PSIA	96 hours	Pass
			192 hours	Pass
Temperature cycling **AEC-Q100 GRADE** O	TCT	−65–150 °C	500 cycles	Pass
			1000 cycles	Pass
			1500 cycles	Pass
			2000 cycles	Pass
HTST **AEC-Q100 GRADE O**	HTST	150 °C	500 hours	Pass
			1000 hours	Pass
			1500 hours	Pass
			2000 hours	Pass

7.9 Electrical Test Considerations

Test considerations are an important factor for embedded technologies. Unit-handling, contact and insertion force, socket design, and unit pad metallization considerations all play key roles in the basic test requirements of a-EASI. Embedded die packages must consider if there are any special requirements for unit handling to ensure minimal yield loss when performing multi-insert testing.

Challenges such as die cracking, dielectric cracking, and the introduction of severe unit warpage are key considerations that can affect package reliability in addition to potential yield loss. Fortunately, ASE has had a lot of experience in handling ultra-thin packages, and this helps to provide a baseline on what to expect with package structures such as a-EASI.

For non-SiP a-EASI units that do not have components on the top side, the pickup and handling is standard, like a thin QFN. When populated with components, some creative tools must be used to ensure proper socket insertion. In some cases, a pretest may be performed before additional component attach at strip/panel level to ensure good embedded die interconnect yields.

Contact insertion force is one that is often varied depending on the resistance level observed when testing. Sometimes increasing the insertion test force may allow for improved contact, which lowers contact resistance. This is seen in BGA units where the BGA solder balls may have a layer of oxide that needs to be broken for proper contact. In the case of a-EASI, most structures are in the LGA format and contract related resistance is not seen as a test challenge.

The design of the test sockets follows similar guidelines as other SiP or thin CSP package platform, where key tolerance of the package is used as a baseline for the socket design. Similar variations as found in substrate-based CSP are designed accordingly. Since most a-EASI package body sizes are relatively small and the pad counts are low, yield loss due to warpage is not a key factor. As the body size grows and the balance of the component/embedded Cu balance changes, this is an area that must be considered.

Overall, the a-EASI test considerations are straightforward and follow most of the prior learning on other thin-based CSP and SiP modules. Being aware of the potential areas of reliability risks due to the test process is always good practice.

7.10 Summary

Data centers and automotive are leading areas driving advances in power solutions to enable innovative designs. Smart factories, smart grid, renewable energy, and communications will all benefit from high power conversion efficiency and superior power density offered by advanced power solutions encompassing traditional silicon and wide band-gap devices. The need to reduce system size and weight while increasing overall efficiency requires power module evolution to smaller, lighter weight packages with higher switching frequencies. Thus, the package parasitics, especially the inductance and resistivity, becomes a critical factor to fully realize the electrical performance of the power module. The increasing current density needs improved thermal dissipation to correlate with the enhanced reliability requirements for industrial and automotive standards. In this chapter, embedded die power modules a-EASI technology has been detailed, which enable the key requirements of next generation power solutions, namely improved package efficiency, better thermal performance, form factor reduction, and integration flexibility. The a-EASI technology is in high-volume manufacturing with > 100 million units shipped and applications across automotive, industrial, and consumer application space.

References

1 Alderman, A., Burgyan, L., Kakizaki, Y., Hama, Y., Nakagawa, H., Böttcher, L. and Löher T. (2015). Current developments in 3D packaging with focus on embedded substrate technologies. *PSMA Technology Report.*

2 Alderman, A., Burgyan, L., Narveson, B. and Parker E. (2015). 3D embedded technology: analyzing its needs and challenges. *IEEE Power Electronics Magazine*, (December), pp. 30–39.

3 Essig, K. (2015). Embedded die substrate solutions for power applications. IMAPS France, Power Electronics Workshop, Session 2, Tours, France.

4 Essig, K., Chiu, C.T., Kuo, J., Chen, P. and Yannou, J.M. (eds). (2016). Higher efficiency power module integrated solutions by chip embedding. *Proceedings of International Symposium on Microelectronics*. Pasadena, USA (10–13 October 2016). USA: IMAPS.

5 Appelt, B. K., Su, B., Lee, D., Yen, U. and Hung M. Eds. (2011). Embedded component substrates moving forward. *IEEE 13th Electronics Packaging Technology Conference Proceedings*. Singapore (7–9 December 2011). USA: IEEE.

6 Wang, T., Chen, X., Lu, G., and Lei, G. (2007). Low temperature sintering with nano-silver paste in die attached interconnection. *Journal of Electronic Materials* 36: 1333–1340.

8

Die Integration Technologies on Advanced Substrates Including Embedding and Cavities

Markus Leitgeb and Christian Vockenberger

AT&S AG, Leoben, Austria

8.1 Introduction

Heterogeneous integration will need a more holistic interconnection technology in the future to form electronic modules at a maximum speed and minimum space and energy consumption. Fan-out packages have recently been introduced in high-end applications to replace stand-alone substrates to lower package height or replace fan-in packages because of the need for more package real estate for inputs and outputs (I/O) or higher reliability. However, the continuing trend toward miniaturization, system cost reduction, and shorter time to market for electronic products drives the need for heterogeneous integration of larger number of components in a single module. Existing epoxy mold compound (EMC)-based fan-out technologies are typically focused on single-component packages or packages with only a few integrated components. Package manufacturing is mainly performed at wafer level with some recent introduction of panel-level fan-out manufacturing [1]. Large panel-based technologies such as printed circuit boards (PCB), substrate-like-PCB (SLP), and/or integrated circuit (IC) substrates can offer the integration of a larger number of components in a single module when combined with chip-first embedding or chip-last assembly using cavity technologies. Moreover, the existing high-volume manufacturing (HVM) PCB and substrate infrastructure can be leveraged to a large extent, which offers attractive potential for volume scaling.

In the electronic industry, many different fields are merging because advanced packaging can overcome challenges in electronic systems architectures [2]. Two major drivers in the electronics industry are electrical performance and mechanical miniaturization. Whereas lines and spaces (L/S) have been getting smaller over the years (current high-density interconnect (HDI) standard is L/S of 40/40 μm

Embedded and Fan-Out Wafer and Panel Level Packaging Technologies for Advanced Application Spaces: High Performance Compute and System-in-Package, First Edition. Edited by Beth Keser and Steffen Kröhnert.

and IC substrate is 9/12 μm), mechanical miniaturization has thus far been mostly limited to decreasing layer count and material thickness.

8.2 Heterogeneous Integration by Use of Embedded Chip Packaging (ECP®)

AT&S has developed heterogeneous packaging based on embedding active and passive components in a laminate-based substrate or PCB. The development of the chip embedding technology has a long history and started with a funded project on chips in polymer in 2000 led by the Fraunhofer Institute IZM (FIZM). The basic research and development (R&D) work for chip embedding was done in the former project titled "Hiding Dies." The first functional demonstrators have been shown in this project, and FIZM patented a face-up embedding technology for chip embedding [3]. With this experience of embedding passive discrete components and the increasing market interest, FIZM and AT&S saw a large opportunity in creating the supply chain for chip embedded applications in the frame of a European consortium. This consortium was willing to develop the chip-embedding technology and the supply chain for the industrialization of embedded applications. The partners of the consortium consist of end users, material suppliers, equipment suppliers, a silicon supplier, a test house, a PCB supplier, and research institutes and are shown in Figure 8.1. It was also important to attract potential players, to create business power and to generate a move to chip-embedding technology. The name of this European-funded industrialization project was HERMES (High density integration by Embedding chips for Reduced size Modules and Electronic Systems).

The selection of the partners in the project focused on the key items to make such a new technology successful. Besides the process of embedding itself, material availability, design and layout, and testing were included.

The embedding of active and passive components offers a wide range of benefits and potential. System miniaturization is one of the top priorities driving the use of embedding technology, which is most interesting for module applications where size constraints are critical and high integration is key. By moving the components to the inner layers, more active and passive components can be placed on the same footprint. A 3D system-in-package (SiP) results in higher functionality and complexity. The improvement of the electrical performance of signals with short copper interconnections reduces parasitics compared to, for example, bonding structures, leading to minimal signal distortions. High mechanical system stability, no soldering or bonding, and highly reliable copper interconnections result in an improved overall reliability, which is a clear benefit of the embedding technology. Furthermore, the components are typically located in the neutral axis

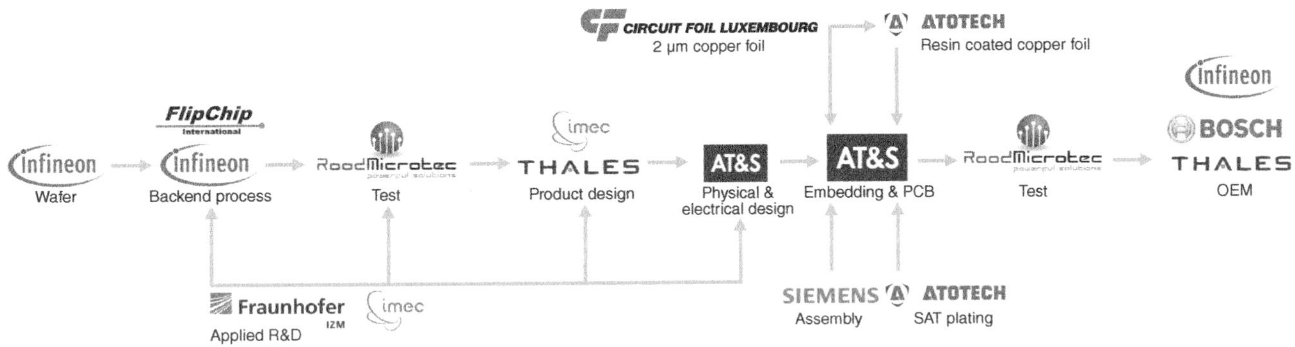

Figure 8.1 HERMES Project consortium structure.

of the PCB; this means the components will see less stress during bending of the PCB, which may occur when dropped.

One of the major challenges of high-performance systems is efficient thermal management in the modules. Copper via or other thermal structures can be contacted directly from the front or back side of the component. Components are located between the PCB conductor layers and not on top of the PCB, and the contacting can be done from either side. The thermal performance is shown in a later example.

During the development of the technology, there was always a focus to stay as close as possible to existing materials, HVM equipment, and robust processes that are already established and well known in PCB manufacturing. Besides the proven robustness of the processes, this approach allows a cost-effective manufacturing and enables fast scalability and expansion of production. On the other hand, the PCB process environment has specific impact on component requirements and design rules, which need to be considered.

8.3 Embedding Process

The initial process steps for the embedding process are identical to PCB manufacturing. First, there is an image transfer to the core (glass cloth with fully cured resin and copper on top and bottom) of the PCB. Subsequently, cavities for the embedded components (EC) are created. These cavities are laser-cut openings in the dielectric material of the core. Laser cutting offers the necessary accuracy and is an established process in HDI PCB manufacturing. In the next step, a temporary carrier is laminated on one side of the core, which will keep the components in place during further process steps. The components will then be placed into the cavities and onto the temporary carrier. This placement is done with standard surface-mount technology (SMT) assembly equipment. This ensures high throughput and the necessary precision. See Figure 8.2 for details.

During the first lamination step, a prepreg will be laminated on top of the core. A prepreg is a sheet of semi-cured glass-reinforced epoxy resin, which is used in PCB manufacturing to create the dielectric layers. During lamination, the resin of the prepreg will fill the gap between the component and the dielectric material of the core. The lamination process is adapted to the properties of the resin, and the resin is cured in the same step. As the component is now fixed by the resin, the temporary carrier can be removed. A second lamination step is done in which a prepreg is attached to the bottom of the core to ensure a symmetric stack up.

The interconnection to the component itself is realized by copper-plated laser vias. Therefore, it is necessary to drill microvias (or μ-vias) with a laser on the

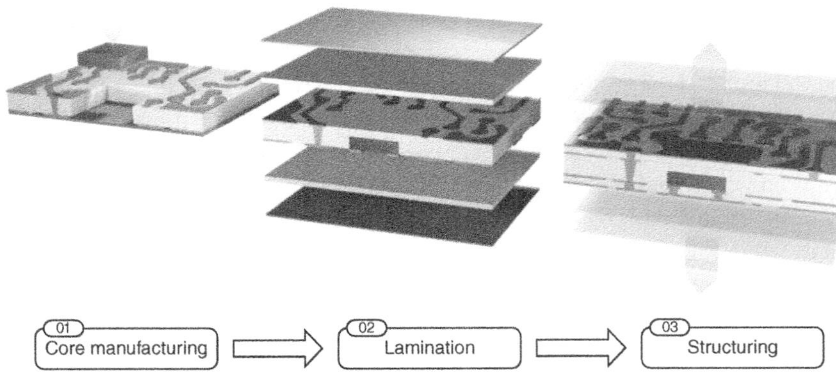

Figure 8.2 Embedded component packaging process flow overview.

component interconnect pads. As the components are hidden in the laminate, a specific registration system has been developed to ensure that the laser hit the component pads accurately. The use of CO_2 laser systems and the copper termination on the embedded device (refer to specific component requirements in the next section) ensure a secure removal of dielectric material and no damage to the component pads. The laser accuracy in combination with the placement tolerance define the possible pad size of the component. This means, a proper registration concept and high-precision equipment are key for driving the design rules.

Further processing is very similar to HDI PCB manufacturing. After having the vias completed, copper plating enables an electrical connection between the PCB circuitry and the EC. The plating process and equipment are identical for laser-drilled HDI PCB; thus, there is no need to develop dedicated solutions. A slight modification and tuning of the plating process is sufficient to fulfill the requirements of the specific product. All remaining processes are the same as for normal PCB production. The semi-finished product, which has been created by the processes described above, can be used in the same ways as a standard core for PCB with the difference that there are electrical components embedded between the copper layers. This means the PCB can be finished as a two-layer or, depending on the application and product requirements, a multilayer PCB. There are no limitations on the PCB stack up. The embedded core can be combined with other cores, or a sequential build-up process can be performed.

A major consideration when handling active and passive components in production is ESD (electrostatic discharge). This fact was considered from the beginning, and all processes and transportation systems have been set up in a way to ensure there will not be an ESD event when running the manufacturing panels through production.

8.4 Component Selection

A key element of the embedded technology are the components that will be integrated into the laminate. The component properties are defined by the intended method of processing. It is obvious that pure mechanical parts, such as connectors or mechanical switches, cannot be embedded into a PCB. Theoretically, all other electronic components could be integrated into the PCB if they fulfill the physical properties. Nevertheless, the component function must be taken into consideration. Sensors and microelectromechanical systems (MEMS) devices, have limited possibilities for embedding, as the process and the encapsulated condition might influence the function of the device itself. The main component requirements are listed in Table 8.1.

After component selection based on function is complete, next the physical properties of the components need to be taken in consideration. Due to integration into the core of the PCB, the thickness of the component must match the thickness of the core itself. PCB cores are only available in certain thicknesses. In addition, there must be enough resin available to fill around the component. In the x and y dimensions, there is also a certain limitation that is driven by the capability of the assembly equipment and by the die-to-package ratio. The die-to-package ratio is an indication for overall warpage of the product and is a critical factor.

A copper termination is needed on the component as a laser-stop. In combination with the plating process, it also ensures a reliable copper-to-copper interconnection. This kind of surface treatment is not a standard surface neither for active nor for passive components, but is available and possible to apply on both component categories. For active components, copper plated directly

Table 8.1 Basic component requirements.

Feature	Passive components	Active components
Type	resistor, capacitor	bare die
Size	0201, 0402	max $8 \times 8\,mm^2$
Thickness	60–300 μm	60–300 μm
Pad surface	only copper	only copper
Copper thickness	min 6 μm	min 6 μm
Pad diameter	min 200 μm	min 200 μm
Thickness tolerance	±10 μm	±10 μm
Packing method	T&R	T&R, wafer

onto an I/O pad or a redistribution layer (RDL) can be used to create the copper metallization. This will also enable the right pad size, which is needed to compensate the registration tolerances. For passive components, a specific process flow is required, which is readily available at the passive component suppliers. In general, the selection of which electronic parts should be embedded is a crucial topic as it defines the overall architecture of the device. An early and open interaction between the customer, the integrator (company that is doing the embedding), and the component supplier is mandatory and absolutely necessary to avoid late changes and even showstoppers in the development process.

8.5 Design Technology

Having a process and the necessary component in place is key for embedding technologies. If it is not possible, embedding in a PCB is not possible either. Also, the ability to virtually design these components using electronic design automation (EDA) tools with embedding capability is required. In order to realize the intended EDA functionality, tool development has been a major task during the deployment of the embedding technology. Originally, it was only possible to place components on the surface of the PCB, and there was no option to place it as an EC inside the substrate. In the early stage of the technology development, AT&S worked closely with specific EDA tool suppliers to develop the tools, enabling customers to adopt the technology.

One of the main difficulties was the different approaches of embedding, each of them having some specialties, which needed to be considered in the layout. Thus, a challenge for the tool suppliers was to develop a one-fits-all solution for embedded die technology. Of course, there are workarounds that allow to physically place a component inside the PCB, but this solution has high risk. It is absolutely necessary to have the component clearly identified as embedded in the PCB because of the design rule check. If the tool does not know that a specific component is placed inside the PCB, it is unable to check whether all necessary constraints are applied in the right way. In the worst case, ECs are placed at the same location as through holes. For surface mount devices, this would not be an issue, but in case of EC, the drill would go through the component and destroy it.

A clear identification of components as embedded is needed to ensure it is recognized to be *in* and not *on* the PCB. This is necessary so the manufacturer will know which components should be embedded and which ones not. It sounds like a simple thing, but overall it also needs to be ensured that there is an automated design and data flow. This means that not only the EDA world needs adapt but also the computer-aided manufacturing (CAM) setup at the PCB supplier needs to be able to automatically handle the information provided by the customer. To be able to have a smooth technology adoption for customers in this regard, AT&S has established a design group, which is focusing on integration of new technologies in existing EDA and CAM tools and providing solutions.

8.6 Testing

Testing PCBs with EC is different from traditional PCB manufacturing. Testing is required for the product qualification and for the in-process testing in mass production. The functional test, applied after reliability stress during product qualification, should be related to the functional tests applied in mass production. For passive components, this is easy, as most PCB test equipment can measure resistance and capacitance. There are two main items that must be taken into consideration. First, the clear identification of the EC, which is related to the topic of automated dataflow. During test program preparation, the nets connected to components have to be treated in a different way. Therefore, it is necessary that the components and related nets are identified automatically and not selected manually, as this is a source of mistakes especially for complex boards. Second, the traces and copper plains in a PCB have a direct impact on the measured value of capacitors and resistors. Executing an in-circuit test (ICT) on surface mount components usually finds that the measured values match the datasheet values of the device itself. For EC, the influence by the PCB is such that the measured value often does not correspond to the datasheet value because of the additional resistance of PCB traces and vias.

For active components, the test is more complex. The easiest test to be implemented is a so-called diode test, with which the interconnection to the die is verified. This test can be performed with standard PCB test equipment, but as it is similar to an open short test, it has certain limitations, and it requires an ESD protection diode implemented at the pin of the device. For single-die packages, where just one component is embedded, the test can be performed straightforward. If more components are embedded and if these components are connected to each other within the PCB, it can be challenging. For surface mount components, it is simple to contact the components or component pads for testing, but for EC this is not possible, not even for passive components. Having resistors in series or capacitors connected to a pin on an active device, the individual components could be tested without the need of additional test points. For embedding components, it is not possible to test the individual components once embedded. Even the nets in between cannot be tested properly without additional effort. If there is a need for it, this must be considered during design phase already, and additional test points must be implemented. For more sophisticated functional testing, dedicated tests and test equipment should be installed, which is typically not a standard setup for PCB manufacturing.

Once a functional test method is defined for mass production, it can also be used for verification after reliability testing. Dedicated standards for reliability testing of embedded devices have been developed, but they are still not fully implemented in the industry yet. Therefore, it is necessary that the qualification tests for individual products are discussed and aligned between customer and manufacturer at an early stage of the project.

8.7 Applications for ECP Technology

The initial design rules for ECP technology were focused on simple products with low I/O count (< 50) and relatively small die. The need for miniaturization is required by applications that have a space constraint, such as mobile applications, wearables, or medical products. Most of the products are small, which means that several thousand units fit on one production panel. It is necessary to target applications that have large enough volume to be able to support panel and lot sizes used in PCB manufacturing. Considering the benefits provided by the technology, such as improved electrical characteristics and heat dissipation, ECP provides an overall good fit for low-power applications such as low-voltage direct current–to–direct current converter (DC/DC) or discrete power devices, as shown in Figure 8.3. These devices are classified as multimarket, so they have a broad field of applications and thus provide the necessary volume to efficiently run an embedded die production.

Table 8.2 shows the quantities shipped by AT&S in the different categories through 2020. Packages consider all products that have just one embedded device. Surface mount components can be added, but it is not mandatory. Discrete power devices would also be associated with this category. Modules have more than one embedded device (active and/or passives) and typically need additional surface mount components. Boards are large form factor main boards and typically just have passive components embedded.

Figure 8.3 Exemplary picture of discrete power devices.

Table 8.2 Quantities of shipped units with embedded components.

	Packages	Modules	Boards
# of shipped units	> 431 million	> 9 million	> 2.3 million
# of assembled components	> 431 million	> 23 million	> 250 million

In PCB processing, it is typically no problem to achieve copper thicknesses of 35 μm and higher. This enables additional applications with higher power. In combination with further development of wide band-gap materials such as Gallium Nitride (GaN), which are switching at higher frequency, embedded die technology is a perfect fit for such applications, not only for discrete devices but also for modules or embedding in board applications. For these kind of applications, it is important to have short loop connections to reduce the parasitic inductances. This drives efficiency improvement and further reduces the size of the overall application. Another benefit of embedded die is the fact that the EC can be contacted from both sides using multiple μ-vias. With copper-filled via technology and the design flexibility of the individual μ-vias, the thermal and electrical resistance can be improved significantly and the heat flow optimized.

The performance increase has been demonstrated using a half bridge module comparing a standard surface mount approach versus embedded die in the module. The module included the following list of components. The schematic is shown in Figure 8.4.

1. Two gallium nitride transistors;
2. High-side bootstrap with SiC diode;
3. SI8271 driver with integrated isolation;
4. LDO to prevent gate drive failures;
5. High-frequency input filter;
6. Thermal measurement device (NTC) for the embedded version;
7. Three power connectors and one signal connector.

For the surface mount version, a four-layer PCB was developed with the GaN transistors mounted on top and all other components on the bottom. Above the transistors, a heat sink has been attached using a thermal interface material (TIM). It is important to mention that the transistor is realized with AT&S ECP. Due to this fact, the performance difference between the two versions is for some parameters not so significant.

The embedded version was also a four-layer build up. The transistors are embedded in the center of the PCB, which leaves the whole surface area for heat sink attach. The transistors have been embedded with active side down in the stack up.

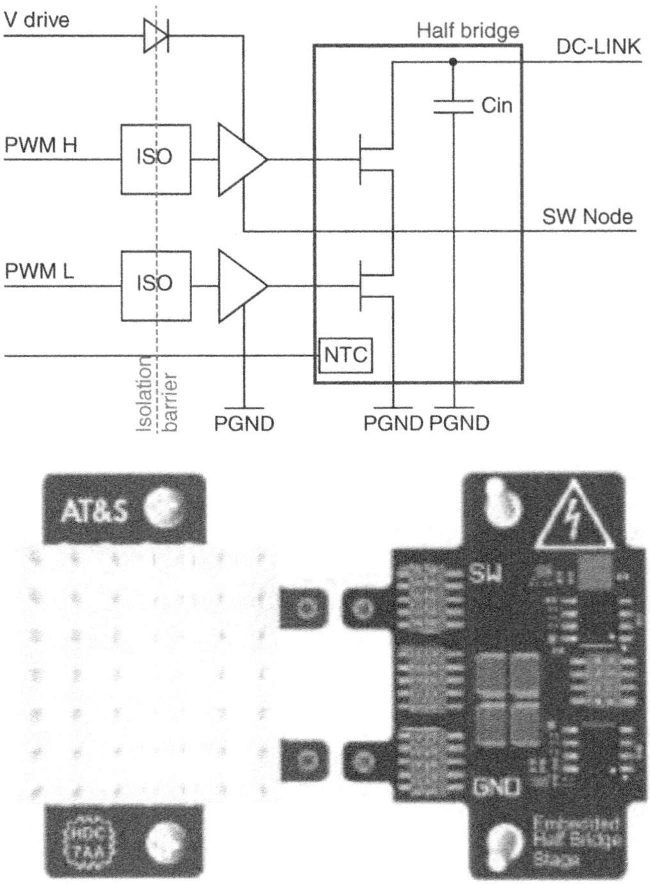

Figure 8.4 Schematic and graphical image with attached heat sink.

The back side of the transistor is connected to layer two in the module by several copper filled μ-vias. A thermal prepreg is used between layer one and layer two to improve the heat dissipation to the heat sink mounted on top. The heat sink has been attached with a thermal adhesive.

Figure 8.5 shows the cross-sectional schematic view of the different construction. It can be seen that the overall thickness of the embedded version is less compared to the SMT version. This is because of a thinner PCB but also because the surface mount transistors and the TIM need more space in the *z*-direction. Figure 8.6 shows a cross-section picture of the PCB with the embedded transistor and the μ-vias contacting both sides of the transistor. Moving the transistors inside the module reduces the distance between the DC link capacitors and the transistors, which results in a shorter commutation loop, as shown in Figure 8.7.

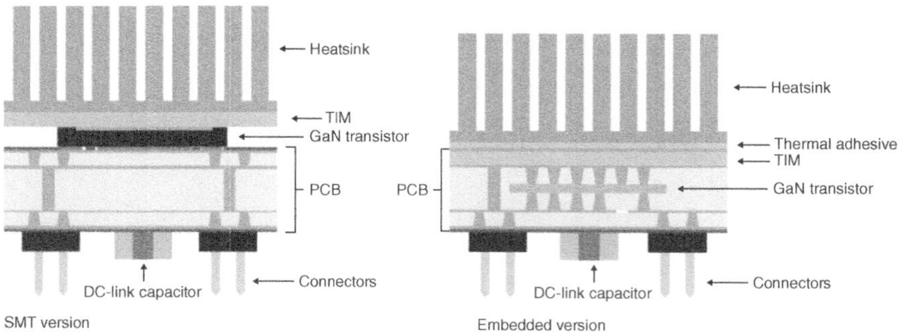

Figure 8.5 Stack up comparing SMT with embedded construction.

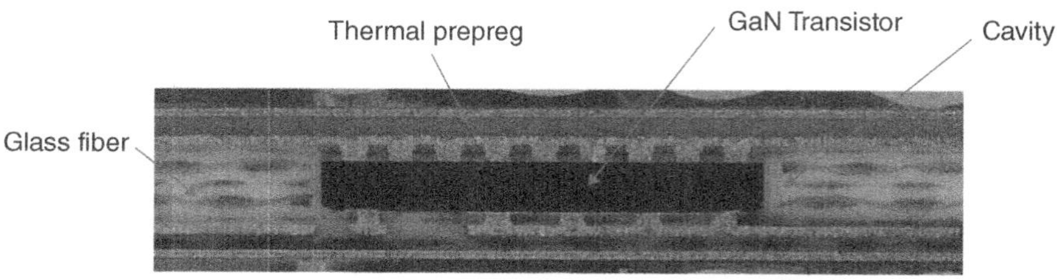

Figure 8.6 Cross section of embedded GaN transistor with double-sided laser via connection.

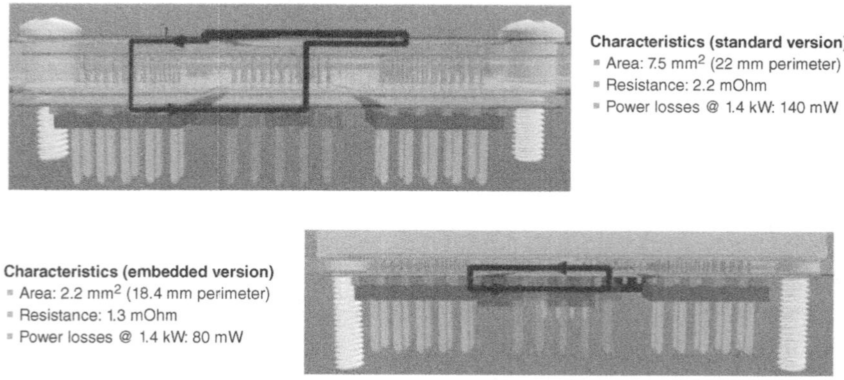

Figure 8.7 Cross-sectional view comparing commutation loop of SMT (top picture) and embedded structure (bottom picture).

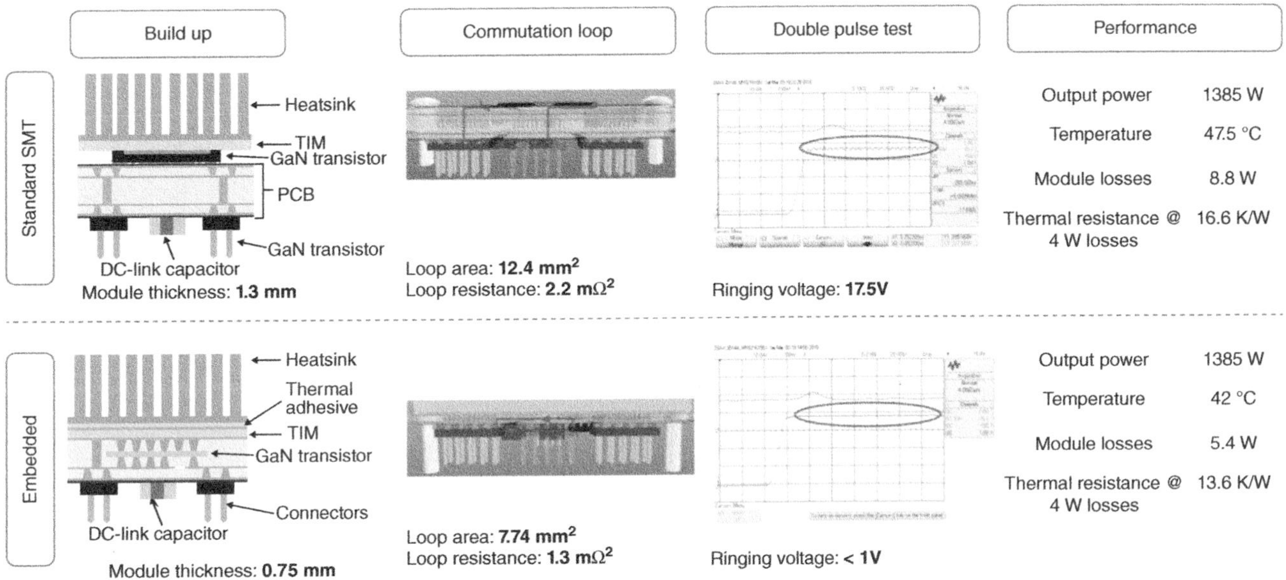

Figure 8.8 Performance results of SMT and embedded structure.

To compare the performance of the two versions, the following tests have been conducted:

1. Double pulse test to evaluate the commutation loop;
2. Efficiency measurement at full load;
3. Apply current ratio for self-heating and compare ambient temperature for thermal performance.

Figure 8.8 shows a summary table of the test results. For the double pulse test, the inductor is charged with 22 A, which is 75% of maximum current load, and a ripple voltage was measured. For the SMT version, the ripple voltage was 17.5 V, which is a good result. The embedded version showed a ripple voltage below 1 V, which is significantly better.

For the efficiency evaluation, the following setting has been used:

Input voltage: 400 V
Output voltage: 200 V
Switching frequency: 100 kHz
Inductor: 128 μH
Load: 0–1.6 kW

The results show that with the embedded version we could reach a maximum efficiency of 99.5% at a temperature of 42 °C. This was again better than the SMT version, which showed a maximum of 99.2% at a temperature of 45 °C.

8.8 Heterogeneous Integration Using Cavities in PCB

Another embedded solution is local height reduction through the introduction of recessed cavities in the PCB (further on called "cavity" or "cavities"). These cavities can be used assembling components and/or device elements to reduce overall PCB z-axis dimension technology needs. These cavities can be also used to improve the thermal management by reducing the thermal path between the hot spot and the heat sink as well as the radio performance of an electronic system by exposing the antenna structure. Cavity approaches in PCB are shown in Figure 8.9.

In recent years, the functionality of such modules has tremendously increased. The enhanced functionality is provided by the heterogeneous integration of components with various functions in advanced miniaturized modules of added value addressing, for instance, power management, advanced data processing, and sensing and actuating. These kinds of electronic modules are also named SiP.

Compared to fan-out wafer-level packages, PCB-based solutions offer specific benefits. First, PCB material provides more stability because of the dielectric material properties such as glass fibers' higher mechanical stability compared to

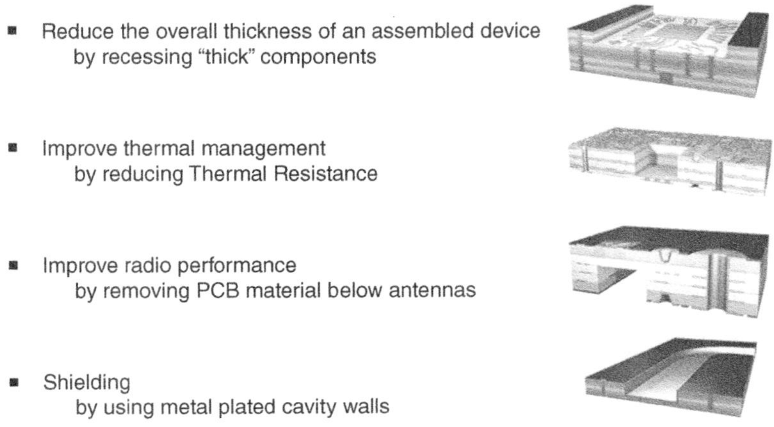

Figure 8.9 Cavities in PCB.

mold-based encapsulation materials. According to the nature of a PCB, several layers can be built up, which can be used for signal routing or integrated shielding. Copper planes and specific via structures can be used to enhance the thermal management of the package and can act as shielding. The copper structures can also be used to create antennas. The possibility of thicker dielectric will result in better the antenna performance. The right mix of different types of materials can help to match the coefficient of thermal expansion to the component for improved reliability. Finally, the large panel format offers cost-effective and high-volume manufacturing. Cavities can be applied to every kind of board in different applications and markets. Known products range from sensor packages and power applications up to mainboards for wearables or even mobile phones (see Figure 8.10).

Cavity forming allows for unlimited flexibility in shape and depth, thereby enabling greater freedom in material selection and PCB design rules. Solderable surfaces and solder mask patterns can also be applied on the cavity layer (see Figure 8.11). The feature relates to defined indents (cavities) in the PCB, which can be used to position electronic components such as capacitors, transistors, and even logic modules lower, thereby giving the assembled PCB a thinner overall structure. Electrical contacting can also take place in the cavity. There is a wide range of base materials (e.g., high speed, high T_g). The main advantages of in-cavity electrical contacting are:

1. No limitations in shape and depth of cavity;
2. Surface finishes in cavities (e.g., ENiG, OSP);
3. Several depths can be achieved in 1 card;
4. Solder mask in cavity.

Figure 8.10 Cavity in mainboard for wearables.

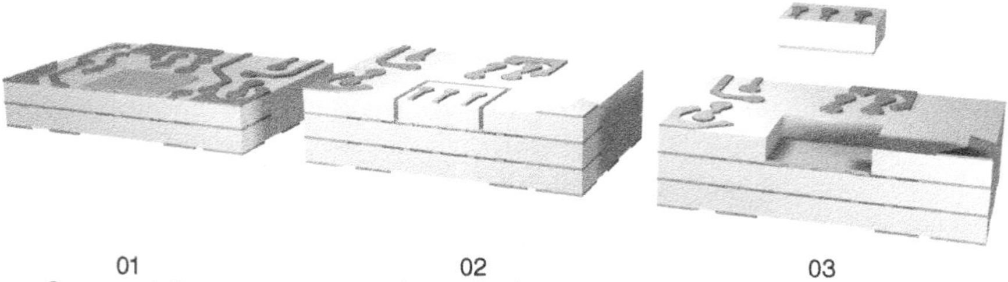

Figure 8.11 Schematic process flow of cavity formation at AT&S.

One main advantage of the cavity technology is that the same design rules apply inside and outside the cavity area. As mentioned above, even solder mask and surface finishes are possible in the cavity area to enable an SMT process.

8.9 Package Performance, Robustness, and Reliability

The reliability results for chip-last and chip-first technologies have been published previously [4, 5]. An example of a chip-last in-cavity package is a 1.1 mm thick PCB with a 10-layer multi-layer Panasonic R1551W material (halogen-reduced epoxy resin-based prepreg) with exception to the outer layers, which contained Mitsui MRG300 RCC-Foil (resin coated copper-foil). This stack-up and production method based on a special technology enables the removal of multiple layers at varying depths. Solder mask and Entek HT (organic surface protection) was used as a surface finish for all solderable surfaces (see Figure 8.12).

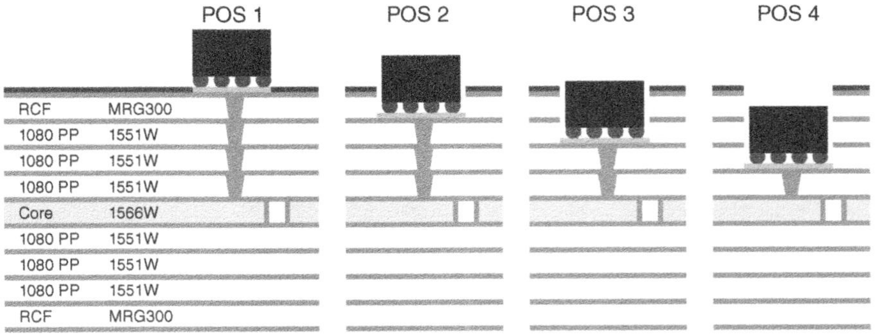

Figure 8.12 Test vehicle build-up with recessed areas (PCB and components are not to scale). POS here refers to position.

Table 8.3 Number of needed cards and components for drop test (DT) and temperature cycle test (TCT).

	DT		TCT	
	Cards	**Components**	**Cards**	**Components**
POS 1	9	5	4	15
POS 2	9	5	4	15
POS 3	9	5	4	15
POS 4	9	5	4	15
Total	**36**	**180**	**16**	**240**

The solder paste used for these trials is a commonly used tin-silver-copper (SAC) type 3 lead-free solder paste. The solder paste printing system used in this trial is a two-part system consisting of a step stencil and a customized squeegee. Four stencils were used in alignment with the four test vehicle variations, all of which were laser cut and electro polished stainless-steel stencils glued in polyester mesh and tensioned in an aluminum frame. The boards have been populated as shown as in Table 8.3.

When comparing the process capability or C_{pk} values of paste volume (see Table 8.4), one finds a clear trend in process capability under these standard conditions, while the test vehicle without cavity (0 μm) displayed a comparatively high C_{pk}, therefore, consistently attaining a sufficient volume. With each reduction in layer count we see sufficient but somewhat diminishing process capability. No major deviations (voids, skips, etc.) were observed with any test vehicles during the one-pass production trials (see Figure 8.13).

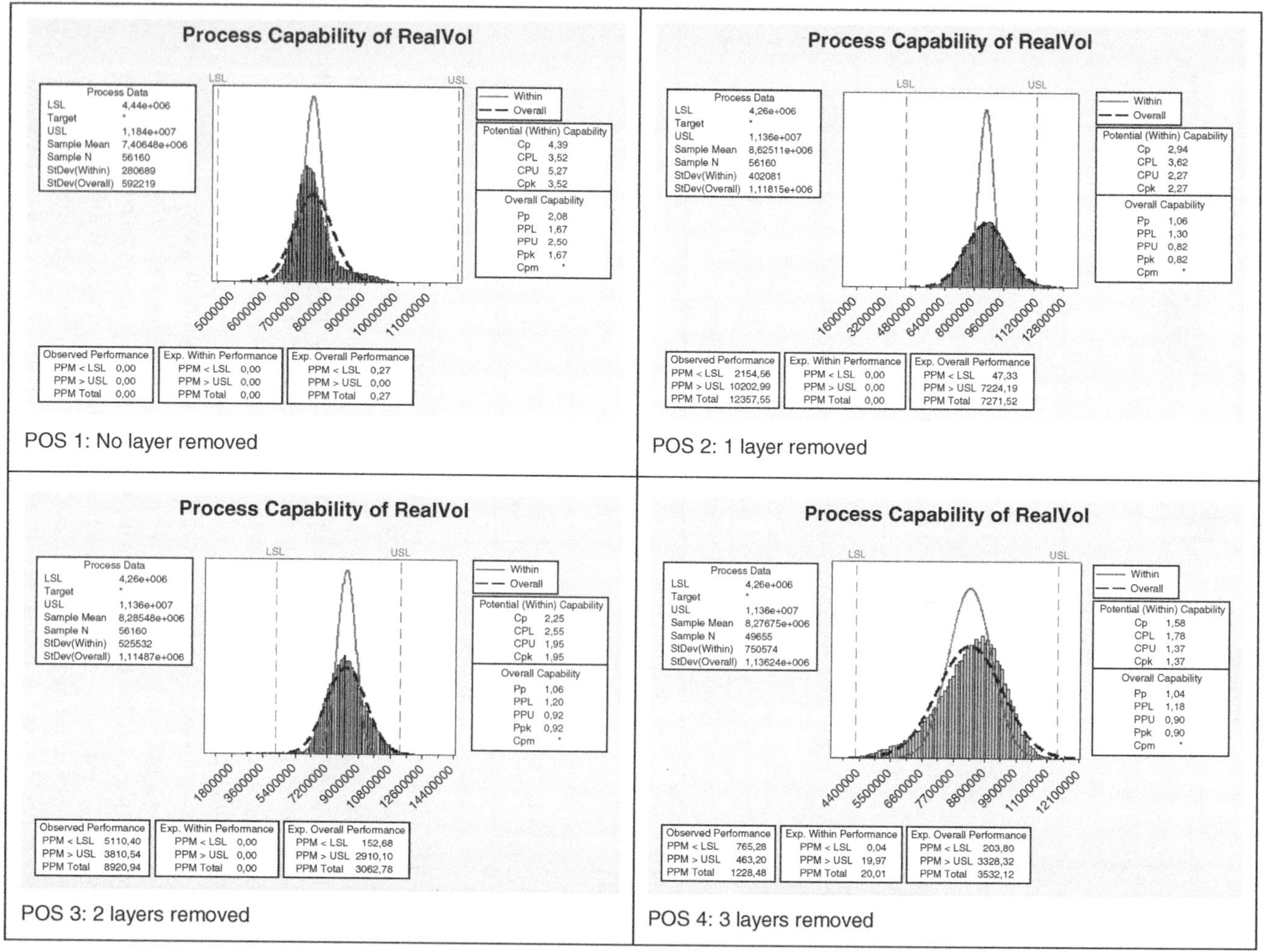

Figure 8.13 Process capability of paste volume for all test vehicles (POS 1, POS 2, POS 3, and POS 4).

Table 8.4 C_{pk} values of paste volume for all test vehicles (POS 1, POS 2, POS 3, and POS 4).

POS	C_{pk}
1	3.53
2	2.27
3	1.95
4	1.37

Figure 8.14 Position of components for drop test.

The board-level drop test specification was based on the JEDEC JESD22-B111. For these trials, five center components were chosen to be assembled because of overall higher exposure to tension during this type of drop test (see Figure 8.14).

The 0 µm test vehicle (no cavity) demonstrated the highest level of performance during drop test with a characteristic life (η) of 527 (see Figure 8.15). For each removed layer, the drop test performance was successively lower.

The earliest failure occurred with the test vehicle POS 3. Upon investigation of the earliest failures, cross sections revealed cracks in the solder near the component pad, that is, no defects were found at PCB solder joint (see Figure 8.16).Test vehicle POS 4 demonstrated, however, a somewhat inferior performance in general when compared with the other test vehicles.

The contour plot shown in Figure 8.17 clearly demonstrates a significant relationship between the cavity performances versus the outer layer (0 µm POS 1) performance. The contour plot graphically illustrates the η and slope (β) of the

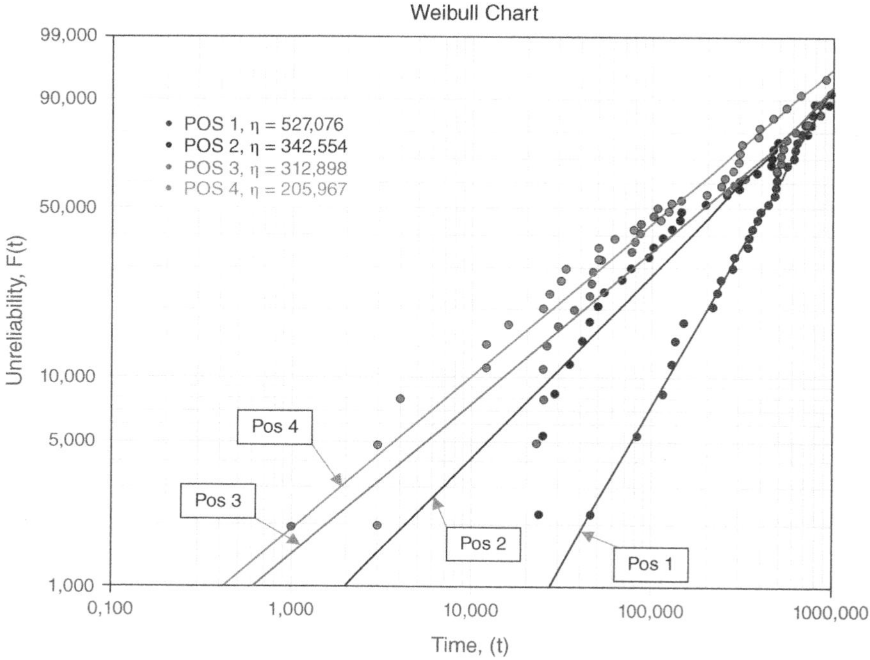

Figure 8.15 Drop test Weibull chart for all test vehicles (POS 1, POS 2, POS 3, and POS 4). Decimals here are notated with commas in the European style.

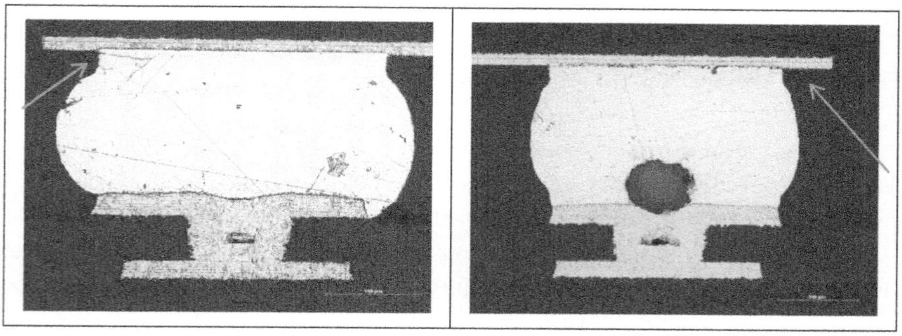

Figure 8.16 Crack in solder near component pad.

Weibull fit. The curve plots show that η and β will fall within the marked areas at the following drop with a probability of 50%. If these plots do not converge, a statistically significant difference exists between the respective population (POS 2, POS 3, and POS 4 versus POS 1). In general, no particularly frequent defect mode was visible during investigation (i.e., cracks in all locations throughout solder joints, bulk solder, and PCB).

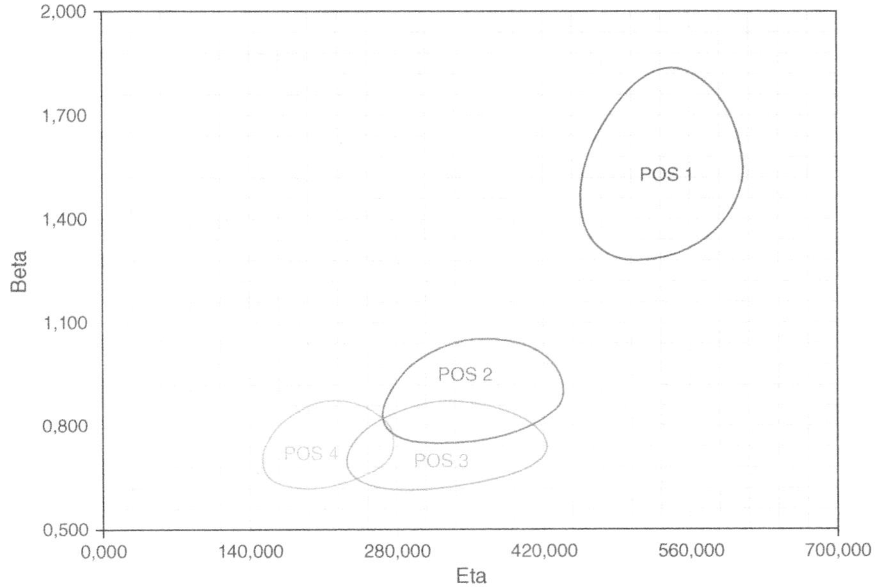

Figure 8.17 Drop test contour plot of characteristic life (Eta) and slope (Beta) plot for all components for all test vehicles (POS 1, POS 2, POS 3, and POS 4). Decimals here are notated with commas in the European style.

The board-level temperature cycle test (TCT) specification was based on the JEDEC JESD22-A104C. Four test vehicles were tested, each with 15 surface-mount device (SMD) daisy chains. Test events were monitored in situ. The boards were subjected to a constant change in temperature in the range of −40 °C to +125 °C. The criteria for passing was 1000 cycles. In all cavity variations, the test vehicle passed 1000 cycles without failure.

It is worth noting in the conclusion of this analysis that the preparation, manufacturing, and availability of materials (i.e., stencil, squeegee, PCB, solder paste, etc.) posed no obstacle to the overall investigation. This fact is important when highlighting our focus on the manufacturability when employing these elements.

A test board, including the SMD and EC on the same vehicle, was designed to ensure that both sets of components were subjected to the same stress elements during testing [5]. The PCB was an 8-layer multi-layer as shown in Figure 8.18. The unused copper on all inner layer was hatched, meaning they were not full copper surfaces. This is standard practice in PCB design to achieve stable thermal and therefore thermo-mechanical performance (i.e., warpage control). The material chosen for the PCB was Panasonic R1551W (halogen-free epoxy resin-based prepreg). This material is standard in HDI applications and is not a material with specialized characteristics for embedding components.

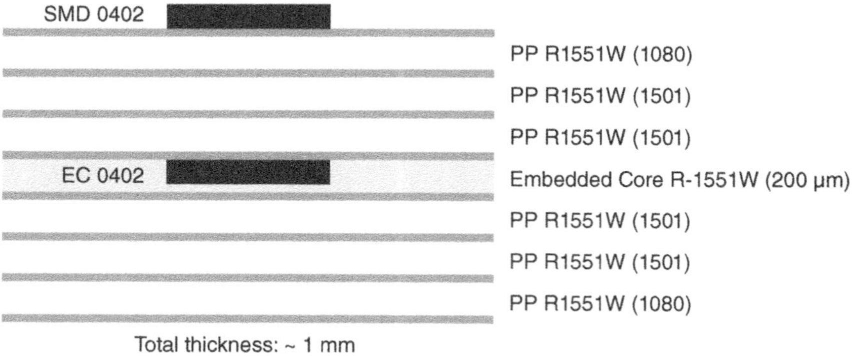

Figure 8.18 Build of an 8-layer PCB with surface mount devices (SMD) and embedded components (EC).

In general, it was decided to create this vehicle using passive components (resistors in this case) due to the simplicity of testing without excessive manufacturing performance variability. Standard 10 ohm 0402 resistor SMD components were chosen for outer layer assembly based on their commonality in both SMT manufacturing and availability at the current EC manufacturing at the facility where the test vehicle was produced. The choice should represent a practical application but is certainly not the only choice that could have been made.

The EC are the size of 0402s in x and y dimensions and the electrical function matches that of the SMD component. The z-axis differs in that it is thinner than the standard SMD 0402. The same component manufacturer was used for both SMD and EC. Because of the nature of the varying methods of achieving connectivity, the terminals of the components also vary. The SMD component is metal grazed (tinned terminals) for the SMT assembly process, whereas the EC has copper terminals.

One of the more obvious characteristics of EC (pertaining to the manufacturing method discussed here) is the fact that they are encased within the PCB, so they are surrounded by the resin and glass fiber of the prepreg. Any external stress would theoretically be distributed within the whole construct and less on the component itself, as opposed to an SMD component and solder joint system, which being on the outer layer is directly exposed a stress source. The central location of the EC is also more neutral when considering mechanical stresses to the planar rigidity of the PCB.

Drop test and TCT was completed to verify reliability. Seventeen out of 18 test vehicles revealed SMD-related defects before 1000 drops, while one test vehicle demonstrated 1000 successful drops without SMD failure. There was one out of 18 vehicles that registered an EC-related failure before 1000 drops. Extensive attempts were made to locate the failure source within the PTH and inner layer traces, but no obvious source could be found.

The boards were subjected to a constant change in temperature in the range of −40 °C to +125 °C. The requirement was to pass 1000 cycles. The total sum of tested daisy chains was 70: 35 SMD and 35 EC. In each daisy chain there were eight components. An event is registered as a failure after demonstrating a resistance change of > 1000 ohm. In all variations (SMD and EC), the test vehicle passed 1000 cycles without failure. No further cross sections or analyses were made because of the lack of any relevant failure.

8.10 Conclusion

Miniaturization of electronic systems drives the need for multi-functional modules. PCB and IC substrate technologies offer a versatile toolbox for small form factor integration of modules (SiP) and large electronic systems with 100 and more components (system-in-board). Chip-last and chip-first technologies can even be combined into one package to optimize the yield/performance ratio, for example, embedding the passive components into the board and assembly active components on a cavity layer. As mentioned in the individual examples, both technologies offer significant benefits for the product itself, such as miniaturization or improvement of electrical and thermal performance, but also volume scaling and efficient process utilization because of the large panel size compared to wafer size.

References

1 "Panel level packaging." (2021). https://www.izm.fraunhofer.de/de/abteilungen/system_integrationinterconnectiontechnologies/arbeitsgebiete/panel-level-packaging.html (accessed 25 May 2021).

2 Mahajan, R., Viswanath, R., Sankman, B. et al. (2018). Intel 2D to 3D package architectures - back to the future. *Proceedings of the International Microelectronics and Packaging Device Packaging Conference.*

3 Jung, E., Landesberger, C., Ostmann, A. (1999). "Method for integrating a chip in a printed board and integrated circuit," German patent DE19954941C2, filed November 16, 1999 and issued November 6, 2003.

4 Leitgeb, M. and Ryder, C.M. (2012). *SMT Manufacturability and reliability in PCB cavities.* https://smtnet.com/library/index.cfm?fuseaction=view_article&article_id=1801&company_id=54036 (accessed 25 May 2021).

5 Ryder, C.M. (2011). Embedded components: a comparative analysis of reliability. https://www.datasheetarchive.com/whats_new/e145f3d44c0527568545f9922f4c3cee.html (accessed 25 May 2021).

9

Advanced Embedded Trace Substrate – A Flexible Alternative to Fan-Out Wafer Level Packaging

Shih Ping Hsu, Byron Hsu, and Adan Chou

Phoenix Pioneer technology (PPt) Co., Ltd., Hsinchu, Taiwan

9.1 Technology Description

9.1.1 C^2iM Technology

C^2iM (copper connection in materials) is a multi-layer integrated circuit (IC) substrate platform with embedded trace and coreless structure. Furthermore, PPt developed the fan-out PLP (panel-level package) technology on the C^2iM platform, called C^2iM-PLP, which uses chip-first face-up method to embed components. Single chip or multiple chip packages can be created with this method. Figure 9.1a shows the traditional embedded trace substrate (ETS) structure used for flip-chip packaging comprised of a dielectric layer containing the glass fiber such as bismaleimide-triazine (BT) material, with interconnections created by laser via [1–3]. Figure 9.1b is the basic structure of C^2iM.The biggest difference between C^2iM and ETS is that C^2iM uses EMC (epoxy molding compound) as the dielectric layer and copper (Cu) pillars as the interconnection via. The advantages of C^2iM are as follows:

(1) Since there is no glass fiber at high frequencies, it has better electrical properties for high-speed applications.
(2) Cu pillar made with photoresist and electroplating has a better cost advantage when the number of vias is large, and it can provide any shape of Cu pillar with more flexible layout space. This is more suitable for fine via pitch needs. Cu pillars made by photoresist and plating can also overcome the high-cost problem of laser via at high via count and the limitation of fine via pitch.
(3) The material composition of C^2iM is simple, mainly based on EMC and copper (Cu); in practice, it can obtain a good reliability capability up to MSL1 (moisture sensitivity level 1).

Embedded and Fan-Out Wafer and Panel Level Packaging Technologies for Advanced Application Spaces: High Performance Compute and System-in-Package, First Edition. Edited by Beth Keser and Steffen Kröhnert.

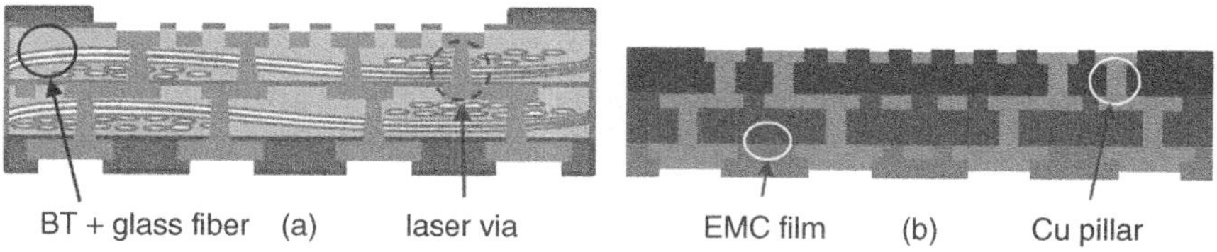

Figure 9.1 (a) ETS substrate. (b) C^{2}iM substrate.

C^{2}iM has been widely used in various electronic components. Package forms include flip-chip chip-scale package (FC-CSP), wire bond ball grid array package (WB-BGA), and coreless flip-chip ball grid array package (FC-BGA). C^{2}iM is in production with power IC for consumer electronics and automotive electronics, AP (application processors), RF (radio frequency), OIS (optical image stabilizer), and FPS (finger print sensor), among many examples.

9.1.2 C^{2}iM-PLP Technology

In order to meet the needs of products with high current, high voltage, high heat dissipation, and high reliability, such as power electronic components for server and automotive, C^{2}iM-PLP adopts thick Cu substrate technology. Compared to FO-WLP (fan-out wafer-level package), C^{2}iM-PLP is lower in cost and can be used in power components, automotive power IC, and other fields. For example, semiconductor power components such as SiC or GaN applied to high-voltage and high-frequency product applications must face the challenges of heat dissipation, electrical performance, and reliability. The unique packaging technology of C^{2}iM-PLP can overcome these difficulties and produce the products that meet customer needs. Figure 9.2 is one of the examples showing C2iM-PLP product image for power switch application.

The development of PPt C^{2}iM-PLP has been completed and has passed customers' electrical performance and reliability verification and will soon be introduced into small volume production.

Figure 9.2 C^{2}iM-PLP.

9.2 Applications and Markets

Embedding the die in thick Cu results in good thermal stability and high reliability due to the connection between die pad and substrate copper traces being copper instead of conventional assembly interconnects such as wire bonds or solder bump joints. C^2iM-PLP can also have better power performance, better heat dissipation, and higher reliability. C^2iM-PLP is suitable for automotive, new energy vehicle, and server markets. Production of diodes for automotive generators, power switches of automotive motor drivers, power regenerators of optical networks, and power modules of servers are currently ongoing. Future target includes advanced driver assistance systems (ADAS) and electric car applications.

9.3 Basic Package Construction

C^2iM-PLP There are two types of C^2iM-PLP. Type A has single redistribution layer (RDL) with thick Cu and type B has multi-layer RDL with components attached like in a 3D system-in-package (SiP) structure. Figure 9.3 shows examples of the two types along with the design features for automotive and power modules. C^2iM-PLP is more focused on automotive and power modules because of performance requirements such as electrical performance, heat dissipation, and form factor.

9.3.1 C^2iM-PLP Experience

PPt offers various package and die dimensions for both type A and type B. For example, package sizes of 3 mm × 3 mm to 8 mm × 8 mm, die sizes of 1 mm × 1 mm to 5 mm × 5 mm, and package heights of 300 to 750 μm made on a panel size of 630 mm × 540 mm for high-volume manufacturing (HVM).

9.3.2 C^2iM-PLP Advantages and Disadvantages Compared to Wirebond Quad Flat No Lead (WB-QFN) and Flip-Chip QFN (FC-QFN) Packages

One advantage of C^2iM-PLP is good current carrying capability and better electromigration (EM) performance [4]. Since the electrical connection of C^2iM-PLP on the chip side uses copper under bump metallurgy (UBM) to connect with package Cu trace and the overall connecting area is increased, it can effectively improve the current carrying capability and reduce EM. The electrical properties of WB-QFN are limited by the cross-sectional area of the Au wire and the length. The electrical

Type A	Type B
One layer RDL with thick Cu	3D and multi-layer RDL with thick Cu
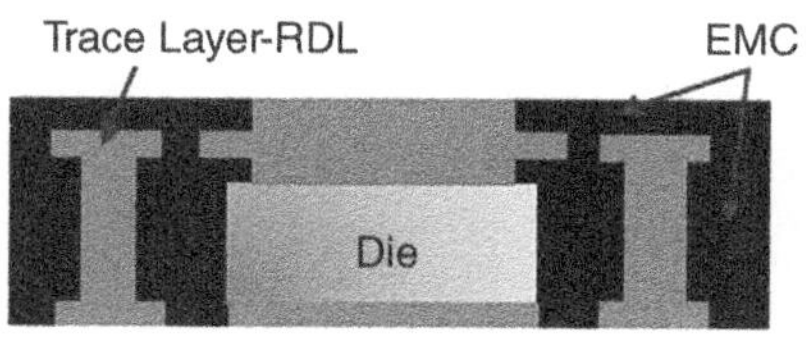	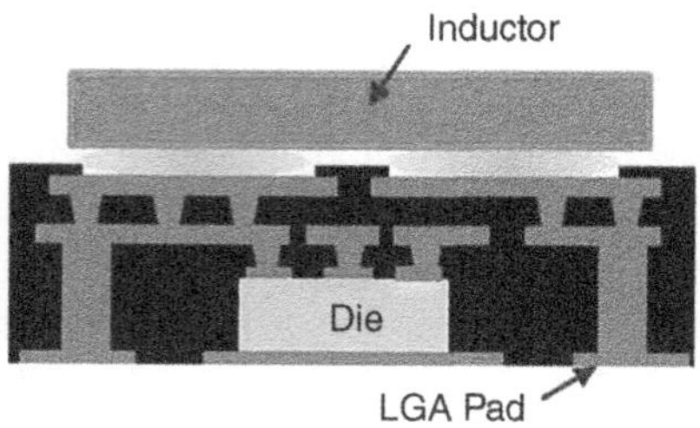
PKG size: 8 × 6.5 mm	PKG size: 5 × 5 mm
PKG thickness: 320 μm	PKG thickness: 370 μm(w/o Inductor)
Die size: 5 × 5 mm	Die size: 2.5 × 3 mm
Feature:	Feature:
PKG size: 8 × 6.5 mm	PKG size: 5 × 5 mm
PKG Thickness: 320 μm	PKG Thickness: 370 μm(w/o Inductor)
Die size: 5 × 5 mm	Die size: 2.5 × 3 mm
Application: Automotive	Application: Power Module

Figure 9.3 C²iM PLP construction.

connection of FC-QFN on the chip side is solder bump interconnects, so it is prone to EM at the interface in high-current applications.

C²iM-PLP has good design flexibility. It is easily routable and can be applied to multi-chip packages such as SiP. Both WB-QFN and FC-QFN are less routable and less flexible compared to C²iM-PLP. Figure 9.4 shows that C²iM allows a routable design for multi-die connection, which conventional QFN cannot meet.

One disadvantage of C²iM-PLP is that it is higher in cost than WB-QFN and FC-QFN, but slightly less than FC-QFN with copper clip products [5]. For a complete list of advantages and disadvantages, see Table 9.1.

9.3.3 C²iM-PLP Advantages and Disadvantages Compared to WLP and FO-WLP

One of the advantages of C²iM-PLP is excellent die protection. The good die protection is a result of the EMC protection on all six sides of a die, which wafer-level

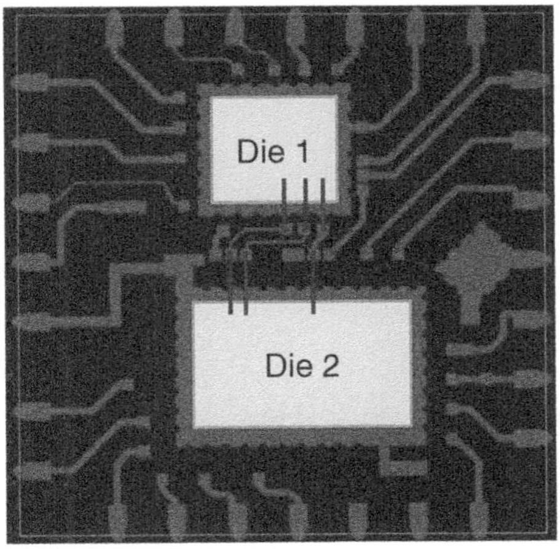

Figure 9.4 C²iM-PLP routable design.

Table 9.1 C2iM-PLP compared to WB-QFN and FC-QFN.

Platform	**WBQFN + Cu clip**	**FCQFN + heat sink**	**C2iM-PLP**
	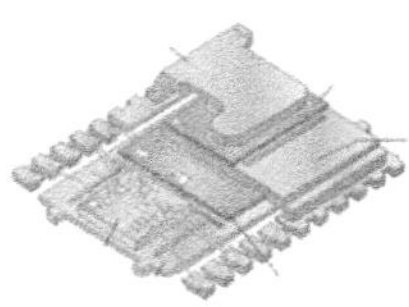		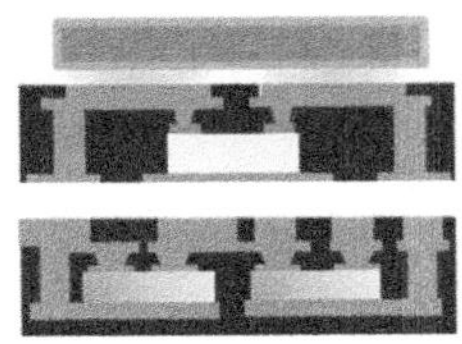
1st level connection	Au or Cu wire Δ	Cu + solder ×	Cu-Cu ⊙
EM current performance	limited by Au/Cu wire loop Δ	FC with solder ○	Cu connect ⊙
Thermal	full Cu pad ⊙	with Adhesive Δ	full exposed pad to PCB ⊙
Design flexibility	not routable ×	not routable ×	routable ⊙
Cost	$$$ ×	$$$ ×	$$ ○

⊙: Excellent; ○: Good; Δ: Normal; ×: Bad

packaging (WLP) and traditional chip-first face-down FO-WLP discussed in other chapters do not have.

Since C²iM-PLP uses substrate technology, it can easily achieve thick copper designs, which can provide good electrical and thermal properties, and can be applied to high-current and high-voltage electronic products. For WLP and FO-WLP to do thick copper traces (e.g. Cu thickness > 40 μm thick), the manufacturing cost will be expensive and the processing is more complex. C²iM-PLP is a process that uses large-size panels, and total cost is cheaper than for FO-WLP.

Table 9.2 C[2]iM-PLP comparing to WLP and FO-WLP.

Platform	WLCSP	FO-WLP	C2iM-WLP

Small form factor	++ ⊙	+ ○	+ ○
I/O count	+ Δ	+++ ⊙	++ ○
Thick Cu(um)	< 10 Δ	< 10 Δ	15–65 ⊙
Die protection By EMC	0 side ×	3 side ○	6 side ⊙
Fine L/S (um)	≤ 10/10 ⊙	≤ 10/10 ⊙	≥ 15/15 ○
EM current performance	1x Δ	1x Δ	> 1.3x (Thick Cu) ⊙
High current application	+ ×	++ Δ	++++ ⊙
Total cost	$ ○	$$$ ×	$$ Δ

⊙: Excellent; ○: Good; Δ: Normal; ×: Bad

Regarding the disadvantage of C[2]iM-PLP, the I/O count capability of C[2]iM-PLP is worse than FO-WLP, and the fine line and space (L/S) capability is also worse than WLP and FO-WLP. For a complete list of advantages and disadvantages, see Table 9.2.

9.3.4 Future Applications

C[2]iM-PLP has high reliability capability, with the use of full EMC and a thick copper package structure and has good electrical and heat dissipation properties. C[2]iM-PLP can effectively reduce the resistance from drain to source when the MOSFET is fully on, which is extremely suitable for power module applications and products such as automotive power modules.

9.3.5 Limitations of C[2]iM-PLP

Today, the applications of C[2]iM-PLP focus on low I/O products. Therefore, for products that require high I/O, fine die pad pitch (< 100 μm) and fine line/space (L/S ≤ 10/10 μm), C[2]iM-PLP is not a good fit. Since the current approach is to use a CO_2 laser drill process, it is difficult to make < 50 μm via diameters to get more offset between die pad and via hole, so the die pad pitch is limited. In order to meet the needs of these high-density products, a new generation of C^{2}iM-PLP technology will be developed.

9.4 Manufacturing Process Flow and BOM

The C^{2}iM-PLP manufacturing uses IC substrate production lines with additional die mount process equipment. The process flow can be found in Figure 9.5. The UBM is formed on the chip, the die is placed face up on the carrier with silver paste, next the photo imaging process and Cu plating creates the Cu pillar, and then molding and grinding reveals the Cu pillar. After laser is used to form a via to the die, the next step uses SAP (semi-additive process) to form the Cu RDL. Finally, the package back-end processes including marking, singulation, test and inspection are completed.

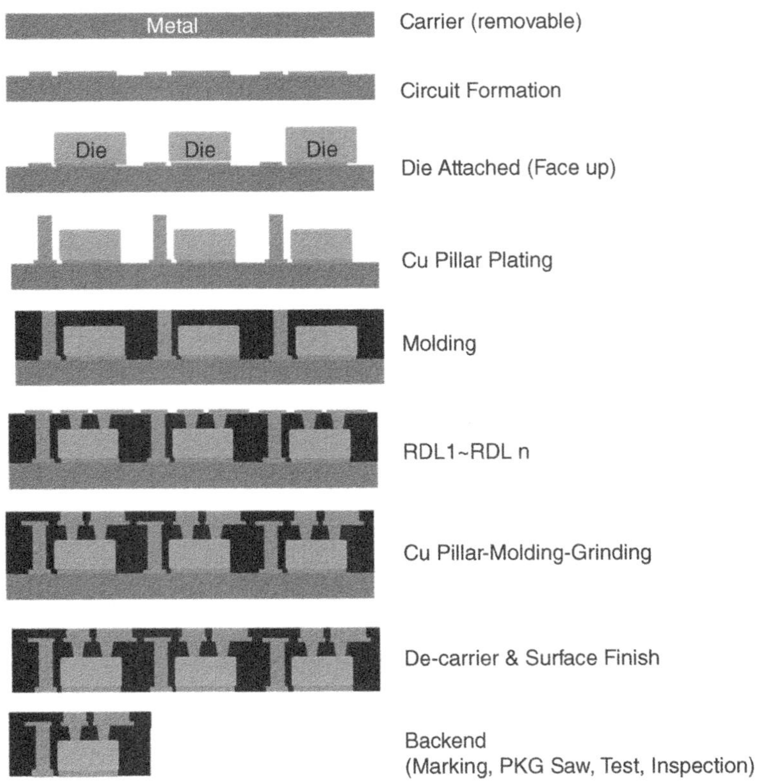

Figure 9.5 C^{2}iM-PLP process flow.

Table 9.3 C^2iM-PLP material properties.

Item		EMC granular type	EMC film type
CTE1 (XY)	ppm/°C	13	7
CTE1 (Z)	ppm/°C	13	7
CTE2 (XY)	ppm/°C	30	21
CTE2 (Z)	ppm/°C	30	21
Tg (TMA)	°C	200	157
Dk	1 MHz	4.2	3.5
	1GHz	4	3.3
Df	1 MHz	0.01	0.007
	1 GHz	0.014	0.006
F. modulus RT	MPa	18 500	8000
F. modulus 260 °C	MPa	2500	S80

The material properties of the bill-of-materials (BOM) of C^2iM-PLP is shown in Table 9.3. The basic material composition of C^2iM-PLP is plated copper and EMC. EMC has two types: one is granular type for embedded die and the other is film type which uses vacuum lamination process CTE mismatch between various packaging materials and the silicon chip is a concern, so lower CTE to prevent warpage needs to be considered. Second, a lower dielectric constant (Df) value for high-frequency products is important. Lastly, the flexural modulus is required to balance the warpage behavior during process.

9.5 Design Features

9.5.1 Package Design Rules

The design features of C^2iM-PLP can be found in Table 9.4. Current capability includes package sizes of 3 mm × 3 mm to 8 mm × 8 mm, die thickness smaller than 300 μm, and I/O count lower than 100. C^2iM-PLP is improving capabilities continuously; the future plan is to be able to do <21 mm × 21 mm package size, die thickness more than 300 μm, higher density products such as an I/O count of 100–300, and ball pitch of 250–350 μm.

9.5.2 Design Rules for Die UBM

The design rules of die UBM are shown in Table 9.5 below. Copper with 8 μm minimum thickness as the UBM metal is needed because of process requirements.

Table 9.4 C²iM-PLP design feature.

Characteristic	C²iM-PLP design feature: Normal	Future
PKG size (mm)	3 × 3 – 8 × 8	21 × 21
Die Thickness (μm)	150–300	125 to > 300
I/O count	2–100	100–300
Pad/ ball pitch (μm)	> 400	250–350
Package height (μm)	260–780	> 780

Table 9.5 Design rules for die UBM.

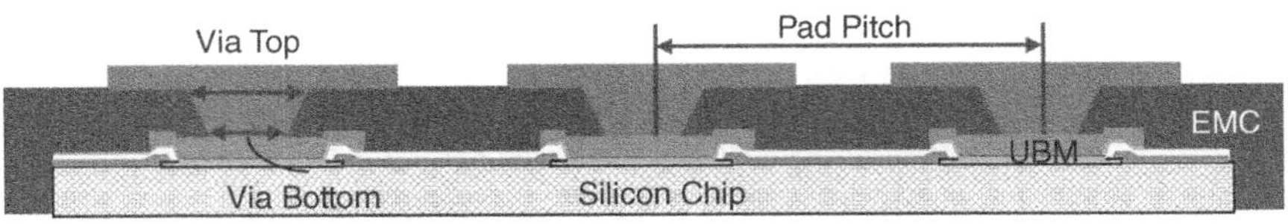

Characteristic	Design (μm): Normal	Advance		Future
Die Pad Pitch (μm)	180	160	150	90
Die Pad Die. (μm)	130	110	100	60
Min. Via Top size (μm)	80	70	60	40
Min. Via Bottom size (μm)	55	50	40	25
UBM Metal Thickness	Copper Min. 8 μm			

C²iM-PLP capability includes 160 μm die pad pitch with 110 μm pad diameter using 70 μm top via diameter currently with plans to shrink design rules to 90 μm die pad pitch in the future.

9.5.3 Design Rules for Die Side by Side

The design rules for die side by side are shown in Table 9.6 below. 300 μm from die to package edge and 600 μm between 2 die are needed currently. 200 μm from die to package edge and 500 μm between 2 die are advanced rules. The future plan is to shrink design rule to 150 μm from die to package edge and 400 μm between 2 die.

Table 9.6 Design rules for die side by side.

		Unit: μm		
Symbol	**Description**	**Normal**	**Advanced**	**Future**
A/B	min. space of die to PKG edge	300	200	150
C	min. space between 2 dies	600	500	400

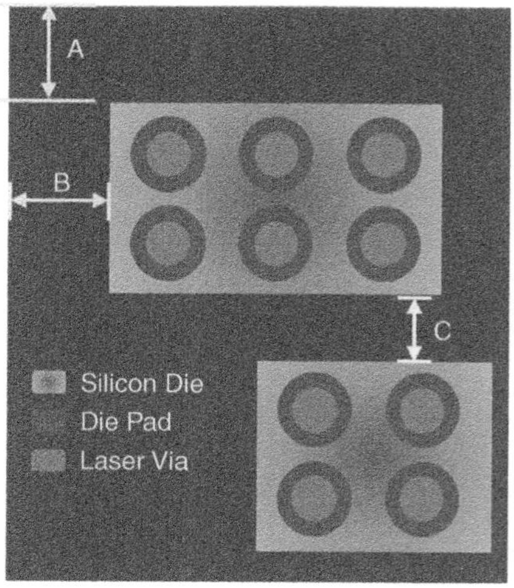

9.5.4 Design Rules for Cu Pillar

C^2iM-PLP Cu pillar design rule can be found in Table 9.7. C^2iM-PLP has to use maximum 200 μm pillar height with 150 μm pillar diameter in the embedded die layer. 280 μm pillar height maximum with 250 μm pillar diameter is planned for the future, with a minimum pillar space of 500 μm. Regarding die edge to RDL distance, 250 μm minimum is the advanced rule with plans to make 200 μm ready. Minimum L/S on RDL of granular EMC is 30/30 μm for advanced rules, but 20/20 μm can be achieved today on EMC film. Future plans include for 15/15 μm on granular EMC and 10/10 μm on film EMC.

Table 9.7 Design rules for Cu pillar (all units in μm).

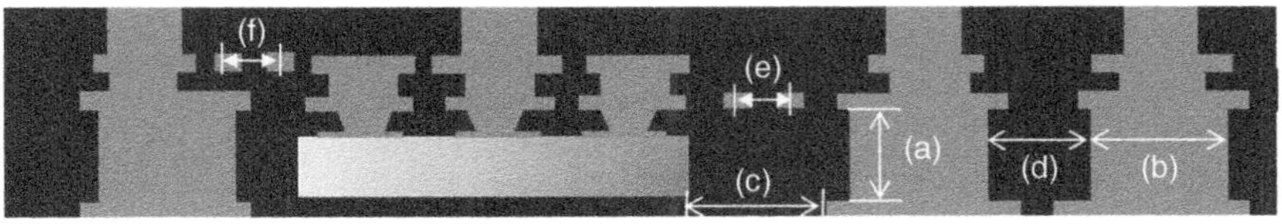

Symbol	Description	Normal	Advanced	Future
(a)	max. pillar height	200	200	280
(b)	min. pillar diameter	200	150	250
(c)	min. die edge to RDL	300	250	200
(d)	pillar space	150	150	500
(e)	min. L/S on RDL 1 (on granular EMC)	50/50	30/30	15/15
(f)	mm. L/S on RDL 2 (on film EMC)	30/30	20/20	10/10

9.6 System Integration Capability

The trend in power modules is toward the integration of multiple functions into a SiP module. Integrating multiple power ICs and passive components on the package can achieve better electrical performance and simplify the system assembly. C^2iM-PLP is very suitable for embedding active and passive components at the same time with integration of the electronic components on a package since C^2iM-PLP can make die to RDL connection on the same plane and use compression molding to overcome the different component thickness variation. C^2iM-PLP can provide SiP solutions and Figure 9.6 shows a SiP concept.

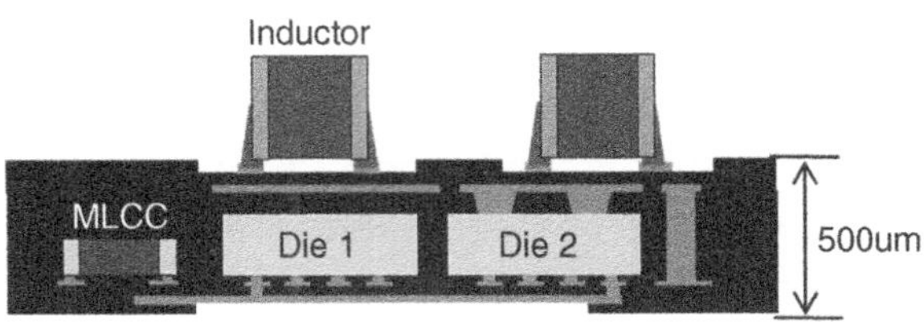

Figure 9.6 C^2iM-PLP SiP structure.

Table 9.8 Manufacturing format—panel size plan.

Process	Manufacturing format	
	Now	Future
Panel size (mm)	540 × 630	smaller
Back-end (Marking /BGA ball/test/saw)	strip 95 mm × 240 mm	quarter panel 290 mm × 240 mm or smaller

9.7 Manufacturing Format and Scalability

The current panel size of C^2iM-PLP is 540 mm × 630 mm (21 in. × 24 in.). In the future, considering the mass production scale and minimizing the risk of the die loss, PLP may fine-tune the panel size to the most appropriate size (see Table 9.8). The reason for the change would be to leverage back-end process equipment capability in the industry. Changing from strip format to quarter panel size can get better utilization rate since, 630 mm × 540 mm panel size does not fit quarter panel size capability of back-end machines.

9.8 Package Performance

9.8.1 Electrical Performance

C^2iM-PLP can increase contact area of die to get higher current carrying capability. See Table 9.9 below for details. C^2iM-PLP can apply a larger opening size and Cu post for thicker Cu to achieve a higher current per ball. See Figure 9.7 for a schematic.

Table 9.9 C^2iM-PLP EM current advantage.

	Interconnect structure	Interconnect area	# of interconnect	Current per interconnect (100 kh @110 °C)	Total current (A)
FCQFN	Cu pillar bump (Sn/Cu interface)	125 μm × 200 μm	5	1.36	6.8
C^2iM-PLP	Cu/Cu interface	70/50 μm Circle	22	0.6	13.2

Figure 9.7 C^2iM-PLP current carrying advantage.

Table 9.10 Thermal performance comparison.

Θja (C/W)	**Laminate PLP**	**C^2iM-PLP**
Structure	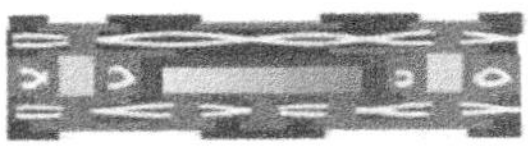	
PKG height	520 μm	370 μm
Simulation	29 C/W	22 C/W
Real testing	19.1 C/W	15.4 C/W
Improvement	POR	19.3%

✓ PKG size: 5 × 5 mm
✓ Die size: 2.5 × 3 mm

9.8.2 Thermal Performance

C^2iM-PLP has good thermal performance. A power module product is used to compare the heat dissipation efficiency. The package features of the C^2iM-PLP are two layers of Cu traces, a package size of 5 mm × 5 mm, and a die size of 2.5 mm × 3.0 mm. The package thickness of the package structure using C^2iM-PLP is thinner than a traditional WB-CSP or FC-CSP package substrate, and the thermal performance (Θja) of the package has a 19% performance improvement over laminate PLP. See Table 9.10 for comparison.

9.9 Robustness and Reliability Data

9.9.1 Automotive Reliability Certification Pass

The C^2iM- PLP with two layers of RDL, package size of 8.6 mm × 6.5 mm, and die size 5 mm × 5 mm is shown in Figure 9.8 and has been certified to the automo-

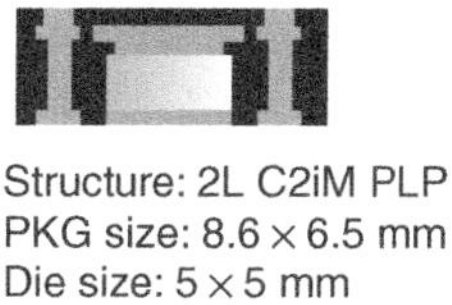

Figure 9.8 C^2iM-PLP two layer.

Table 9.11 Reliability test result for automotive.

Item		Criteria	Sample size	Result
1	High temperature storage	Ta = 175 °C. 1000 hours	77 pcs	pass
2	Power cycle-DC	Tc = −50–160 °C, 50 A. 25 000 cycles	77 pcs	pass
3	Power cycle-AC	Tc = −50–175 °C, 150 A, 25 000 cycles	77 pcs	pass

tive AEC Q100 grade 0 reliability specification for all relevant tests. The results of reliability certification are shown in Table 9.11.

9.9.2 Board Level Reliability Verification Pass

The C^2iM-PLP with one layer of RDL, a package size of 6.6 mm × 6.6 mm, a die size of 6 mm × 6 mm, and a ball pitch of 0.5 mm passed board level testing. The test condition for component level was MSL1 + temperature cycle test (TCT) of −40 to 125 °C for 1000 cycles with a sample size of 32 units. All units passed the board level TCT. Test standard used for board level TCT was IPC9701 TC3 (−40 to 125 °C).

9.10 Electrical Test Considerations

Large panel electrical test is more suitable for high-volume production and for cost savings. Currently, the volume production of open/short (O/S) test can only be done in strip or unit form. The test equipment for the panel process is very expensive since it is not yet fully mature and has not yet been used in production. Large panel test can reduce handling and cycle time including alignment, pick and place, and loading and unloading times compared with unit or strip format testing so that total running cost is better than current approach.

Electrical test methods including in process O/S testing, package O/S, and final test for quarter panel test or panel test will be required in the future. Test equipment suppliers need to develop lower cost machines, and PLP suppliers must cooperate with equipment suppliers to develop test equipment that meets the needs of the panel process.

9.11 Summary

C^2iM-PLP technology is good for many applications especially for power module, high voltage, and high-current requirements; it is very suitable for third-generation semiconductors such as SiC or GaN chip products. C^2iM-PLP has already qualified by automotive products and has been in high-volume production since Q1'2021.

References

1 Chen, E., You, J., Lan, A., and Liao, M. (2014). Structure reliability and characterization for FC package w/Embedded Trace coreless Substrate. *Proceedings of the 16th Electronics Packaging Technology Conference (EPTC)*. IEEE.
2 Tang, T., Lan, A., Tsai, J., Lin, S., Ho, D., and You, J. (2015). Challenges of flip chip packaging with embedded fine line and multi-layer coreless substrate. *Proceedings of the 65th Electronic Components and Technology Conference (ECTC)*. IEEE.
3 Hsieh, M., Cho, N. and Kang, K. (2017). Development of thin flip chip package with low cost substrate technology. *Proceedings of the 12th International Microsystems, Packaging, Assembly and Circuits Technology Conference (IMPACT)*. IEEE.
4 Wei, C.C., Yu, C.H., Tung, C.H. et al. (2011). Comparison of the electromigration behaviors between micro-bumps and C4 solder bumps. *Proceedings of the 61st Electronic Components and Technology Conference (ECTC)*. IEEE.
5 Lwin K.K., Tubillo, C.E., Panumard, T.J.D. et al. (2016). Copper Clip Package for high performance MOSFETs and its optimization. *Proceedings of the 18th Electronics Packaging Technology Conference*. IEEE.

10

Flexible Hybrid Electronics Using Fan-Out Wafer-Level Packaging

Subramanian S. Iyer and Arsalan Alam

Center for Heterogeneous Integration and Performance Scaling, Samueli School of Engineering, University of California, Los Angeles, CA, USA

10.1 Introduction

Flexible hybrid electronics (FHE) refers to a packaging construct that is physically flexible in contrast to more conventional packaging, which tends to be rigid. Physically flexible electronics can be made conformable to irregular surfaces ranging from the human body, both outside and inside the body, to curved surfaces such as the surfaces presented by a car or aircraft or curved displays. In many cases, the conformal packages may be applied, removed, and reapplied several times over their lifetime. Figure 10.1a [1], b [2], and c [3] show some typical FHE devices that are available in the market today. FHE revenues are still small, approaching about \$200 M in 2020 but are projected to increase to over \$3 B [4] over the next decade or so. While the initial development was focused on large-area solar cells [5] and flexible displays using organic light-emitting diodes (OLED) and thin film transistors (TFT) [6], manufacturing scaleup has been challenging. Today, by far, a dominant driving force is medical electronics, where conformable wellness devices can monitor a variety of body parameters such as temperature, blood pressure, blood components including oxygen, sugar, and so forth, heart rate, and muscle activity almost continuously for health, therapeutic, and wellness considerations. But there are also much simpler albeit very high-volume applications such as RFID tags and smart packaging, which can drive huge volumes if they hit the right price point. A special issue of the *Proceedings of the IEEE* [7], (vol. 103, issue 4) was published in 2015 and describes the technologies that have been developed as well as some of the drawbacks and challenges of these approaches.

Embedded and Fan-Out Wafer and Panel Level Packaging Technologies for Advanced Application Spaces: High Performance Compute and System-in-Package, First Edition. Edited by Beth Keser and Steffen Kröhnert.

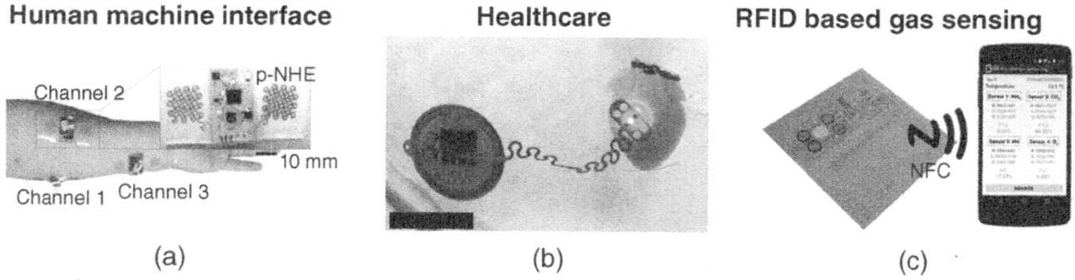

Figure 10.1 A few examples of commercial FHE systems: (a) human-machine interface [1] (b) health care [2], and (c) RFID-based gas sensing [3].

At a very basic level, the concept of flexibility needs elaboration. There are three main attributes: bendability [1], stretchability [2], and twistability [3]. These are illustrated in Figure 10.2. Consider a membrane. We define flexibility in terms of the bending radius, R. The smaller R, the more bendable the material is. Of course, the three attributes outlined above are not independent. A membrane of finite thickness will stretch on the outer side and squeeze on the inner side, as shown in Figure 10.2. However, it is commonly understood that stretchability refers to elongation along a direction in a plane. Twistability encompasses simultaneous bendability along multiple axes and consequently some stretching as well. Another key attribute is repeatability without any degradation in mechanical or electrical properties. Flexibility requires that the membrane can revert to its original shape without any residual distortions. In a composite material, such as a membrane with die and wiring, the stresses are distributed among these components dictated by the material parameters such as Young's modulus and geometric factors such as thickness and lateral dimensions.

The typical stress-strain relation for a material of elastic modulus E is shown schematically in Figure 10.3 [8]. Beyond the yield stress, which is typically $\sim E/100$, the linear relationship breaks down. If the material is brittle (true for most semiconductors, glass, and even organic laminates), the material may break catastrophically at stresses that are above the yield stress. However, ductile metals (curve A) will elongate with increasing stress, though their cross-section will narrow. This process is often called necking. At some point, the necking is

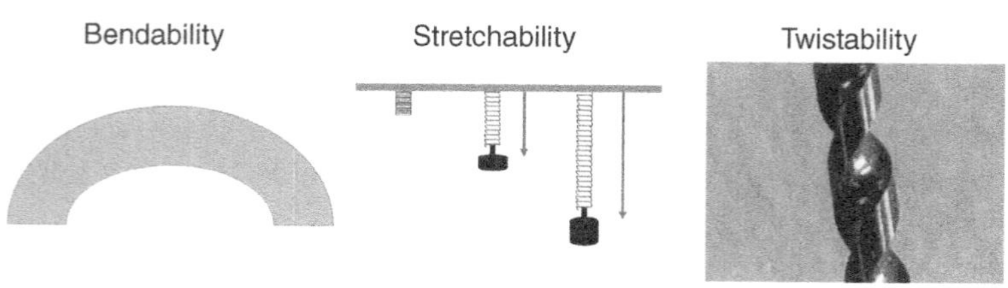

Figure 10.2 Bendability, stretchability, and twistability.

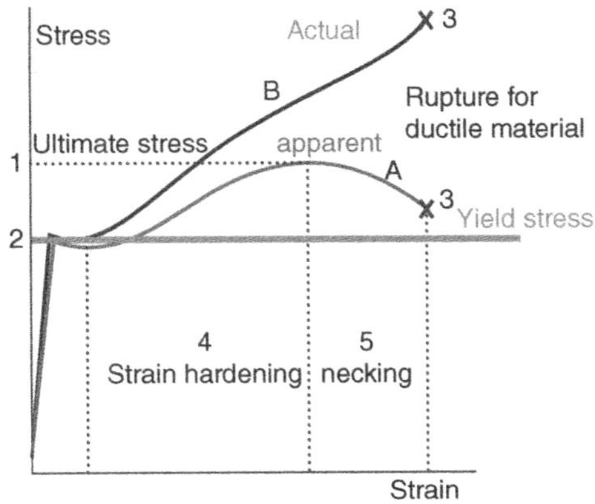

Figure 10.3 A qualitative view of the stress-strain relationship for a typical material. [8].

large enough to rupture the specimen. This point is the ultimate stress point. Elastomeric materials with low Young's modulus and low glass transition temperatures do not break easily but deform plastically, and an important characteristic is the so-called elongation at break. The larger this parameter is, the more stretchable and, therefore, more flexible the material is. These materials are usually polymers (such as rubber), and their unstressed and stressed configurations are shown in Figure 10.4. An additional property shared by many elastomers is viscoelasticity, where the stress-strain curve in the elastic region exhibits hysteresis. However, this strain is time dependent, and the membrane eventually returns close to its original shape. However, the deformation is not elastic but dissipative, that is, some of the work applied to deform the material is lost to heat, and there is a rearrangement of molecules in the material. This, in turn, can lead to long-term wear-out.

When considering FHE, one must bear in mind that there are multiple components of the system: The integrated circuit (IC) or sensor die usually made of silicon, microelectromechanical system (MEMS) and III–Vs, passive devices, batteries as well as conductive interconnects (usually metallic), fluidic

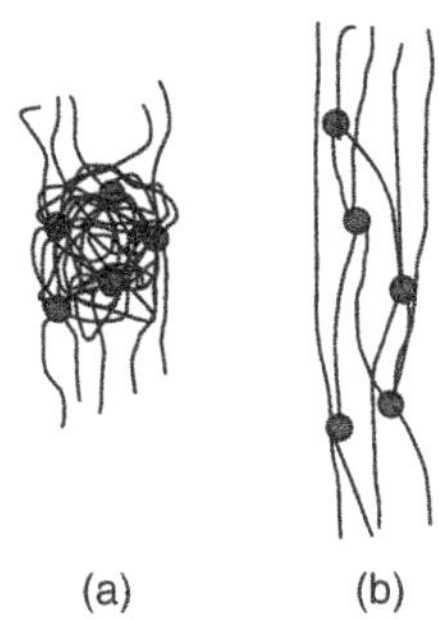

Figure 10.4 Schematic of an elastomeric material in the (a) neutral state and (b) after stretching.

interconnects, other embedded structures, and of course the substrate. To be flexible, all these components need to be flexible to more or less the same degree. This is obviously not possible as these different materials have vastly different flexibility properties. There is another consideration here. In general, all materials exhibit increased flexibility when thinned. Consider a substrate (laminate or printed circuit board-PCB) with another membrane or film attached to it (this could be a die or a conductive interconnect). This is understood through Stoney's equation [9],

$$\sigma^{(f)} = \frac{E_s h_s^2 \kappa}{6h_f(1 - \nu_s)}.$$

Here, here $\sigma^{(f)}$ is the stress in the film, E_s is Young's modulus, ν_s is the Poisson's ratio of the substrate, h_s and h_f are the thicknesses of the substrate and film, respectively, and κ is the curvature ($1/r$ for bending around a cylinder of radius r) of the composite system.

If further we just consider a single material—say the die or the metallization—and assume the substrate is extremely flexible, the strain in thin film is given by $\varepsilon = h_f/2r$ (recall E_f [Young's modulus] $= \sigma_f/\varepsilon_f$). This structure can be considered flexible as long as the stress σ_f is below the yield stress for brittle materials such as silicon or glass ($\sim E_f/100$) or necking-mediated rupture for ductile materials (such as wiring). For a silicon die, the limiting strain, before fracture, is $\varepsilon_{\text{limiting}} \sim 0.007$ (0.7%). This can be translated into a maximum bending radius $r = t/2\varepsilon_{\text{limiting}}$, which is plotted in Figure 10.5. Thus, to bend a silicon die to a radius of 5 mm, one

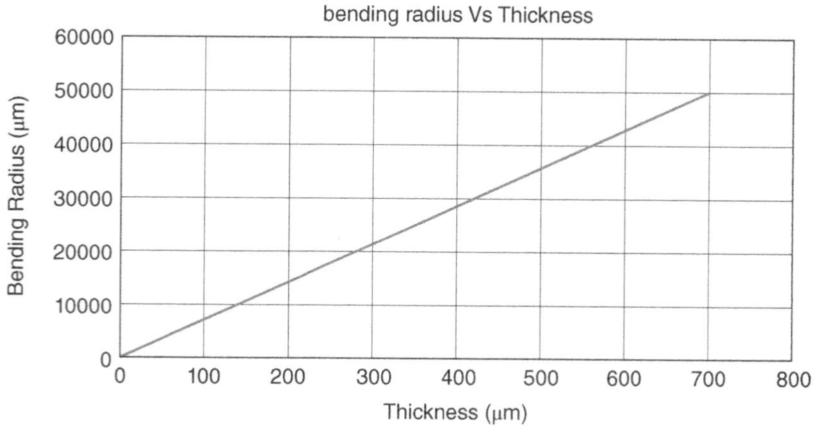

Figure 10.5 Maximum bending radius of a silicon die as a function thickness before catastrophic failure.

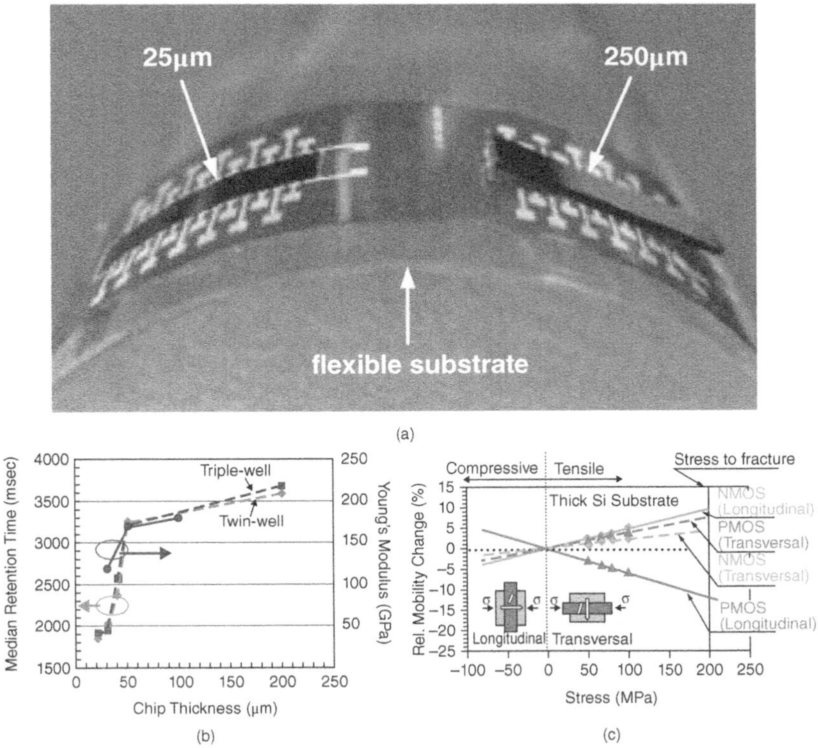

Figure 10.6 (a) Delamination of thick die from a flexible substrate on bending compared to a thinned die [10]. (b) Thickness dependent increase of Young's modulus as a function of die thickness [11] (c) Effect of stress on mobility in CMOS devices [12].

would need to thin it to about 70 μm. One approach to FHE is to surface mount such thinned die to a flexible plastic substrate such as Kapton (polyimide) or PTE (polyethylene terephthalate) that has printed wiring on it using a conductive adhesive or solder. Depending on the thickness and the strength of the surface mount conductive adhesive, when bent, a thicker die can delaminate from the substrate, as shown in Figure 10.6a [10]. Appropriately thinned die will remain conformal as long as the stress is below the yield stress. Burghartz et al. [11] describe the thinning process consideration and some applications of these thinned die mainly for flexibility. However, thinned dies present several problems: Handling and assembling super-thin die is difficult and a big yield detractor; there is also some evidence that thinning can cause degradation of the electronic and functional properties for DRAMs [13] (retention time degradation) as well as

CMOS devices [12] (mobility degradation). Additionally, there is some evidence of Young's modulus reduction as silicon is thinned [9], which in turn can result in lower yield stress.

The handling issue is often resolved by a first-level chip-scale package, sometimes called semiconductor-on-polymer chip-scale-packaging [14], as shown in Figure 10.7, where these die are sub-assembled in an encapsulant that provides some protection to the thinned die. However, the sub-assembly is thicker and heavier and does not bend as much as the individual die but does provide a solution. In a manner of speaking, this is an embodiment of classical hierarchical packaging applied to physically flexible systems.

As far as the interconnect (wiring) is concerned, the current approach is to print the wires and mating pads using a conductive epoxy material. Typically, conductive particles—a few nanometer to a few micrometer in size—are embedded in an epoxy that is dispensed onto the flexible substrate on which the bare die and CSPs are attached. A full description of the different methods of interconnect printing is beyond the scope of this chapter, so please refer to the literature for excellent reviews of this topic. Figure 10.8 shows such inkjet-printed wires and electrodes on a flexible Kapton substrate [15] after optimization. Inkjet printing has several limitations: the width of the printed wires are typically hundreds of micrometers wide, although, with care, these can be brought down to a width of 10s of micrometers. However, doing so reduces the wire thickness, and this increases the sheet resistance as well. Since printed wires rely on the tenuous connections of multiple nano- or micro-conductive particles, the best conductivity achievable is typically at least five times worse than the bulk material. This degrades significantly as the thickness is reduced. Narrower lines also exhibit more line-edge-roughness, which in turn increases variance. Furthermore, narrower lines tend to be even thinner, exacerbating this issue, although this can be circumvented by multiple passes. Generally, though, fine wire printing is a slow process, and while adequate for small volume and prototyping, it cannot be scaled up. Scaling up is usually

Figure 10.7 A flexible chip-scale package (CSP) that can be used as a component for a flexible assembly [14]. These CSPs allow for limited flexibility and greatly ease the handling concerns of thin die. Flexibility is affected, though much less, as the encapsulant is thin and flexible. This is the equivalent of the first-level laminate in conventional PCB packaging.

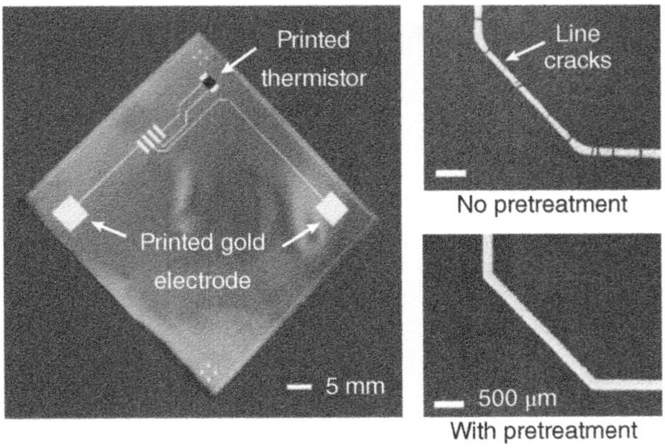

Figure 10.8 Printed wiring and electrodes [15].

done with screen printing, but as in the use of screen printing for PCB interconnects, there are limitations on linewidth, pitch, and conductivity. Finally, repeated bending of these printed wires can lead to the generation of discontinuities. The increased wire resistance degrades performance and can even cause catastrophic failure.

To summarize, FHE so far has drawn heavily from conventional packaging technology but has worked diligently to make the structures more physically flexible by thinning the substrates and die significantly to improve flexibility. Although the flexibility offered by these methods is limited, FHE has made significant inroads in the application space described earlier in this section. In what follows, we explore what we call FHE2.0, where we attempt two improvements: significantly more flexibility, including the possibility of folding, and higher performance interconnects. We will start with a brief review of recent trends in packaging and see how those trends can influence FHE.

10.2 Recent Trends in Packaging

Microelectronics has made tremendous progress over the last several decades, adhering to what is popularly called Moore's law. One measure of Moore's law is the scaling factor of minimum features on a silicon IC. This trend is shown in blue in Figure 10.9 [16] and shows over a 1000-fold decrease in minimum feature size, corresponding to a million-fold transistor density improvement, and a corresponding reduction in power per function as well as a corresponding reduction of cost and price per function. Packaging till recently, however, did not scale as is seen in Figure 10.9. For example, in 1967, when flip-chip bonding

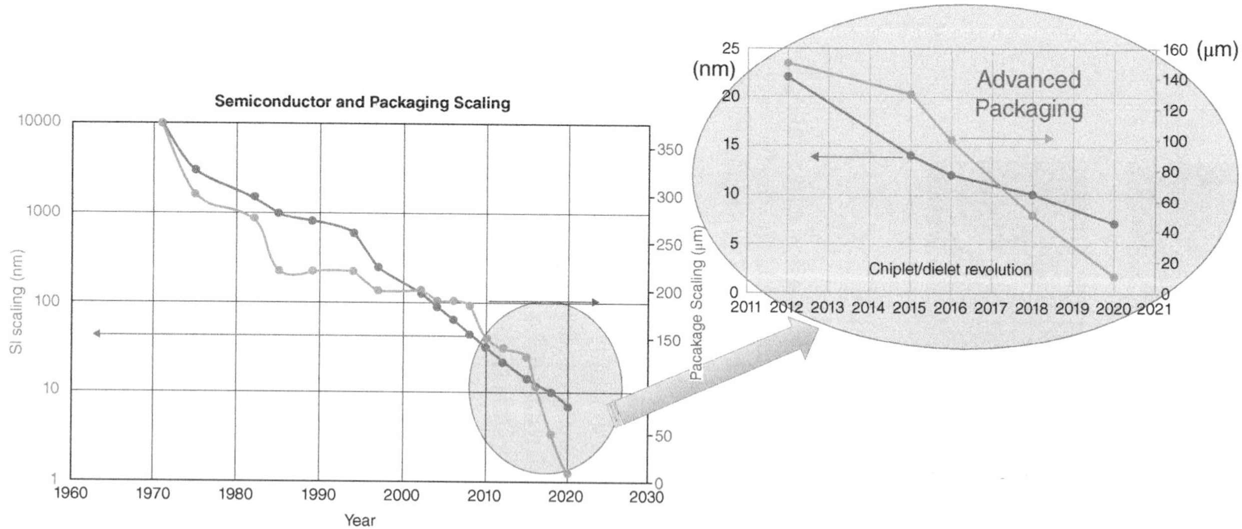

Figure 10.9 Scaling trends for CMOS features as well as package features. Package scaling has lagged significantly compared to Si scaling though the adoption of silicon-like technology has accelerated the scaling somewhat [16].

was first introduced, the bump pitch was 400 μm. Even today, the bump pitch to laminates has scaled to about 130 μm, while ball grid array pitches and trace pitches on laminates and PCBs have not fared better. However, in the last few years, we have seen an acceleration of these metrics, as shown in the inset in Figure 10.9. Note that the silicon scale is in nanometers while the packaging scale is in micrometers. There are two key factors that have mediated this acceleration: the first one is the adoption of silicon-like processing materials and methods to achieve scaling, including silicon interposers, and the second one, importantly, is fan-out wafer-level packaging.

Why is it important to scale packaging? Packaging dimensions determine the size of the system, especially since the scale is 10–100 times larger than chip dimensions. Power too is a major consideration. Communication power between chips accounts for 30–40% of total system power. So for SWaP (size, weight, and power) as well as cost, scaling the package has advantages. The key parameters that affect SWaP are dominated by packaging metrics, and scaling the package has a greater impact on SWaP than mere silicon scaling. For FHE, form factor and power play a critical role. Most FHE devices are mobile and dependent on battery power. As such, scaled FHE packages will benefit immensely from scaling.

Another aspect of advanced packaging that has received much attention in recent years is heterogeneous integration (HI). This term needs some clarification. Most packaging constructs do, in fact, achieve HI via the integration of diverse packaged chips on an extended substrate such as a PC board. So HI, in general and in itself, is not new. However, in the context of advanced packaging, HI refers to the integration of bare die on a first-level package. This could be organic, ceramic, or silicon interposer. The key feature that distinguishes it from classical or conventional packaging is the pitch of the connections between the bare die and the substrate, the number of connections between the interconnected bare die, the size of the die, and hence a significant simplification in the communication protocols of inter-die signaling. It is generally accepted that if the bump pitch is $<\tilde{}50$ μm, the inter-die spacing is $<\tilde{}2$ mm, and the trace pitch (wiring between the die) $<\tilde{}5$ μm, we are in the regime of advanced packaging.

Finally, chiplets and dielets are another feature of advanced packaging. A complex system or large chip design is fragmented into smaller entities called chiplets and then instantiated in silicon as dielets. These dielets are then re-integrated intimately at fine pitch (bump and trace) as well as short inter-die spacing (as described in the previous paragraph) to synthesize a subsystem or module. This can be further assembled on a PCB or in the case of wafer-scale systems can represent the entire system itself [17, 18].

Both these features of advanced rigid substrate packaging may be applied to FHE and can result in improved SWaP parameters well beyond what is achievable in conventional FHE. This, in turn, will open up new applications that leverage HI, including the use of higher-performance, lower-power dielets, higher-performance interconnects, and the integration of AI engines in these applications. This last possibility is important as we develop more energy-efficient edge computing techniques, including in-memory analog computing for edge applications. Computing at the edge refers to computing at the stand-alone mobile device, for example, a biomedical patch. This allows for simplified processing of the data accumulated by the device locally and a reduction of the amount of data that must be sent to the cloud for further processing or storage. This requires the use of advanced low-power processor die and memory locally along with tighter interconnect pitches.

In what follows, we describe FO-WLP as a viable method to achieve such scaling. Other chapters in this book describe FO-WLP in great detail, so we shall only address the unique features of FO-WLP as applied to FHE. What is important to note is that FO-WLP can implement these important features of advanced packaging, namely fine pitch and closely spaced dielets.

10.3 FHE Using FO-WLP - FlexTrate™

The taxonomy of fan-out wafer level packaging (FO-WLP) has been described elsewhere in this book and is summarized in Figures 10.10–10.12. The process most suited for FHE is the die-first face-down approach. This type of FHE called FlexTrate™.

The process is described below:

1. **Design considerations:** generally, the system consists of several chips (that may include sensors) that need to be interconnected at one or more wiring levels. A floor plan is made consistent with the design manual for the

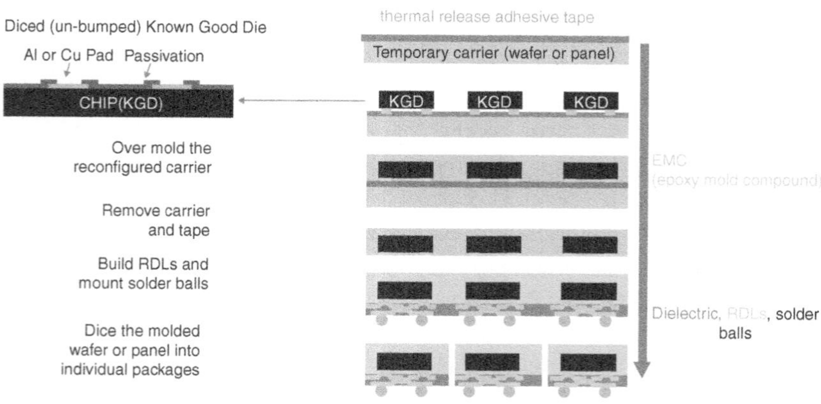

Figure 10.10 FO-WLP process flow of die-first (face down).

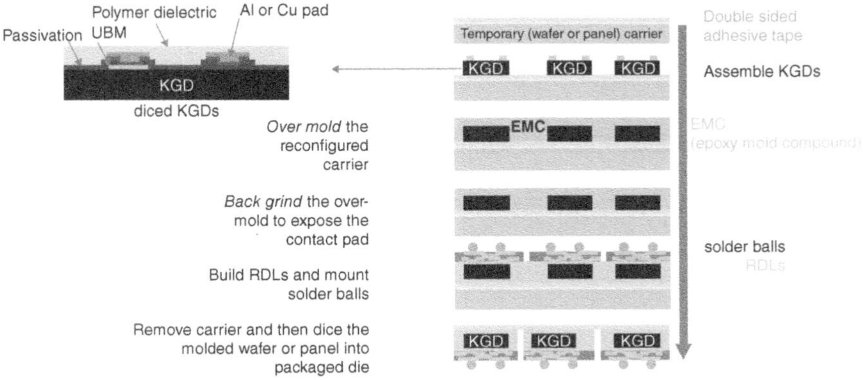

Figure 10.11 FO-WLP Process flow of Die-first (face up).

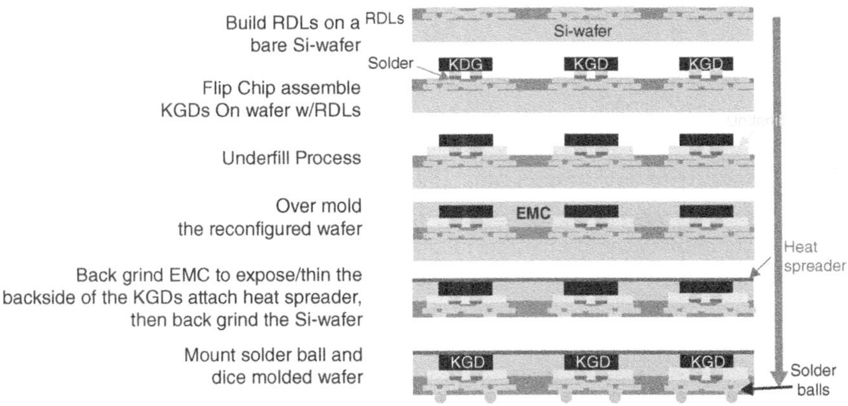

Figure 10.12 FO-WLP Process flow of Die-Last (face up).

technology. This floor plan comprehends the electrical requirements of the system with special attention to die alignment, signal and power integrity, power delivery and storage (e.g., batteries, supercapacitors, wireless power transfer etc.) The design is checked for Design rule violations using the design rule checker (DRC) and node-to-node interconnect integrity, often called a layout-to-schematic (LVS) check. In this aspect, there is no substantial difference from conventional packaging and chip design procedures except that the level of heterogeneity is significant and may require special tools.

2. **Preparation of the first temporary handler:** The first handler forms a temporary substrate on which the die are assembled face down (circuit side facing substrate). A substrate placement map is prepared, and appropriate alignment marks are etched onto the handler. The alignment marks are usually made of

a high-contrast metal that is visible through the adhesive layer. Usually, the temporary handler is reused multiple times. The handler may be silicon or glass. The latter has the advantage of being transparent, and this can allow for optical alignment.

3. **Adhesive application:** The temporary adhesive layer may either be spun-on or be applied as a composite tape. Composite tapes are either single- or double-coated, as shown schematically in Figure 10.13 [19]. The thermal release layer is an important feature of the adhesive. The release layer material consists of a large number of microscopic micron-size pellets. The materials in these pellets expand significantly around the release temperature causing the adhesive to delaminate. Tape manufacturers tune the release temperature, and a wide variety of tapes are available. In our case, we use a tape that releases at about 90 °C. Unfortunately, the tape surface and this release mechanism cause the surface to be rough, and this roughness transfers in part to the molding compound surface. In case this is an issue, for example, in display applications, a planarizing layer will need to be applied.
4. **Dielet placement:** Known good bare dielets of varying thickness (typically $<\tilde{}300\,\mu m$ thick) can be assembled on the temporary adhesive using flip-chip placement or bonding. The die assembly on the adhesive is done using a conventional die-to-substrate bonder that can achieve alignment accuracies in the order of 1 μm today. We expect this alignment accuracy to improve with improved tooling capability. The alignment can then be performed by looking at the alignment marks through the adhesive. A dual side IR camera as part of the die-to-wafer bonder can automatically detect complementary alignments on the flipped die and the substrate below and perform accurate placement. In our case, a commercially sold K&S APAMA tool is used for this purpose.
5. **Substrate reconstitution:** After the accurate placement of the die on the temporary adhesive of the first handler, a flexible molding compound is dispensed on the substrate and then compression molded using a second

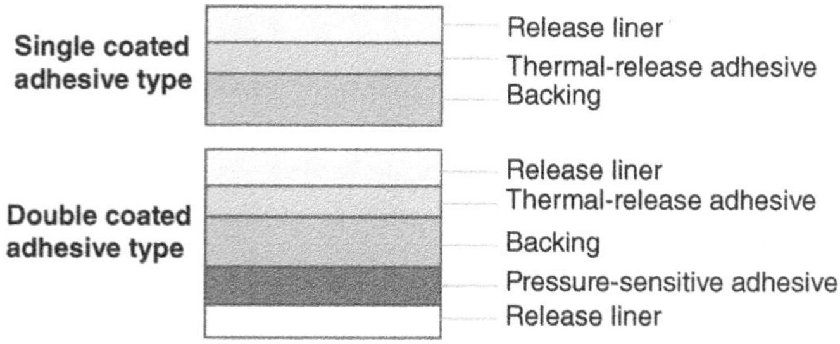

Figure 10.13 Schematic of single- and double-coated temperature release adhesive tape.

handler with its own temporary adhesive. Typically, the structure of this adhesive is similar to the first adhesive used (see step 3), with one important difference. The release temperature for the second adhesive is higher than that of the first adhesive—by typically more than 40 °C. This allows the first handler to be removed by thermal means without affecting the second handler. Pressure is applied on top of the second handler using a standard weight or some other technique. The thickness of the overall substrate is determined by a Teflon ring that is concentrically placed at the wafer edge. The Teflon ring can be customized to different thicknesses and shapes to get the required substrate form factors. The flexible molding compound is then cured. In our case, the molding compound that is used is PDMS (poly di-methyl siloxane), from Silastic (MDX4-4210), which is a flexible, biocompatible, viscoelastic polymer. The PDMS is typically cured at room temperature for 24 hours. Die shift is a common phenomenon during the compression molding that occurs due to the curing-induced drag forces on the assembled die and the shrinkage factor of the post-cured substrate [20–23]. Using a low Young's modulus material such as PDMS cured at very low temperatures allows us to significantly reduce this die shift. We have currently demonstrated die shifts of < 6 μm across a 100 mm wafer [24]. We will discuss die shift considerations later in this chapter. After substrate reconstitution, the first handler is debonded using a suitable technique: thermal release, chemical release, or UV irradiation. The removal of this handler exposes the PDMS surface with embedded die ready for metallization.

6. **Deposition of stress buffer layers:** The exposed substrate (reconstituted) surface has an inherent roughness due to the roughness of the temporary adhesive. We have found that typical photoresists may not be directly applied to the exposed PDMS. Doing so causes the resist to crack upon curing, rendering it unusable. To avoid this, a stress buffer layers composite is first deposited on top of the reconstituted substrate before the fabrication of the redistribution layers (RDL). A conformally coated layer of parylene-C, a biocompatible poly-para-xylylene, using a chemical vapor deposition (CVD) technique on top of the PDMS substrate is used as a stress buffer layer. A spin-coated layer of SU-8 on top of this parylene-C layer is used as a further planarization- and stress-mediating layer. These stress buffer layers can be optimized to get the required flexibility and sample planarity by adjusting the deposited film thicknesses [25, 26]. Resist may then be spun on and patterned.
7. **Deposition of the SU-8 corrugations:** We have mentioned earlier that there is a significant shrinkage of the PDMS upon curing. However, since the PDMS is well adhered to the second handler, it does not shrink but extends. As a result, the PDMS is under tensile stress. If we were to metallize the surface at this point, the metallic line would be unstressed. However, upon release

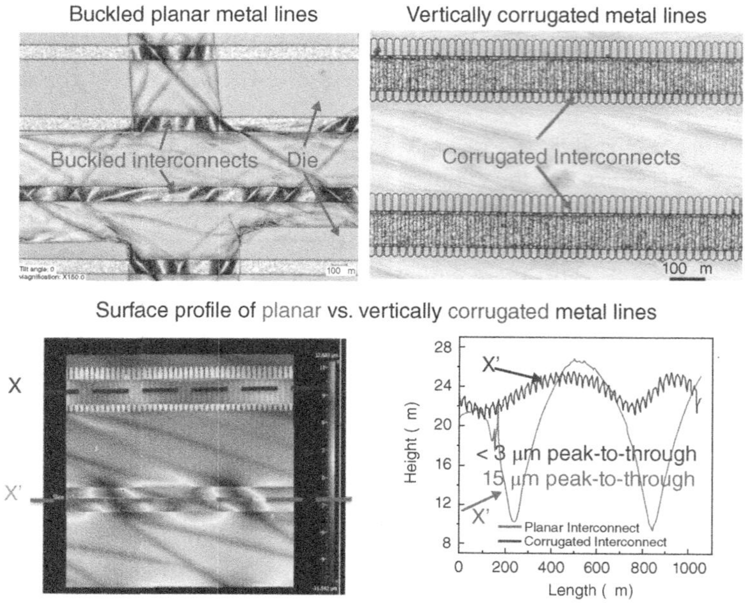

Figure 10.14 Buckling in planar metal lines. Vertically corrugated metal lines are less buckled than planar metal lines [24].

from the second handler, the stress would be relieved and the PDMS would contract, putting the previously unstressed metal line under compressive stress. A line under compressive stress will be unstable and tend to buckle. We have observed that this is exactly what happens to the metal lines. This buckling leads to large and somewhat spatially random undulations in the wire, as shown in Figure 10.14 [24, 27–29].

From basic solid mechanics, it is known that the critical buckling stress of a metal wire is directly proportional to the Young's modulus of the metal and inversely proportional to the square of the wire length [30, 31]. Long wires on the substrate would hence easily buckle as the bending-induced stresses exceed the critical buckling stress. Our studies with 40 μm wide copper wires on FlexTrate™ showed that this bucking amplitude could exceed 15 μm causing possible delamination or breakage of the wire. One solution is to break up the long wire into a whole bunch of shorter wires so as to increase the critical buckling stress of each wire segment, causing an overall more mechanically reliable wire. On FlexTrate™, we take this approach by vertically corrugating the metal wires so that the effective length of the wire in plane with the substrate is reduced to the corrugation pitch, thus significantly increasing the critical bucking stress and preventing the wire buckling. In fact, we demonstrated that by vertically corrugating our metal lines, we could reduce the bucking amplitude by a factor of 5x, significantly boosting the mechanical reliability of the wire. The vertical corrugations are made on the substrate using photo-patternable SU-8. The corrugation height is typically in the order of around 5 μm. After the corrugations are fabricated, the copper metal lines are semi-additively electroplated on top of these corrugations, as shown in Figure 10.14 [24, 31, 32].

8. **RDL fabrication:** After the formation of the vertical corrugation structures using SU-8 on the substrate, a seed layer of Ti/Cu is sputtered, followed by the semi-additive electroplating of copper to form the first metallization layer. The sample is then coated with a layer of SU-8 or KMSF 1000 by MicroChem, which acts as an isolation dielectric. Lithography is used to define via holes in the dielectric. Semi-additive electroplating is then performed to form the second layer of metallization. The use of two levels of metallization provides sufficient routing capability for most modern consumer medical devices. Currently, we have demonstrated the ability to fabricate <40 μm pitch interconnects on FlexTrate™. The ability to push down the interconnect pitches to sub 10 μm has also been explored. This ultimately comes down to the ability to perform reliable lithography at these fine dimensions on a non-conventional transparent organic substrate. Optical techniques such as the use of an anti-reflection layer and substrate aware mask patterns should enable the interconnect pitches to be pushed down to <10 μm. Figure 10.15 shows an array of 200 die interconnected at 40 μm pad pitch [32]. Figure 10.16 shows a two-level metallization scheme on Flextrate™ [33].
9. **Passivation of the substrate and final release:** After the metallization layers have been fabricated, the next step is the passivation of the substrate to both protect the metallization from corrosion and to enable biocompatibility of the overall fabricated sample. On FlexTrate™, we have explored the use of a 2 μm thick layer of CVD deposited parylene-C to enable corrosion and humidity resistance and improve the biocompatibility of the substrate. The parylene-C is not completely capable of preventing moisture from reaching the interconnects and causing corrosion. Hence further multilayers of inorganic moisture barriers such as aluminum oxide and silicon dioxide are to be deposited on top.
10. **Final release of the passivated substrate:** The passivated substrate is then finally released from the second handler using an appropriate debonding process (thermal, chemical, or optical). In our FlexTrate™, the thermal release tape on handler 2 has a release temperature of around 170 °C and can hence be debonded by heating the sample to this temperature. A final flexible FlexTrate™ with multiple interconnected die is shown in Figure 10.17 [34].

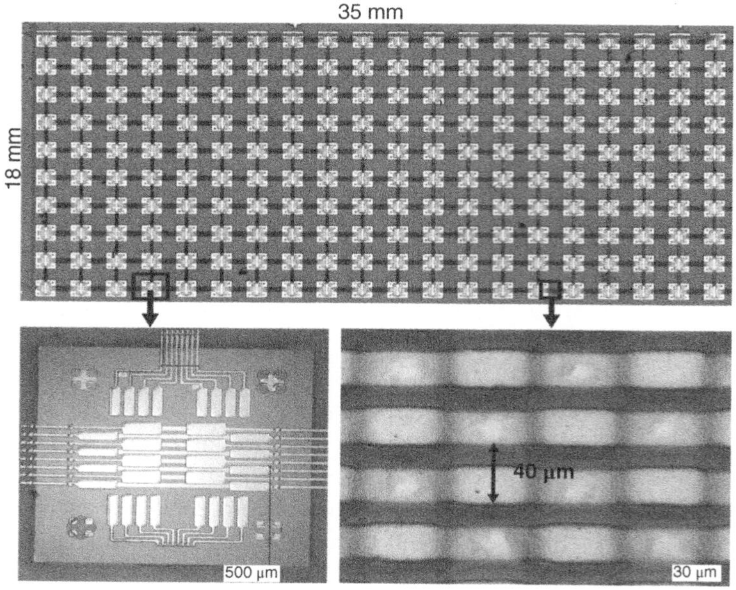

Figure 10.15 Image of 200 die interconnected at 40 µm pad pitch on FlexTrate™ [32].

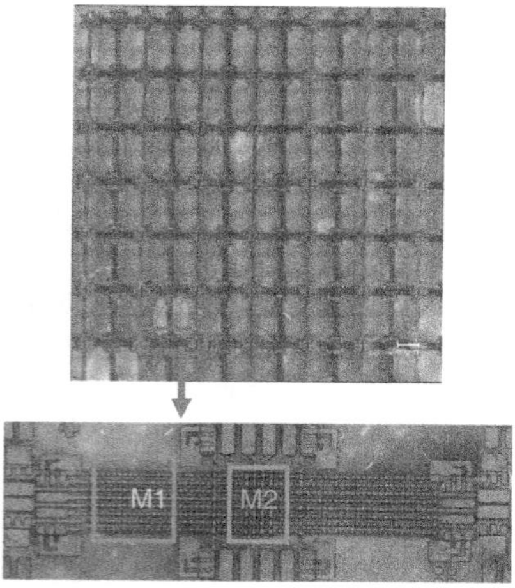

Figure 10.16 Image of two-level metallization at 40 µm pitch on FlexTrate™ [33].

Die shift is a common occurrence in wafer-level fan-out packaging. As shown in Figure 10.18, die can shift in multiple ways [26]. In the z-direction, the die can "pop up." This is invariably due to the die embedded in the adhesive layer when they are aligned and placed on the first handler. There is a trade-off in that we would like the die to adhere well to the adhesive, which requires some pressure to be applied

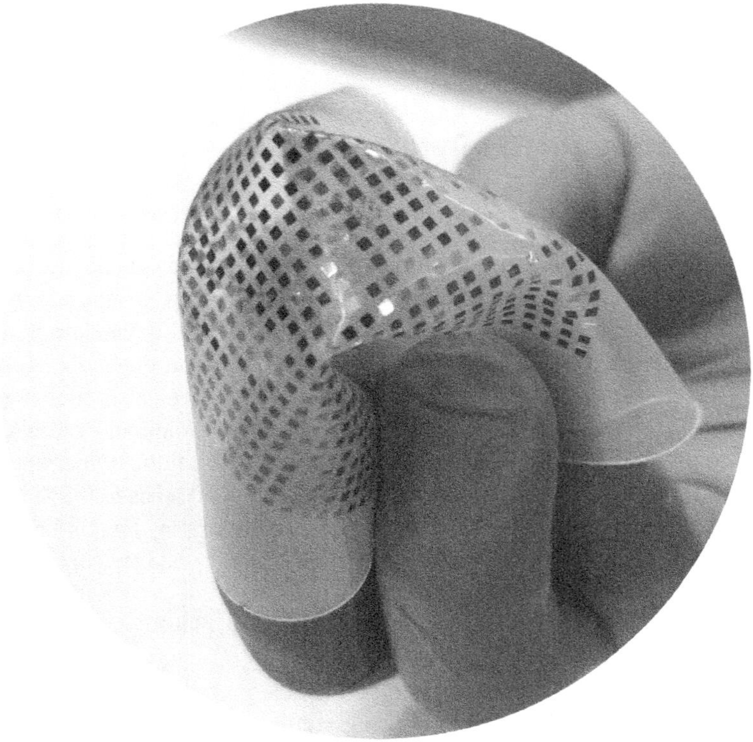

Figure 10.17 Image of FlexTrate™ with multiple interconnected die [34].

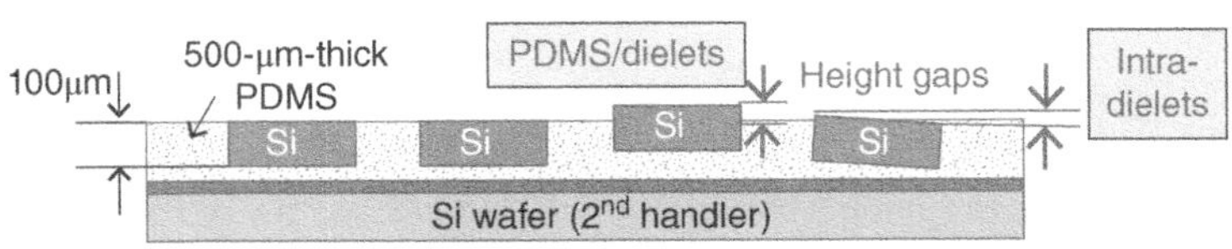

Figure 10.18 Die in FlexTrate™ can shift in multiple ways [26].

during placement. However, too much pressure will cause the die to embed partially into the adhesive leading to pop-up. If the pop-up is too large, the topography will cause challenges in lithography and line resolution and step coverage. By careful optimization, this pop-up can be minimized to < 1 μm. As the PDMS cures, the die experience stresses. These can cause the die to shift in the *x*- and *y*-direction and can also cause the die to rotate and tilt as shown in Figure 10.18. The x-y shift is dominated by the shrinkage of the PDMS and has a radial dependency. Die shift consideration is described in detail elsewhere [24], and we have been able to contain them to about 6 μm in both directions. Die tilt too has been reduced to negligible values.

10.4 Applications on FlexTrate™

Some of the applications demonstrated on FlexTrate™ are as follows:

1. **Foldable display:** The passivated multiple commercially available InGaN LEDs (DA 1000, Cree Inc., 1 mm × 1 mm × 0.335 mm) and in-house fabricated Si die (1 mm × 1 mm × 0.2 mm) are integrated to demonstrate a seven-segment foldable display (37 mm x 52 mm x 0.5 mm) on FlexTrate™, as shown in Figure 10.19 [32]. The LEDs are dipped in aqua-regia solution for ~300 seconds to remove the solder termination from the anode and cathode pads before placement on the first handler. The Si die have fiducial marks to help with ±1 μm alignment accuracy for different alignment steps to fabricate the foldable display. Each segment of the seven-segment display has six LEDs connected in parallel through 40 μm pitch vertically corrugated interconnects. The LEDs are powered by a power supply operating at a current limit of 200 mA to display "UCLA CHIPS," where each letter is displayed one at a time on the foldable display, as shown in Figure 10.19.
   Additionally, Figure 10.19 illustrates that the LEDs remain illuminated as the display is manually folded to a 0 mm bending radius. The display is shown to

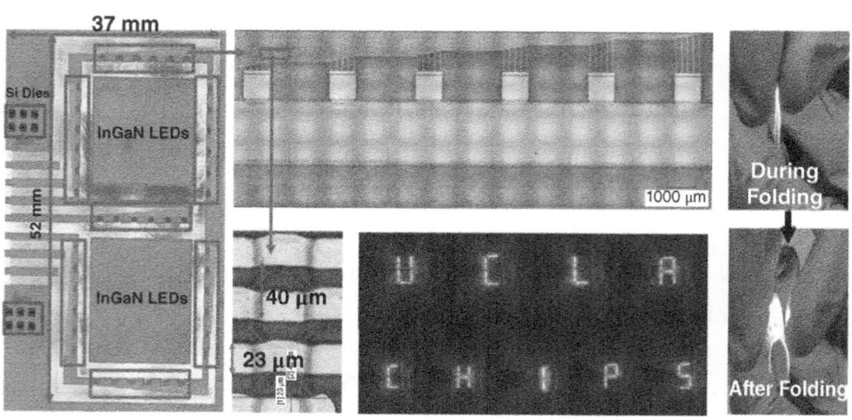

Figure 10.19 Fabricated foldable display on FlexTrate™ with multiple LEDs connected with 40 μm pitch vertically corrugated interconnects. The LEDs of the display remain illuminated as the display is being folded [32].

be reliably bent to a 1 mm bending radius for 1000 bending cycles without any performance degradation, including no delamination of any metal lines and die, demonstrating that the package is extremely reliable even when subjected to cyclic bending at extreme bending radii. In the future, this system can be scaled up to produce high-density, full-color, highly flexible displays.

2. **Flexible inorganic μLED display:** In this work, UCLA has demonstrated HI of arrays of blue GaN μLEDs that have < 100 μm dimensions with Si die on FlexTrate™, as shown in Figure 10.20 [35]. Diode-pumped solid-state (DPSS) laser source is used to release the GaN LEDs from the sapphire substrates through the laser lift-off (LLO) process. Plating Ni, which acts as a stress buffer, on top of the LEDs allows achieving ~100% yield during the LLO process [36].

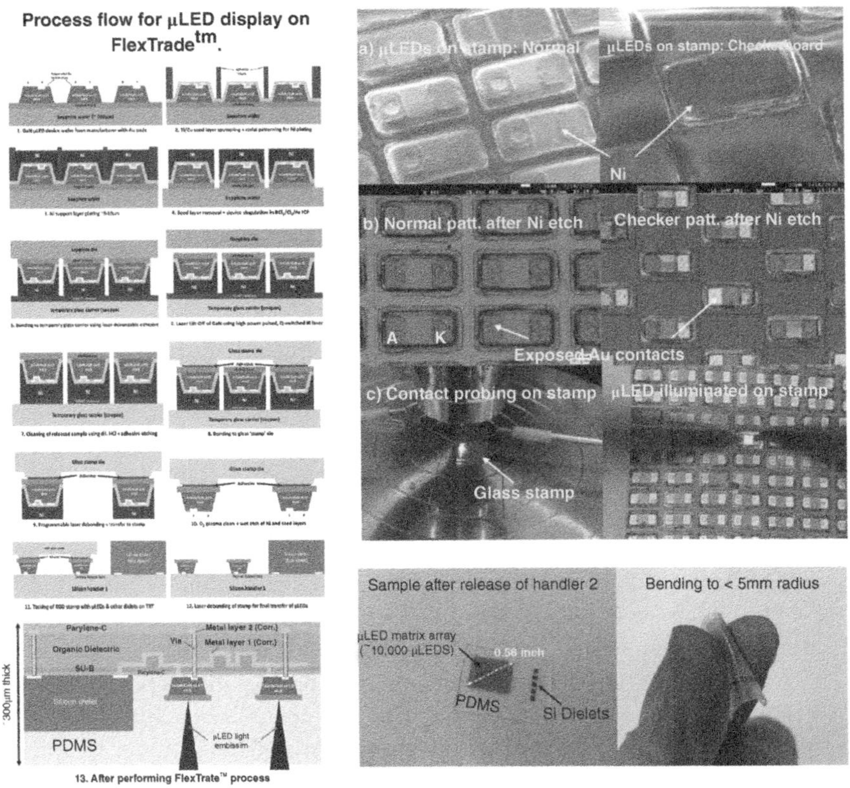

Figure 10.20 Process flow for fabricating μLED display on FlexTrate™ along with images of the sample at different processing steps [35].

The released LEDs are embedded in the FlexTrate™'s PDMS substrate with > 99% yield using a novel mass transfer technique involving adhesive bonding. The pixel resolution of ~200 PPI is demonstrated by the transfer of the GaN μLEDs on FlexTrate™. Throughout the transfer process of the μLEDs to FlexTrate™, no degradation in the LED performance is observed. The μLED display on FlexTrate™ can be reliably bent to a 5 mm bending radius without any performance degradation. Such μLED display can be used in high-performance and high-resolution flexible wearable and implantable systems.

3. **Wireless brain implant system:** A wirelessly powered subdermal brain implant system is demonstrated on FlexTrate™. This system includes a coil powered wirelessly by an external coil through resonant magnetic coupling at 13.56 MHz, a diode, multiple decoupling deep trench capacitors, a low drop out regulator, and a green LED (AlGaInP), as shown in Figure 10.21 [37]. All the electronic components are in bare die form and are integrated, without any solder, by 40 μm pitch vertically corrugated interconnects. The LED is shown to be illuminated at 15% power transfer efficiency in body medium conditions when the external coil powers the implant from a coupling distance of 10 mm. There is no significant change in the implant performance as it is bent to a 5 mm bending radius for 1000 bending cycles. Overall, the implant is just ~535 μm in thickness and 20 mm in diameter. Such a flexible implant can be used for wireless optogenetics applications [38].
4. **Personal Environment Monitors:** A MEMS gas sensor is integrated on FlexTrate™ for the purpose of personal environmental monitoring, as shown in Figure 10.22 [39]. The MEMS gas sensor, fabricated at IISC Bangalore, includes interdigitated electrodes along with the metal-oxide (ZnO)–based sensing material and a microheater [40]. An additional polymer layer is used to protect the sensitive membrane of the gas sensor that increases the yield of sensor integration in FlexTrate™ from 10% to > 90%. Additionally, this system includes an integrated analog-to-digital converter, trans-impedance amplifier, and Bluetooth system that enables wireless communication from the FlexTrate™ to the outside world, such as a smartphone. All these components are connected by 40 μm pitch interconnects that help integrate the overall system in small form factor by enabling high-density integration. As shown in Figure 10.22, both the temperature of the microheater and the current in the metal oxide increase with the heater voltage, confirming the proper functioning of the MEMS gas sensor post integration in FlexTrate™. Moreover, the recovery time and the response time of the sensor at 5 V heater voltage are 300 and 50 seconds, respectively, without any baseline drift. Overall, the system is lightweight, flexible, inexpensive, and easy to wear.

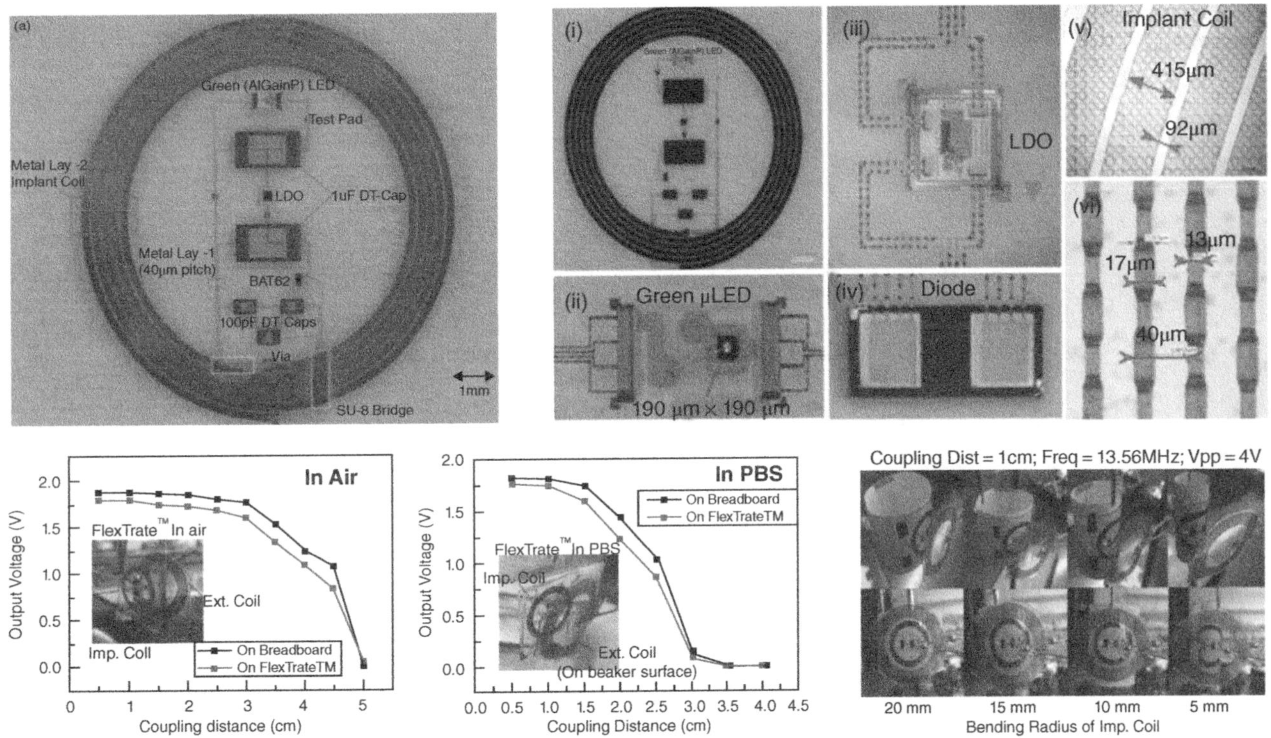

Figure 10.21 Image of the fabricated wireless brain implant system on FlexTrate™. The implant is reliability bendable to a 5 mm bending radius [37].

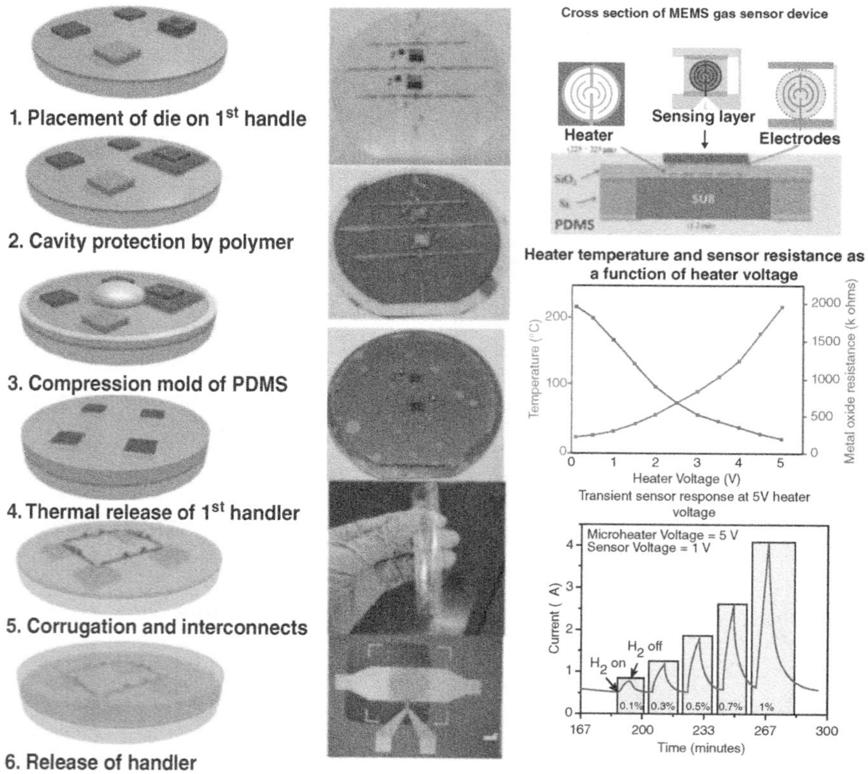

Figure 10.22 Personal environment monitors on FlexTrate™. The motivation of this work is to monitor local pollution through a wearable gas sensor system [39].

5. **FlexsEMG: Multi-channel sEMG system:** A surface electromyography (sEMG) is demonstrated on FlexTrate™, called FlexsEMG, as shown in Figure 10.23 [41–44]. It consists of heterogeneous solderless integration of passives, a commercially available sEMG data processing and acquisition die (RHD 2216), wireless data transfer system over Bluetooth (EYSHSNZWZ), and up to 20 vertically corrugated Au capped Cu electrodes of 6 mm diameter and 12 mm inter-electrode spacing, and one reference electrode, all connected by 40 μm pitch interconnects. A rechargeable 70 mAh battery and a transparent adhesive are externally added to the FlexsEMG system to help power and attach the FlexsEMG system to the human body, respectively. The FlexsEMG system is highly flexible, just 5 g in weight, which makes it easy to wear, and 75 mm × 45 mm × 1 mm in size. The sEMG electrodes on the FlexsEMG system are able to record as good muscle signals as the commercially available

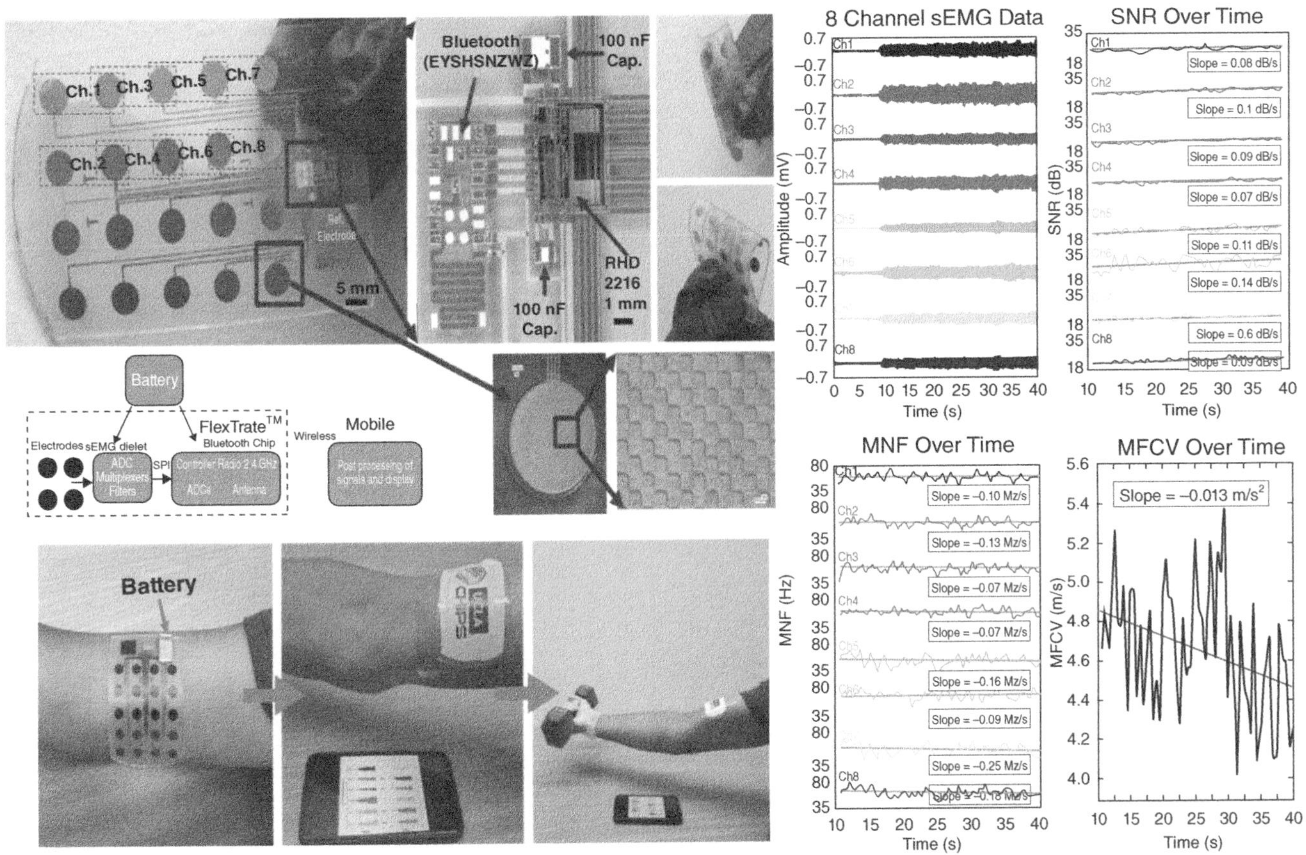

Figure 10.23 Multi-channel sEMG system on FlexTrate™, called FlexsEMG. The FlexEMG system has all the electronics integrated, including multiple electrodes, ADCs, signal filters, multiplexers, passives, Bluetooth communication, etc., and can record muscle signals from 8-channels [41].

Ag/AgCl electrodes while being > 5X smaller in size [33]. Important muscle parameters such as mean frequency (MNF) and muscle fiber conduction velocity (MFCV) that help in the study of many muscular disorders can be evaluated with the help of the 8 channels of this device [45–48]. Such a wireless multi-channel sEMG system can be used to assist in spine surgery monitoring to reduce the number of wires attached to the patient undergoing surgery as well as for patient rehabilitation and general physiological monitoring [49–52].

Acknowledgments

The authors would like to thank the UCLA CHIPS group, especially Tak Fukushima (Tohoku University), Goutham Ezhilarasu, and Amir Hanna (Keysight Technologies). The work reported here was supported NBMC, FlexTech, SRC, DARPA, and the UCLA Chips Consortium.

References

1 Kwon, Y.T., Kim, Y.S., Kwon, S. et al. (2020). All-printed nanomembrane wireless bioelectronics using a biocompatible solderable graphene for multimodal human-machine interfaces. *Nature Communications* 11: 3450. https://doi.org/10.1038/s41467-020-17288-0.

2 Gutruf, P., Yin, R.T., Lee, K.B. et al. (2019). Wireless, battery-free, fully implantable multimodal and multisite pacemakers for applications in small animal models. *Nature Communications* 10: 5742. https://doi.org/10.1038/s41467-019-13637-w.

3 Escobedo, P., Erenas, M.M., López-Ruiz, N. et al. (2017). Flexible passive near field communication tag for multigas sensing. *Analytical Chemistry 2017* 89 (3): 1697–1703. https://doi.org/10.1021/acs.analchem.6b03901.

4 IDTechEx Ltd (2020). *Flexible Hybrid Electronics 2020–2030: Applications, Challenges, Innovations and Forecasts*. IDTechEx Ltd.

5 Li, Y., Meng, L., Yang, Y. et al. (2016). High-efficiency robust perovskite solar cells on ultrathin flexible substrates. *Nature Communications* 7: 10214. https://doi.org/10.1038/ncomms10214.

6 O'Rourke, S.M., Loy, D.E., Moyer C. et al. (2008). Direct fabrication of a-Si:H thin film transistor arrays on plastic and metal foils for flexible displays. *Proc. 26th Army Sci. Conf.*, Orlando, USA (December 2008).

7 Proceedings of the IEEE (2015). 103(4).

8 Richfield, D. https://commons.wikimedia.org/w/index.php?curid=6039110. (accessed 27 May 2021).

9 Stoney, G.G. (1909). The tension of metallic films deposited by electrolysis. *Proceedings of the Royal Society of London A* 82: 172–175.

10 Marinov, V. (2016). The IC side of the flexible hybrid electronics technology. Presented at the SEMICON West. (July 2016).

11 Burghartz, J.N., Appel, W., Harendt, C. et al. (2009). Ultra-thin chips and related applications, a new paradigm in silicon technology. *Proceedings of the European Solid State Device Research Conference*. 29–36. doi: 10.1109/ESSDERC.2009.5331441.

12 Chen, Y., Wu, W., Liu, C.J. et al. (2014). Simultaneous optimization of analog circuits with reliability and variability for applications on flexible electronics. *IEEE Transactions on Computer-Aided Design of Integrated Circuits and Systems* 33 (1): 24–35. https://doi.org/10.1109/TCAD.2013.2282757.

13 Lee, K., Tanikawa, S., Murugesan, M. et al. (2013). Degradation of memory retention characteristics in DRAM chip by Si thinning for 3-D integration. *IEEE Electron Device Letters* 34 (8): 1038–1040. https://doi.org/10.1109/LED.2013.2265336.

14 Hackler, D., Wilson, D., Prack, E. (2019). Semiconductor on polymer chip scale packaging. *Chip Scale Review*, (July–August) pp. 5–6.

15 Khan, Y., Garg, M., Gui, Q. et al. (2016). Flexible hybrid electronics: direct interfacing of soft and hard electronics for wearable health monitoring. *Advanced Functional Materials* 26 (47): 8764–8775. https://doi.org/10.1002/adfm.201603763.

16 Jangam, S. and Iyer, S.S. (2021). Silicon-interconnect fabric for fine-pitch ($\leq$10 μm) heterogeneous integration. *IEEE Transactions on Components, Packaging and Manufacturing Technology* 11 (5): 727–738. https://doi.org/10.1109/TCPMT.2021.3075219.

17 Iyer, S.S., Jangam, S., and Vaisband, B. (2019). Silicon interconnect fabric: a versatile heterogeneous integration platform for AI systems. *IBM Journal of Research and Development* 63 (6): 5:1–5:16. https://doi.org/10.1147/JRD.2019.2940427.

18 Pal, S., Petrisko, D., Tomei, M. et al. (2019). Architecting waferscale processors-a GPU case study. *IEEE International Symposium on High Performance Computer Architecture (HPCA)* (February 2019). IEEE press.

19 Nitto. https://www.nitto.com/us/en/products/e_parts/electronic001 (accessed May 27, 2021).

20 Han, Y., Ding, M.Z., Lin B. et al. (2016). Comprehensive Investigation of Die Shift in Compression Molding Process for 12 Inch Fan-Out Wafer Level Packaging. *IEEE 66th Electronic Components and Technology Conference (ECTC)*, Las Vegas, USA (May 31–June 3, 2016). IEEE press.

21 Che, F., Ho, D., Ding, M.Z. et al. (2016). Study on Process Induced Wafer Level Warpage of Fan-Out Wafer Level Packaging. *IEEE 66th Electronic Components* *and Technology Conference (ECTC)*, Las Vegas, USA (May 31–June 3, 2016). IEEE press.

22 Sharma, G., Kumar, A., Rao, V.S. et al. (2011). Solutions strategies for die shift problem in wafer level compression molding. *IEEE Transactions on Components, Packaging and Manufacturing Technology* 1 (4): 502–509.

23 Rao, V.S., Chong, C.T., Ho, D. et al. (2017). Process and reliability of large fan-out wafer level package based package-on-package. *IEEE 67th Electronic Components and Technology Conference (ECTC)*, Orlando, USA (May 30–June 2, 2017). IEEE press.

24 Hanna, A., Alam, A., Fukushima, T. et al. (2018). Extremely flexible (1 mm bending radius) biocompatible heterogeneous fan-out wafer-level platform with the lowest reported die-shift (<6 μm) and reliable flexible Cu-based interconnects. *IEEE 68th Electronic Components and Technology Conference (ECTC)*, San Diego, USA (May 29–June 1, 2018). IEEE press.

25 Fukushima, T., Alam, A., Wan, Z. et al. (2017) 'FlexTrateTM' — scaled heterogeneous integration on flexible biocompatible substrates using FOWLP. *IEEE 67th Electronic Components and Technology Conference (ECTC)*, Orlando, USA (May 30–June 2, 2017). IEEE press.

26 Fukushima, T., Alam, A., Hanna, A. et al. (2018). Flexible hybrid electronics technology using die-first FOWLP for high-performance and scalable heterogeneous system integration. *IEEE Transactions on Components, Packaging and Manufacturing Technology* 8 (10): 1738–1746. https://doi.org/10.1109/TCPMT.2018.2871603.

27 Wu, D., Xie, H., Yin, Y. et al. (2013). Micro-scale delaminating and buckling of thin film on soft substrate. *Journal of Micromechanics and Microengineering* 23: 035040. https://doi.org/10.1088/0960-1317/23/3/035040.

28 Mei, H., Pang, Y., Im, S.H. et al. (2008). Fracture, delamination, and buckling of elastic thin films on compliant substrates. *11th Intersociety Conference on Thermal and Thermomechanical Phenomena in Electronic Systems*, Orlando, Florida (May 28–31, 2008). IEEE Press.

29 Bowden, N., Brittain, S., Evans, A.G. et al. (1998). Spontaneous formation of ordered structures in thin films of metals supported on an elastomeric polymer. *Nature* 393: 146–149. https://doi.org/10.1038/30193.

30 Guo, Z. and Tan, L. (2009). *Fundamentals and Applications of Nanomaterials*. Artech House.

31 Iyer, S.S., Alam, A., Hanna, A. Fukushima, T. (2018). Flexible and stretchable interconnects for flexible systems. International application # PCT/US2019/015840. US Patent App. 16/965,934, filed January 31, 2018, and published March 11, 2021.

32 Alam, A., Hanna, A., Irwin, R. et al. (2019). Heterogeneous integration of a fan-out wafer-level packaging based foldable display on elastomeric substrate.

IEEE 69th Electronic Components and Technology Conference (ECTC), Las Vegas, USA (May 28–31, 2019). IEEE press.

33 Alam, A., Molter, M., Gaonkar, B. et al. (2020). A high spatial resolution surface electromyography (sEMG) system using fan-out wafer-level packaging on FlexTrate™. *IEEE 70th Electronic Components and Technology Conference (ECTC)*, Vista, USA (May 26–29, 2020). IEEE press.

34 Alam, A. (2021). Development of FlexTrateTM and demonstration of flexible heterogeneously integrated low form-factor wireless multi-channel surface electromyography (sEMG) device. Order No. 28321669. University of California, Los Angeles.

35 Ezhilarasu, G., Paranjpe, A., Lee, J. et al. (2020). A "Heterogeneously integrated, high resolution and flexible inorganic μLED display using fan-out wafer-level packaging." *IEEE 70th Electronic Components and Technology Conference (ECTC)*, Vista, USA (May 26–29, 2020). IEEE press.

36 Ezhilarasu, G., Hanna, A., Paranjpe, A. et al. (2019). High yield precision transfer and assembly of GaN μLEDs using laser assisted micro transfer printing. *IEEE 69th Electronic Components and Technology Conference (ECTC)*, Las Vegas, USA (May 28–31, 2019). IEEE press.

37 Ezhilarasu, G., Hanna, A., Irwin, R. et al. (2018). A Flexible, heterogeneously integrated wireless powered system for bio-implantable applications using fan-out wafer-level packaging. *IEEE International Electron Devices Meeting (IEDM)*, San Francisco, USA (December 1–5, 2018). IEEE press.

38 Ordaz, J.D., Wu, W., and Xu, X.M. (2017). Optogenetics and its application in neural degeneration and regeneration. *Neural Regeneration Research* 12 (8): 1197–1209. https://doi.org/10.4103/1673-5374.213532.

39 Benedict, S., Nagarajan, A., Kumar, T. et al. (2020). Heterogenous Integration of MEMS Gas Sensor using FOWLP: Personal Environment Monitors. *IEEE 70th Electronic Components and Technology Conference (ECTC)*, Vista, USA (May 26–29, 2020). IEEE press.

40 Prajapati, C.S., Soman, R., Rudraswamy, S.B. et al. (2017). Single chip gas sensor array for air quality monitoring. *Journal of Microelectromechanical Systems* 26 (2): 433–439. https://doi.org/10.1109/JMEMS.2017.2657788.

41 Alam, A., Molter, M., Kapoor, A. et al. (2021). Flexible heterogeneously integrated low form factor wireless multi-channel surface electromyography (sEMG) device. *IEEE 71st Electronic Components and Technology Conference (ECTC)*, USA (1 June-4 July 2021). IEEE press.

42 Duchateau, J. and Enoka, R.M. (2011). Human motor unit recordings: origins and insight into the integrated motor system. *Brain Research* 1409: 42–61. https://doi.org/10.1016/j.brainres.2011.06.011.

43 Heckman, C.J. and Enoka, R.M. (2012). Motor unit. *Comprehensive Physiology* 2 (4): 2629–2682. https://doi.org/10.1002/cphy.c100087.

44 Al-Mulla, M.R., Sepulveda, F., and Colley, M. (2011). A review of non-invasive techniques to detect and predict localised muscle fatigue. *Sensors (Basel)* 11 (4): 3545–3594. https://doi.org/10.3390/s110403545.

45 Drost, G., Stegeman, D.F., van Engelen, B.G. et al. (2006). Clinical applications of high-density surface EMG: a systematic review. *Journal of Electromyography and Kinesiology* 16 (6): 586–602. https://doi.org/10.1016/j.jelekin.2006.09.005.

46 Stegeman, D.F., Kleine, B.U., Lapatki, B.G. et al. (2012). High-density surface EMG: techniques and applications at a motor unit level. *Biocybernetics and Biomedical Engineering* 32 (3): 3–27. https://doi.org/10.1016/S0208-5216(12)70039-6.

47 Zwarts, M.J. and Stegeman, D.F. (2003). Multichannel surface EMG: basic aspects and clinical utility. *Muscle & Nerve* 28 (1): 1–17. https://doi.org/10.1002/mus.10358.

48 Yao, B., Zhang, X., Li, S. et al. (2015). Analysis of linear electrode array EMG for assessment of hemiparetic biceps brachii muscles. *Frontiers in Human Neuroscience* 9: 569. https://doi.org/10.3389/fnhum.2015.00569.

49 Stecker, M.M. (2012). A review of intraoperative monitoring for spinal surgery. *Surgical Neurology International* 3 (Suppl 3): S174–S187. https://doi.org/10.4103/2152-7806.98579.

50 Mays, J., Rampy, P., Sucato, D. et al. (2016). A wireless system improves reliability of intraoperative monitoring recordings. *IEEE Topical Conference on Biomedical Wireless Technologies, Networks, and Sensing Systems (BioWireleSS)*, Austin, USA, (January 24–27, 2016). IEEE press.

51 Sen-Gupta, E., Wright, D.E., Caccese, J.W. et al. (2019). A pivotal study to validate the performance of a novel wearable sensor and system for biometric monitoring in clinical and remote environments. *Digital Biomarkers* 3 (1): 1–13. https://doi.org/10.1159/000493642.

52 Kim, Y-S., Mahmood, M. Kwon, S. et al. (2019). Robust human-machine interfaces enabled by a skin-like, electromyogram sensing system. *Proc. SPIE, Nano-, Bio-, Info-Tech Sensors and 3D Systems III*. https://doi.org/10.1117/12.2513782.

11

Polylithic Integrated Circuits using 2.5D and 3D Heterogeneous Integration: Electrical and Thermal Design Considerations and Demonstrations

Ting Zheng, Ankit Kaul, Sreejith Kochupurackal Rajan, and Muhannad S. Bakir

School of Electrical and Computer Engineering; Georgia Institute of Technology, Atlanta, Georgia

11.1 Introduction

An explosion of data-centric computation is driving the need for heterogeneous integration (HI) and advanced packaging. Driving the exponential performance trend of high-end server CPUs has historically required increasing core counts and die sizes, among other factors. This traditionally meant working close to the reticle limit, which is becoming increasingly costly [1]. Moreover, the cost of advanced technologies such as 7 nm and smaller is increasing [2] while not all functions of a system-on-a-chip (SoC) benefit from the leading-edge technology (for example, I/O circuits). These challenges have pushed integrated circuit implementation towards modular designs. This includes Xilinx's "Everest" [3], AMD's chiplet-based "Rome" and "Matisse" SoCs [2], as well as Intel's bridge-based [4] and TSV-based 3D [5] integration. One objective of most advanced integration schemes is to connect disaggregated chips to match the functionality of monolithic SoCs. However, one of the biggest bottlenecks for such systems is the interdie bandwidth density (BWD), which can cause functional blocks to be idle during data transfer [6], leading to lower system performance.

Advanced forms of multi-die HI schemes such as 2.5D and 3D integration can address the need for higher bandwidth and lower energy in chip-to-chip signal interconnections to enable opportunities in high performance and low-power computing. Silicon interposer-based 2.5D integration of field-programmable gate arrays (FPGA) achieves an aggregate BWD in excess of 400 Gb/s [3]. The 3D processor-on-memory integration using through silicon vias (TSVs) exhibits a maximum memory BWD of 510.4 Gb/s at 277 MHz [7]. Monolithic 3D integration is another promising option, which achieves even higher BWD than TSV-based 3D

Embedded and Fan-Out Wafer and Panel Level Packaging Technologies for Advanced Application Spaces: High Performance Compute and System-in-Package, First Edition. Edited by Beth Keser and Steffen Kröhnert.

integration resulting from the utilization of shorter and denser nanoscale vertical vias [8]. Moreover, as noted earlier, there has been recent interest in multi-die packages using bridge-chip technology, including embedded multi-interconnect bridge technology [4] and heterogeneous interconnection stitching technology [9] to enable 2.5D microsystems. In its simplest form, bridge-chip technology utilizes a silicon die with high-density interconnects for inter-die communication. The performance metrics of these 2.5D integration technologies are comparable to interposer-based 2.5D solutions, but many other benefits are offered, including the elimination of TSVs.

In this chapter, we first discuss chiplet and interposer scale interconnection using stitch-chips, or bridge-chips. The discussion includes both electrical design considerations and experimental demonstrations. Next, thermal design considerations in 2.5D and 3D integrated circuits (ICs) are presented. Furthermore, microfluidic cooling in single-die and 2.5D ICs is demonstrated and benchmarked using active circuits (28 nm and 14 nm silicon). Lastly, TSV integration in microfluidic cooled chiplets for 3D integration is presented.

11.2 Heterogeneous Interconnect Stitching Technology (HIST)

To enable heterogeneous integration with monolithic-like performance and low-loss dense signal interconnections, a polylithic integration technology called heterogeneous interconnect stitching technology (HIST) was proposed and is shown in Figure 11.1. The technology enables both 2.5D and 3D integration by fine-pitch and multi-height compressible microinterconnects (CMIs).

In Figure 11.1, 2.5D integration is achieved by connecting multiple "anchor chips" through stitch-chips that are placed between the anchor chips and the package substrate. The anchor chips here emulate the actual chiplets assembled in a system. Fine-pitch CMIs, as off-chip I/Os, interface the anchor chips and the stitch-chips. The fine-pitch CMIs and fine-pitch traces on the stitch-chips provide high-bandwidth signal interconnections between the anchor chips. Furthermore, this technology has the potential to achieve direct interconnection between an anchor chip and a silicon photonic integrated circuit (PIC) with direct fiber assembly, as shown in Figure 11.1.

To provide 3D-like integration, "surface-embedded chips" are integrated underneath the anchor chips and are interconnected with face-to-face bonding using multi-height and multi-pitch CMIs [9]. Multi-height CMIs, instead of conventional solder bumps, can provide signal interconnections between the anchor chips and stitch-chips/surface-embedded chips, as shown in Figure 11.1. In addition, in Figure 11.1, CMIs, which connect anchor chips and package

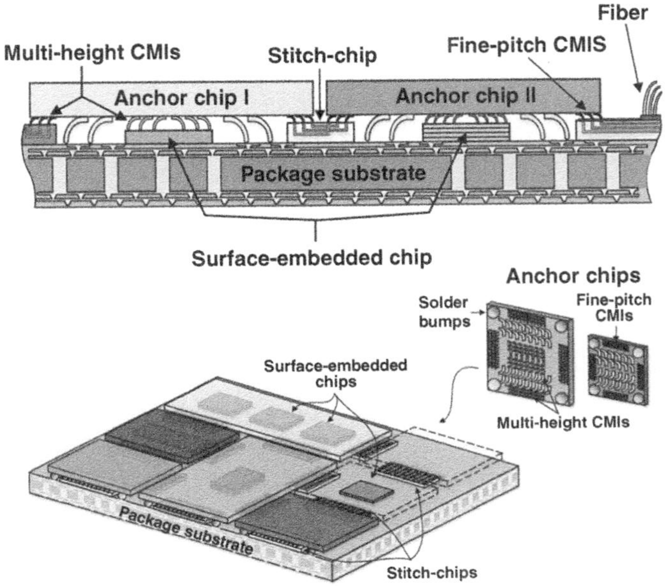

Figure 11.1 Schematic of envisioned HIST for polylithic integration [9].

substrate directly, can provide power delivery with pads and redistribution layers (RDLs) on the package substrate. Mechanical bonding between the anchor chips and the package substrate is enabled by using large solder bumps on each of the four die corners to enable the reworkability of the assembled die [9].

As a critical technology enabler, CMIs possess several benefits. Unlike conventional solder bumps, CMIs are elastically compressible so that they can compensate for possible off-chip interconnection distance disparities resulting from chip thickness differences, which enables both 2.5D and 3D face-to-face interconnections in one platform [9]. CMIs, as temporary interconnections, can improve system yield by facilitating die replacement/rework. In [10], CMIs are reported to improve the thermo-mechanical reliability of an assembled system. In addition, CMIs provide the flexibility in the materials of die and substrates to be stitched together irrespective of their CTE mismatch (e.g. silicon, glass, organic, and compound semiconductor) [9].

To enable high-bandwidth signal interconnections between anchor chips, fine-pitch CMIs are required. An array of 30 μm by 30 μm pitch CMIs are fabricated, as shown in Figure 11.2. These gold-coated NiW CMIs are approximately 9.5 μm tall, 25 μm long, and 2.5 μm thick. Note that the CMIs' geometry is lithographically defined in the demonstrated fabrication process [9] and they can be further scaled to finer pitch. Based on the indentation test results for these

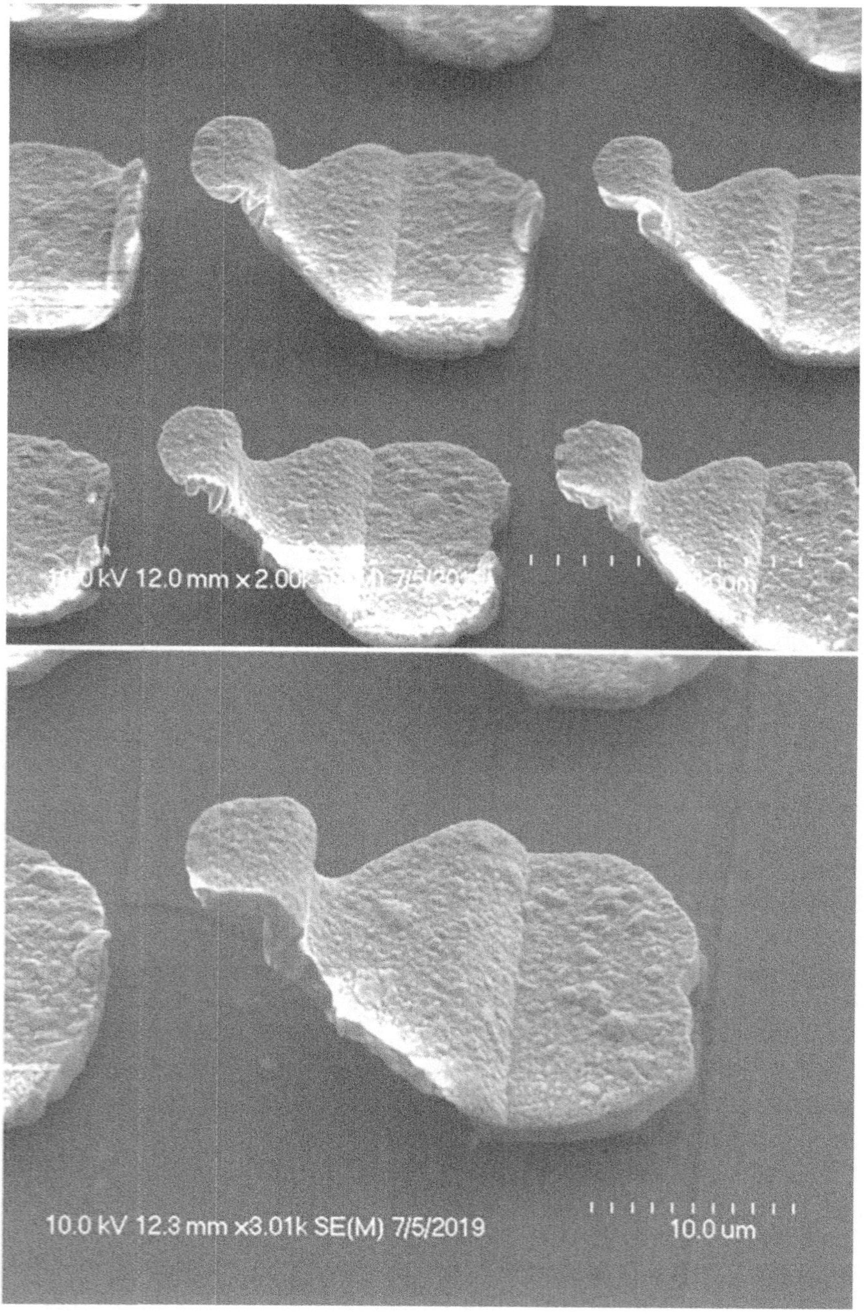

Figure 11.2 SEM images of fine-pitch CMIs [9].

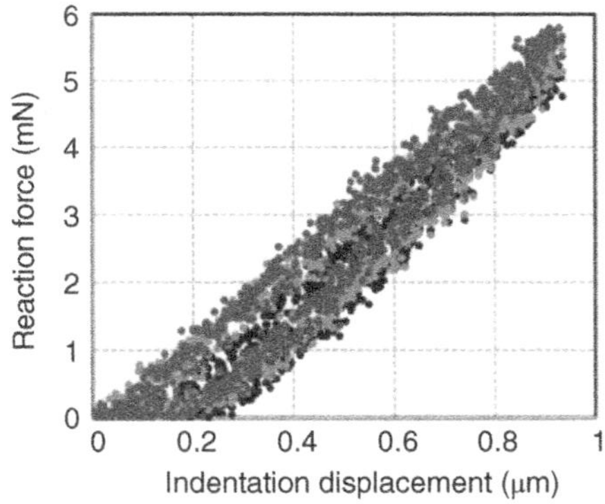

Figure 11.3 Indentation test of fine-pitch CMIs [9].

fine-pitch CMIs, as shown in Figure 11.3, their average measured compliance is around 0.16 mm/N. The compliance can be engineered to a higher value by adjusting the CMIs' thickness [9].

The multi-height gold-coated NiW CMIs' fabrication and characterization are demonstrated in [11]. As shown in Figure 11.4, the heights of the fabricated CMIs are 75 μm, 55 μm, and 30 μm, and their in-line pitch is 150 μm. The indentation tests are performed and reported in [12]: the 75 μm, 55 μm, and 30 μm tall CMIs are measured to have 12.21 mm/N, 8.82 mm/N, and 3.91 mm/N compliance, respectively; their respective vertical elastic deformations are 40 μm, 35 μm, and

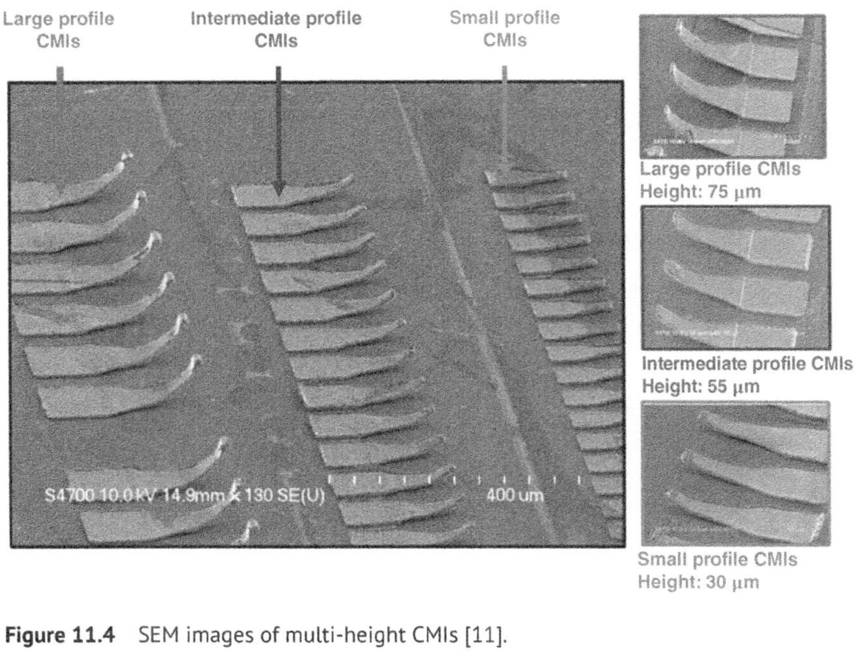

Figure 11.4 SEM images of multi-height CMIs [11].

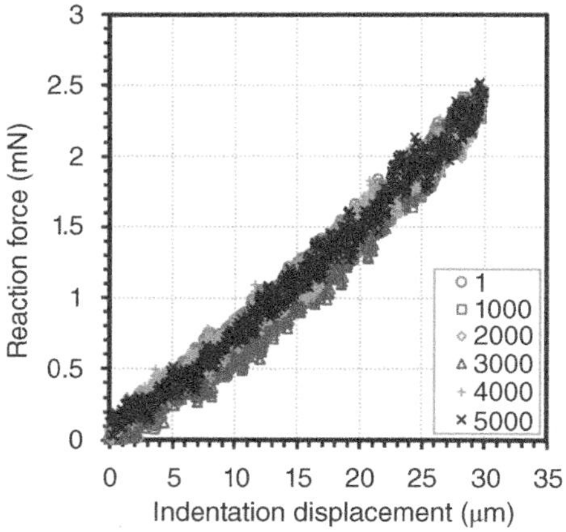

Figure 11.5 5000 cycles indentation test for 75 μm tall CMI [11].

20 μm. To achieve uniform bonding forces during assembly, the compliance of the multi-height CMIs can be optimized easily to have the similar value by adjusting top-view shape and thickness [12]. As for the lifetime fatigue reliability of the fabricated CMIs, a 75 μm tall CMI is indented consecutively for 5000 cycles. Results shown in Figure 11.5 demonstrate that there is no apparent plastic deformation after 5000 indentation cycles. As for the electrical performance, the average measured resistance of the 75 μm, 55 μm, and 30 μm tall CMIs are 67.17 mΩ, 55.18 mΩ, and 48.13 mΩ, respectively [12]. In [13], ~13 μm tall 20 μm pitch Cu pillar exhibits approximately 40 to 50 mΩ with/without underfill. Such multi-height CMIs can be utilized to integrate the surface-embedded chips having various thickness, as shown in Figure 11.1. In addition, the height of the fabricated multi-height CMIs can be easily modified by adjusting the lithography process [11].

For RF/mmWAVE applications, HIST can provide low-loss and low-parasitic signal transitions between heterogeneous chiplets through stitch-chips, as shown in Figure 11.6. The stitch-chips are made by low-loss dielectric material such as fused-silica (dielectric constant: 3.9, loss tangent: 0.0002) in this example.

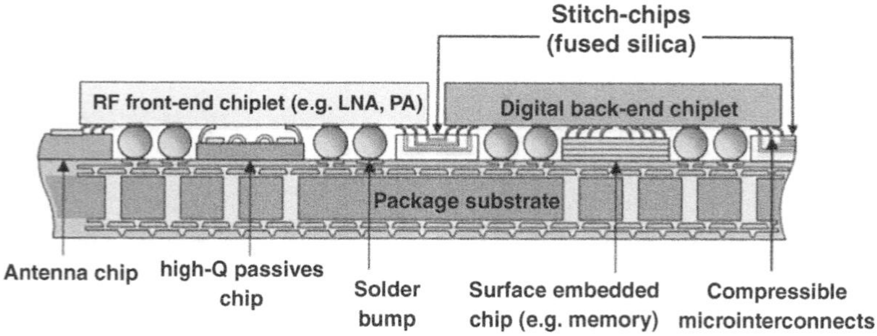

Figure 11.6 HIST for RF/mmWAVE applications [14].

The stitch-chips are envisioned to have short and impedance-matched channels, which consist of coplanar waveguides (CPWs) and gold-coated NiW CMIs. The assembly process with CMIs does not require underfill, which results in a huge reduction of the channels' parasitic capacitance. In addition, HIST provides a flexible approach to separate passives from lossy silicon chiplets and integrate them onto low-loss surface-embedded chips to enhance the system performance [14].

An accurate RF characterization up to 30 GHz is reported in [14] for the signal channels on the stitch-chips. In [14], 200 μm-pitch 500 μm-long stitch-chip channels (500 μm-long CPW and CMIs) and 200 μm-pitch 1 mm-long stitch-chip channels are measured up to 30 GHz to extract S-parameters, as shown in Figure 11.7. The data suggests that the stitch-chip channels enable low-loss signal transitions, as 500 μm-long channels possess less than 0.4 dB insertion loss up to

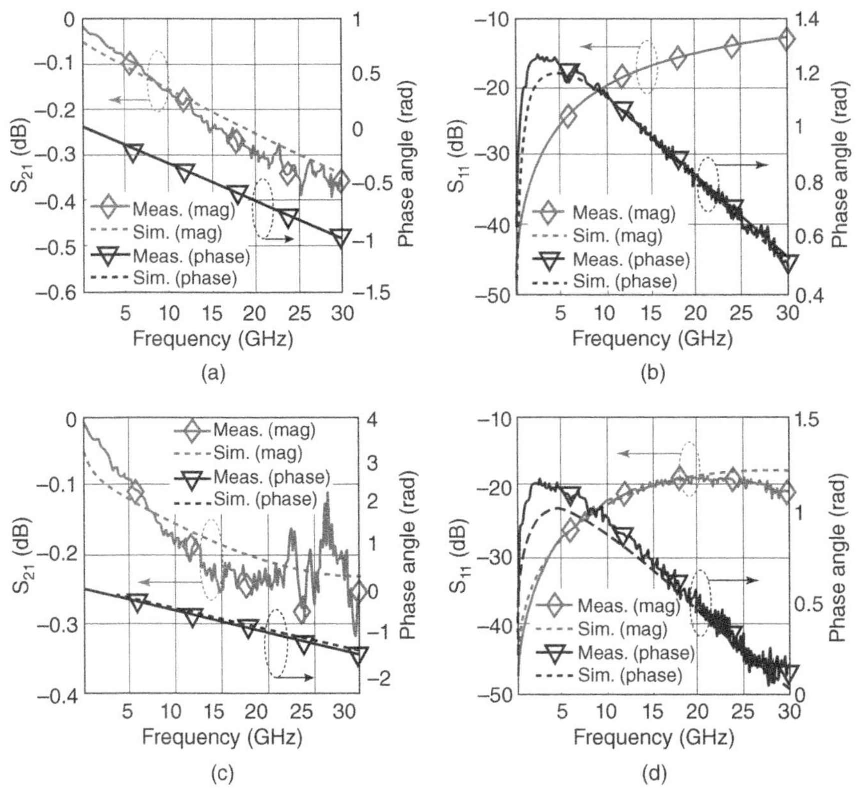

Figure 11.7 Measured S-parameters of channels on stitch-chips: (a) Insertion loss of 500 μm-long stitch-chip channel, (b) return loss of 500 μm-long stitch-chip channel, (c) insertion loss of 1000 μm-long stitch-chip channel, and (d) return loss of 1000 μm-long stitch-chip channel [14].

30 GHz thanks to the used fused-silica stitch-chip substrate (dielectric constant: 3.9, loss tangent: 0.0002).

With respect to loss performance, such low-loss stitch-chip channels surpass conventional Si-interposer channels for RF/mmWAVE applications, such as 5G wireless (28 GHz bandwidth). In [14], a signal channel on CMOS-grade Si (10 S/m conductivity) with identical dimensions as the measured stitch-chip channel is simulated to represent a Si-interposer channel. At 28 GHz, the 1 mm-long stitch-chip channel shows superior insertion loss (0.2 dB) while the 1 mm-long comparable Si-interposer channel exhibits much larger insertion loss (3.78 dB).

For time-domain characterization, eye-diagram simulation is performed with the S-parameters of the 1 mm-long stitch-chip channels and 1 mm-long Si-interposer channels in [14]. The results are shown in Figure 11.8. The low-loss stitch-chip channel increases eye height by 49.4% and reduces rise/fall time (3.3 ps vs. 8.3 ps for the Si-interposer channel), which implies lower signal latency and channel parasitics. A clear eye opening is observed for the 1 mm-long stitch-chip channel at 50 Gbps without any equalization methods applied.

CMIs' RF performance is also reported in [14] for the first time. Using an L-2L de-embedding technique, the measured S-parameters of ground-signal-ground

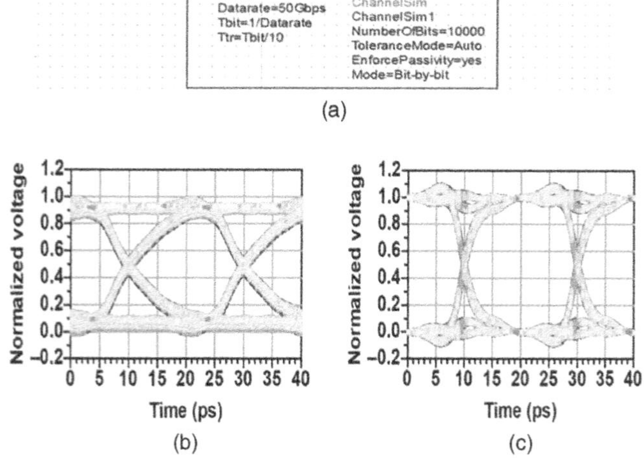

Figure 11.8 Eye diagram simulation: (a) ADS simulation schematic, (b) Si-interposer channel with 50 fF ESD capacitors at 50 Gbps data rate, and (c) stitch-chip channel with 50 fF ESD capacitors at 50 Gbps data rate [14].

CMIs are extracted up to 30 GHz. Their insertion loss is less than 0.17 dB, and the return loss is low (−16.4 dB). The results indicate that the CMIs' loss is negligible up to 30 GHz.

The stitch-chip concept discussed above can also be applied at the interposer level. Silicon interposers provide high-data rate and low-energy communication between chiplets within a single interposer. Interposer-to-interposer communication is routed through the motherboard traditionally, which results in a very significant loss of energy efficiency and high-bandwidth connectivity. To this end, we have previously demonstrated [15] a novel large-scale silicon platform consisting of multiple interposer tiles, which are essentially silicon interposers, that are assembled directly on the printed wiring board (PWB) in a tile-like pattern (Figure 11.9). This approach enables system integration with multiple interposers having size below reticle size limit. Electrical interconnections between the tiles are provided through silicon bridges, which are mounted on top and span two or more interposer tiles. The tiles and the bridges are electrically interconnected using flexible interconnects, which are developed to allow repeatable and low resistance contact with electrical pads [15]. Positive self-alignment structures (PSASs) and inverted pyramid pits passively self-align the interposer tiles with the PWB and the interposer tiles with the silicon bridges. This demonstration is a first step toward the goal of a platform that provides high bandwidth and low-energy per bit (EPB) interconnections among interposer tiles. The key features of this demonstration are: (1) accurate alignment of interposer tiles, bridges, and PWB and (2) electrical interconnectivity among tiles through the bridges. Ultimately, nanophotonics would be integrated into the platform for interconnect lengths that offer better bandwidth density and EPB. Figure 11.10 illustrates a fully assembled

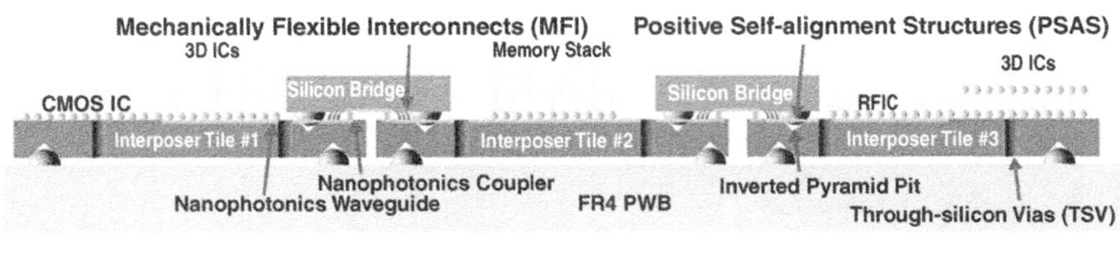

Figure 11.9 Direct mounting of silicon interposers on FR4 PWB [15].

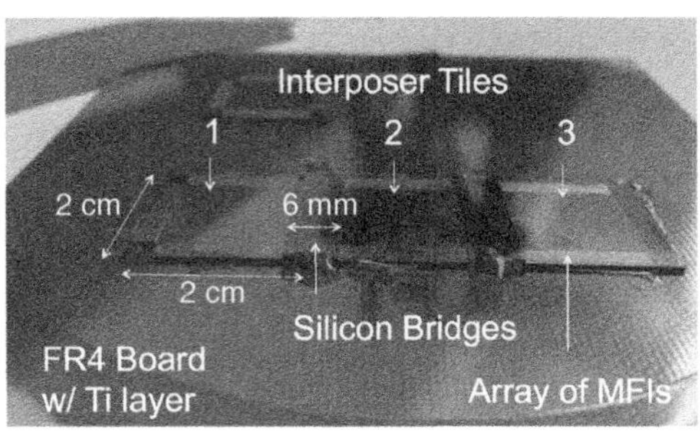

Figure 11.10 Three interposer tiles mounted directly on PWB and interconnected using silicon bridges/stitch-chips [15].

silicon interposer titles on a PWB and interconnected using high-bandwidth density silicon bridge-chips [15].

11.3 Thermal Evaluation of 2.5D Integration Using Bridge-Chip Technology

As multi-die packaging enables using more high-performance chips in a package, thermal challenges are expected to increase. Air cooling is expected to become ineffective at cooling such systems, keeping much of the silicon dark (either idle or significantly underclocked) [16]. Thermal coupling from high-power chips to low-power chips will also lower the overall system performance [17]. Therefore, despite the potential electrical benefits provided by these emerging integration architectures, there exist significant thermal issues.

Thermal analysis and optimization have been extensively conducted for interposer-based 2.5D integration [18, 19], as well as TSV-based [20, 21] and monolithic 3D IC integrations [22, 23]. However, there are no thermal modeling efforts focused on 2.5D bridge-chip-based interconnection platforms. Therefore, in this section we first explore the thermal attributes of bridge-chip-based 2.5D ICs and benchmark with other 2.5D and 3D solutions. Second, we conduct a deep-dive look at bridge-chip-based 2.5D integration and evaluate thermal performance as a function of various technology parameters, such as thermal interface material (TIM) properties, die thickness, and die spacing. These studies will help the community understand the thermal limits and challenges facing bridge-chip-based integration technologies. Next we propose a back-end-of-line (BEOL)-embedded integration scheme, which can help avoid long wire lengths and large pad sizes leading to improved EPB and lower chip-to-chip delay.

11.3.1 2.5D and 3D Benchmark Architectures

11.3.1.1 2.5D Integration

Figure 11.11 shows three 2.5D integration technologies with different chip-to-chip interconnection techniques. Figure 11.11a (1) represents a bridge-based integration where a Si bridge is embedded within an organic package substrate. Dense interconnects on the Si bridge along with fine-pitch μ-bumps are used for die-to-die interconnection. Figure 11.11a (2) represents a traditional interposer-based integration. The third approach, shown in Figure 11.11a (3), de-embeds the Si bridge chip and places it between the active chips and the package.

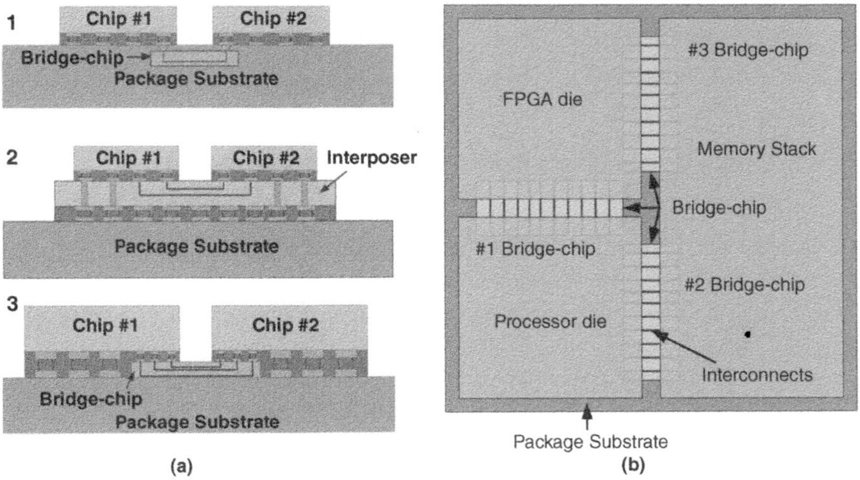

Figure 11.11 2.5D chip stack using (a1) bridge-chip technology and (a2) interposer technology. (a3) Non-embedded bridge-chip using multi-height microbumps technology. (b) Illustration of the envisioned FPGA-CPU-Memory 2.5D chip stack using bridge-chip technology (top view). [24]

In this section, we focus on a CPU-FPGA-Memory 2.5D microsystem as a test vehicle for the bridge-chip technology; thus, all benchmarks are based on this chip set (Figure 11.11b).

11.3.1.2 3D Integration

Figure 11.12 represents two types of 3D integration architectures [24]. TSV-based 3D is shown in Figure 11.12a, where solder-capped copper pillars (or μ-bumps) and through-silicon vias (TSV) are utilized to establish vertical interconnection between two or more known good dice. Figure 11.12b represents monolithic 3D integration, where two or more active device layers and interconnects are sequentially processed using standard lithography tools.

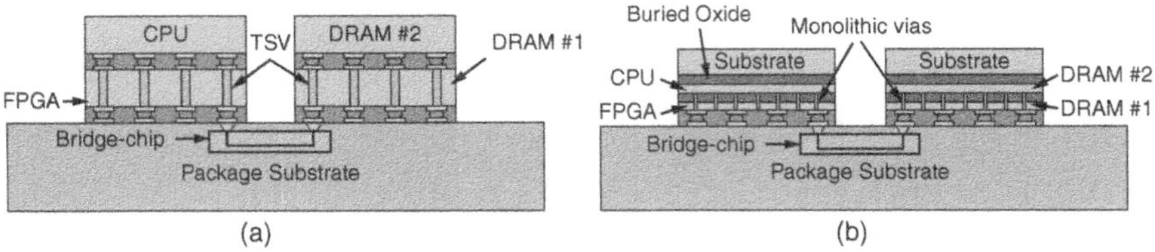

Figure 11.12 3D chip stack. (a) TSV-based. (b) Monolithic nanoscale via-based. [24]

In both cases, we consider a CPU-FPGA-DRAM integrated microsystem. We assume two 3D stacks: a CPU-on-FPGA stack (CPU placed on top for thermal consideration), and a DRAM chip stack.

11.3.2 Thermal Modeling and Specifications

The thermal framework used in this section is reported in [25]. For lower computational costs, non-conformal meshing methods are implemented in the x-, y-, and z-directions. The steady-state and transient-state models are validated against ANSYS, and the maximum relative error in junction temperature (T_j) rise is less than 7% [24, 26].

Figure 11.13 shows a 2.5D integrated system with an air-cooled heat sink. The 3D IC configurations considered are similar except that the chips are stacked. For thermal modeling, the heat sink and the PWB are abstracted as primary and secondary cooling boundaries, respectively. We model both boundaries using a uniform convection coefficient, which is applied to the top surface of the heat

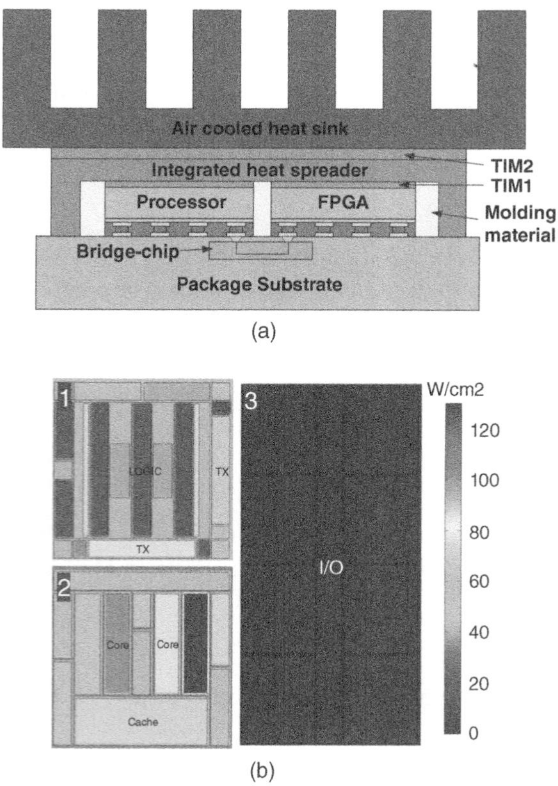

Figure 11.13 (a) 2.5D integration using bridge-chip technology with detailed layer information. (b) Power density maps of each die: (1) FPGA die, 44.8 W. (2) Processor die, 74.49 W. (3) DRAM die (cell circuit), 5.65 W for cell circuit. [24]

spreader and to the bottom surface of the package substrate, respectively. Specific details of layer thicknesses, material properties, geometry parameters, boundary conditions, μ-bumps, TSVs, and power density maps are available in [24].

11.3.3 Comparison of Different 2.5D Integration Schemes

The three 2.5D IC cases are configured with an air-cooled heat sink (Figure 11.13a). The chip layout and power maps are shown in Figures 11.11b and 11.13b, respectively. Unless stated otherwise, all thermal analyses are steady-state results using the maximum power maps shown in Figure 11.13b for a worst-case estimate.

The thermal profiles of each die in all cases is shown in Figure 11.14. All cases exhibit a similar thermal profile as most of the heat is conducted through the attached air cooled heat sink on top (97.17%, 97.19%, and 98.18%, for the interposer, non-embedded bridge-chip, and bridge chip cases, respectively). Nevertheless, a small difference in the T_j results from different secondary heat paths in each case. Additionally, a clear lateral thermal coupling is observed between different dice in all cases due to conduction via heat spreader.

11.3.4 Thermal Comparison between 2.5D and 3D Integration

The power density of a vertical 3D stack will be larger than that of its 2.5D equivalent configurations. Consequently, thermal challenges can be more difficult to address. In this section, we explore two types of 3D integration approaches and compare the thermal performance with bridge-chip-based 2.5D integration.

Figure 11.15b lists the maximum T_j of bridge chip-based 2.5D and the two 3D IC cases. These results show that 2.5D integration has better thermal characteristics than the two 3D IC cases. Figure 11.15a shows the thermal profiles of the CPU and FPGA dice in the two considered 3D IC cases. Compared to the 2.5D bridge-chip

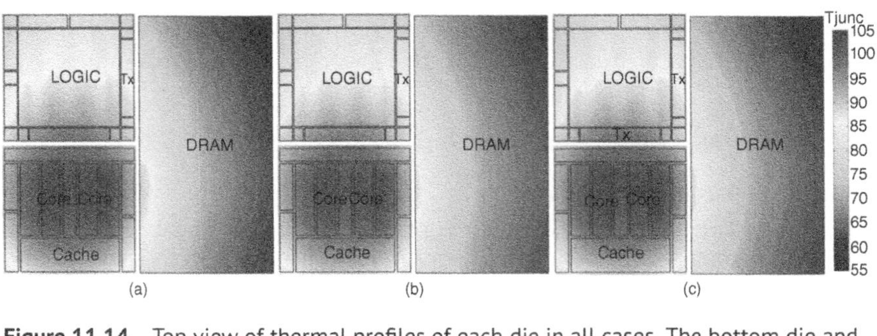

Figure 11.14 Top view of thermal profiles of each die in all cases. The bottom die and the hottest die of the DRAM stack are plotted. (a) Embedded bridge chip, T_{max} : 104.92 °C. (b) Interposer, T_{max} : 102.8 °C. (c) Non-embedded bridge-chip, T_{max} : 104.23 °C. [24]

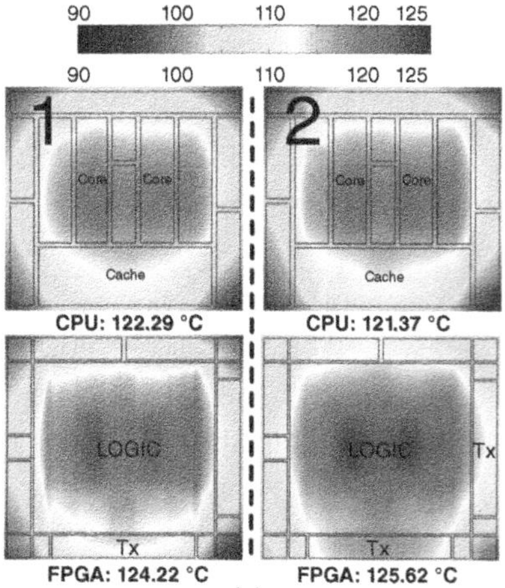

(a)

Unit: °C	CPU		FPGA		DRAM[1]	
	T_{max}	T_{min}	T_{max}	T_{min}	T_{max}	T_{min}
Bridge-chip	104.92	83.08	98.28	75.02	89.17	60.01
Monolithic	122.29	93.61	124.22	94.25	96.25	63.57
TSV	121.37	94.64	125.62	98.94	98.18	66.57

1 For DRAM, we show the maximum temperature of the bottom die in the stack (the hottest die).

(b)

Figure 11.15 (a) Thermal profile of each die in the 3D stack cases. (1) Monolithic 3D. (2) TSV-based 3D. (b) Thermal comparison of bridge-chip 2.5D and 3D integration. [24]

case shown in Figure 11.14(a), the 3D cases have stronger thermal coupling, leading to the thermal profiles of both dice exhibiting a mirror image of each other. The monolithic 3D case has a smaller active layer thickness, which leads to worse spreading than in the TSV-based 3D IC case, as shown in Figure 11.12. However, the thermal resistance from the FPGA to the heat sink is slightly smaller than the TSV-based 3D IC case, resulting in a lower maximum temperature.

11.3.5 Thermal Study of Bridge-Chip 2.5D Integration

11.3.5.1 Impact of TIM conductivity

We consider two TIM layers: The first is present between the heat spreader and each die (TIM1), and the second is applied between the heat spreader and the heat sink (TIM2), as shown in Figure 11.13a. To investigate the impact of TIM properties, we sweep the thermal conductivity of TIM1 and TIM2 from 0.9 to 400 W/°C-m (TIM1 and TIM2 are assumed to be the same material). From the results (Figure 11.16a), there is a crossing point in the thermal conductivity at

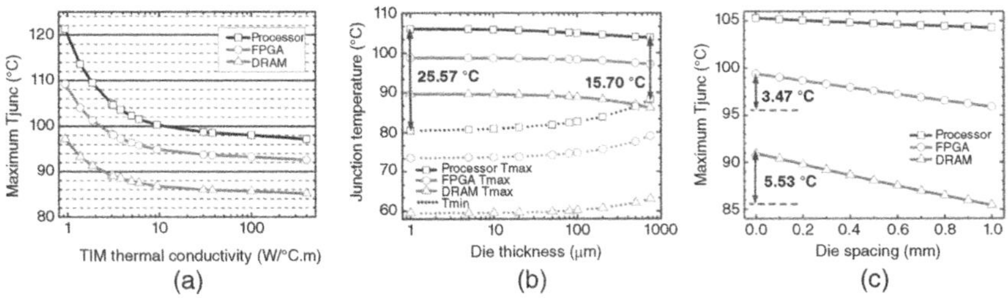

Figure 11.16 (a) Thermal profile of each die (die thickness is 1 μm). The heat spreading is confined, and the block outlines are clearly observed from the thermal maps. (b) Impact of die thickness scaling. The dotted line plots T_{min} of each die. (c) Impact of die spacing: as the die spacing increases, T_j decreases. [24]

approximately 3 W/°C-m, beyond which better TIM conductivity does not provide significant benefits. Similarly, changing the TIM thickness leads to comparable results [24].

11.3.5.2 Die Thickness

Bulk thickness is important for heat spreading to reduce localized hotspot temperatures. As die thickness decreases, the lateral thermal resistance increases and heat spreading decreases, as observed in Figure 11.16b.

11.3.5.3 Die Spacing

Figure 11.16c shows that with increasing die spacing (keeping heat spreader as the same size), the T_j of each die decreases. However, the rate of temperature reduction is larger for the die with smaller power, which implies that the low-power die is more vulnerable to thermal coupling. Since DRAM performance degrades in the extended temperature range (above 85 °C) [27], it is prudent to take advantage of this effect and carefully design the spacing between the DRAM die and the other high-power chips. However, the trade-offs between thermal considerations and off-chip interconnection metrics, such as BWD/EPB and integration density, also need to be considered.

11.3.6 Polylithic 3D Integration

To bridge the gap in connectivity and heterogeneity between TSV-based and monolithic 3D ICs, a BEOL-embedded integration scheme is proposed [28]. This scheme (Figure 11.17) represents a densely integrated system divided into multiple device tiers where custom chiplets, such as I/O drivers and RF front-ends are embedded into the back-end of a base tier (application processor) with a monolithic memory tier, e.g. RRAM. The proposed 3D seamless off-chip

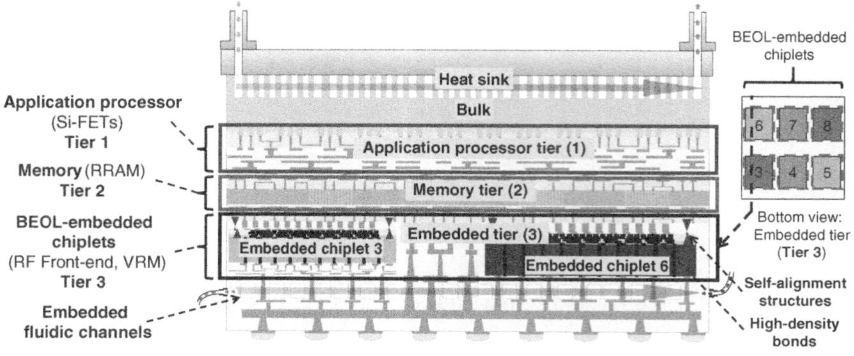

Figure 11.17 Proposed 3D seamless off-chip connectivity (3D SoC+) concept: BEOL-embedded chiplet integration. [29]

connectivity (SoC+) concept aims to combine the best of both TSV-based 3D and monolithic 3D, including extreme efficient signaling and large BWD. This scheme can be enabled by high-density interconnections and high-accuracy self-alignment techniques.

11.4 Monolithic Microfluidic Cooling of High-Power Electronics

As discussed in the previous sections, traditional cooling systems are being pushed to their limit with the increase in power densities and aggregate package power associated with modern computer systems.The increased thermal challenges associated with the advanced dense heterogeneous integration schemes such as 2.5D and 3D are exacerbating these issues. Microfluidic cooling, first demonstrated by Tuckerman and Pease [30], has the potential to solve this cooling challenge for high-power and high-performance integrated circuits. Also, adapting this approach for 2.5D and 3D systems can help mitigate issues such as thermal coupling between the chiplets.

Further, with a liquid coolant, such as water, which has a volumetric heat capacity that is more than 3000× higher than that of air (close to standard temperature and pressure), the necessary volume for heat exchange and fluid delivery can be dramatically decreased. This, coupled with the high surface area-to-volume ratio structures that can be generated by advanced silicon micro-fabrication techniques can help create ultra-low form factor and high efficiency thermal management solutions, which can help increase the compute element stacking density in data-center kind of applications.

In this section, we explore the various real-world performance benefits that we get from monolithic microfluidic cooling of active compute chips. Results from a single die FPGA test-bed will be used for this discussion. Subsequently, similar design and silicon micro-fabrication techniques would be used in conjunction with advanced additive manufacturing techniques to create a microfluidic heat sink for 2.5D packages. Finally, we will discuss the electrical implications of extrapolating these techniques to 3D ICs.

11.4.1 Experimental Demonstration and Characterization on Single Die Systems

The most common method of cooling microelectronics has long been an air-cooled heat sink (ACHS) mounted on top of the packaged integrated circuit, as shown in Figure 11.18a. Efficacy is limited with this solution and can be improved in several ways through the use of microfluidic cooling. First, due to the small heat sink dimensions and properties of the liquid coolants, much lower convective thermal resistances can be achieved with microfluidic cooling when compared with direct air cooling. In addition, in the traditional configuration, shown in Figure 11.18a, there is a large conductive thermal resistance due to the large distance and, more importantly, several material interfaces through which heat must be conducted in order to reach the heat sink. In order to improve thermal resistance at these interfaces, two levels of thermal interface material (TIM) are used, but these interfaces still remain a major bottleneck in total junction-to-ambient thermal resistance. By etching the heat sink directly into the silicon die, conductive thermal resistance between the heat source and heat sink is minimized.

A considerable amount of literature exists on the design and evaluation of micropin-fin-based monolithic microfluidic cooling (Figure 11.18b). While most of these were done on passive silicon with resistive heaters, we focus our discussions in this section on an implementation with active CMOS. An Altera

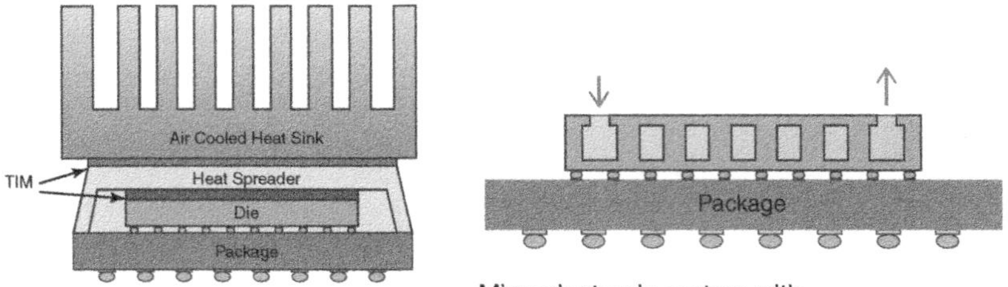

Traditional Microelectronics system.
(a)

Microelectronic system with monolithically integrated microfluidic heat sink (MFHS).
(b)

Figure 11.18 Thermal resistance reduction in monolithic microfluidic heat sinks. [31]

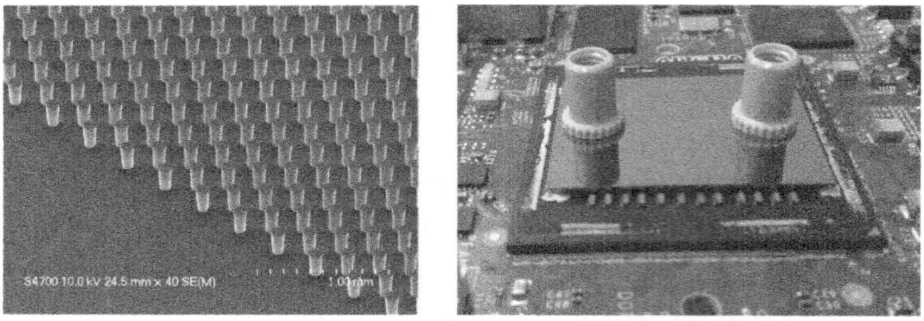

SEM image of etched micropin-fins.

(a)

Processed FPGA soldered to development board with silicon cap and nanoports for fluid delivery.

(b)

Figure 11.19 Test-bed fabrication results. [31]

Stratix V field-programmable gate array (FPGA), built in a 28-nm process, was postprocessed to integrate a micropin-fin heat sink directly into the back of the flip-chip bonded silicon die, a few hundred micrometers from the active circuitry. This microfluidic-cooled FPGA was then tested with deionized water as a coolant at several flow rates and inlet temperatures.All testing was performed with Altera Stratix V DSP development boards. A comparison was made with a stock board with the default air-cooled heat sink. Testing was performed at flow rates ranging from 0.15 to 3.0 mL/s and with inlet temperatures ranging from 21 °C to 50 °C.

For this demonstration, the fabrication process was adapted to work with a prepackaged flip-chip bonded die. The FPGA BGA was initially desoldered from the development board. After de-lidding the die to remove the heat spreader and TIM, the package was then mounted on to a carrier wafer with cool grease. The device was subsequently patterned using photolithography, and a Bosch process was used to etch silicon micropin-fins with vertical sidewalls up to an average depth of 240 μm. The etch was optimized to minimize tapering, where the micropin-fin base is narrower than the top, which could reduce fin efficiency and thermal performance by limiting the cross-sectional area through which heat can be conducted up the micropin-fins. Inlet and outlet plena were formed in the same etching step by etching regions on either side of the micropin-fin array without any micropin-fins. An SEM image of the etched micropin-fins is shown in Figure 11.19a. A separate silicon lid was fabricated with an inlet and an outlet port. The lid was first tacked on to the top of the etched FPGA with high-temperature epoxy, in order to provide a smooth surface for resoldering to the development board. After soldering, the lid was permanently secured with epoxy and nanoports were attached to deliver coolant. A photograph of the resoldered FPGA with nanoports can be seen in Figure 11.19b.

The FPGA was loaded with a custom pulse compression algorithm designed to mimic common DSP-style use cases of FPGAs. The FPGA was tested in an

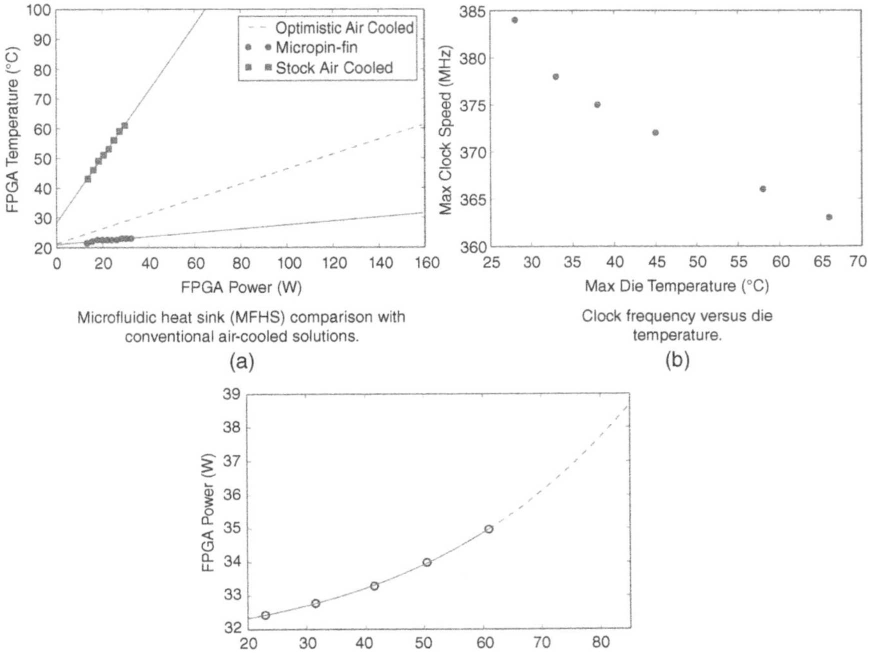

Microfluidic heat sink (MFHS) comparison with conventional air-cooled solutions.
(a)

Clock frequency versus die temperature.
(b)

Die power versus die temperature.
(c)

Figure 11.20 Measurement results. [31]

open-loop system, with deionized water as a coolant. The measurement results are shown in Figure 11.20. An average junction-to-inlet thermal resistance of 0.07 °C/W was achieved at a flow rate of 3.0 mL/s and pressure drop of 97 kPa. This corresponds to a much lower operational temperature when compared to air-cooled solutions. Further, the lower operational temperatures allow for higher clock speeds (Figure 11.20b), as well as lower leakage power (Figure 11.20c). The FPGA was also cooled with inlet water temperatures up to 50 °C, enabling high efficiency through heat exchange directly to ambient air, or waste heat reuse [32].

11.4.2 Microfluidic Cooling of 2.5D Devices: Experimental Demonstration

Interconnect density for tightly integrating multiple chiplets is driving the industry toward 2.5D integration. While this tight integration improves performance, it also increases the challenge of heat extraction by increasing aggregate package powers and introducing thermal crosstalk between the adjacent dice as shown earlier.

The silicon micro-fabrication approach used in the previous test case can also be applied to 2.5D devices as a potential solution. Advances in 3D printing techniques could be leveraged to create ultra-low profile fluid delivery structures, which, in conjunction with the low volume requirements of silicon-based coolers can help create a compact solution. In this section, we describe a silicon micropin-fin heat sink designed and fabricated to cool a Stratix 10 GX field-programmable gate array (FPGA) consisting of five heterogeneous dice. The micropin-fin geometry is locally varied to match the heat flux of the underlying dice. The silicon micropin-fin heat sink is embedded in a 3D printed plastic piece that seals the top of the silicon micropin fins and connects the heat sink to inlet and outlet tubing. A conceptual cross-sectional diagram of the heat sink concept can be seen in Figure 11.21a.

An initial heterogeneous micropin-fin design was created by beginning with a high aspect ratio micropin-fin geometry to populate the regions of the heat sink directly over the transceiver dice, while a second micropin-fin design was used to

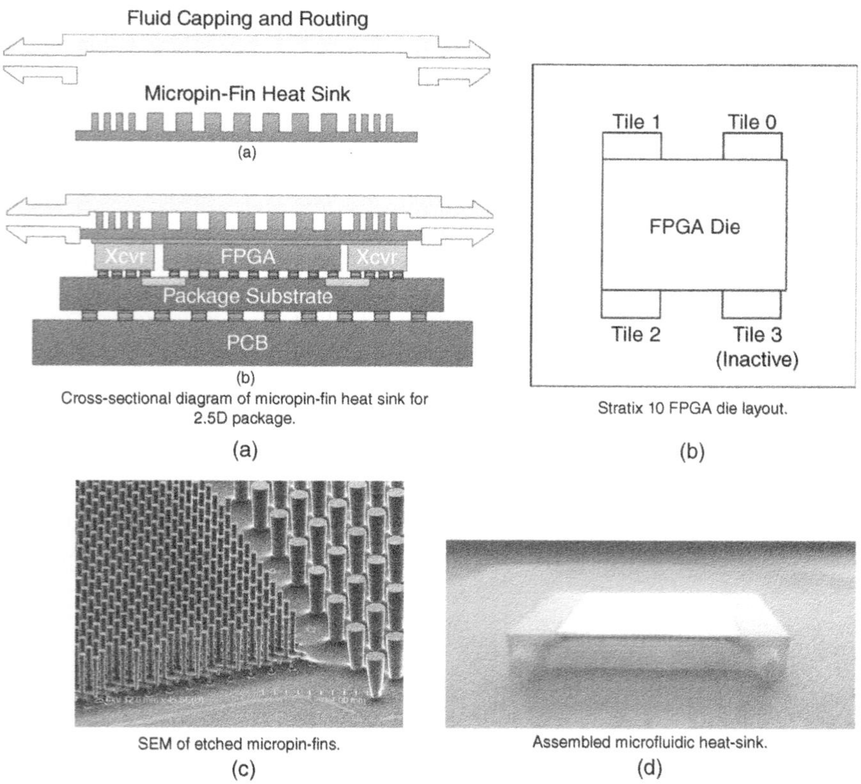

Cross-sectional diagram of micropin-fin heat sink for 2.5D package.
(a)

Stratix 10 FPGA die layout.
(b)

SEM of etched micropin-fins.
(c)

Assembled microfluidic heat-sink.
(d)

Figure 11.21 Heterogeneous micropin-fin heat-sink for 2.5D. [33]

Die	Air	Microfluidic
FPGA	0.46 °C/W	0.057 °C/W
Transciever Title 0	0.48 °C/W	0.003 °C/W
Transciever Title 1	0.43 °C/W	0.004 °C/W
Transciever Title 2	0.45 °C/W	0.045 °C/W

Thermal coupling between dice: change in die Temperatures as a function of FPGA die power

(a)

Die	Air	Liquid	Air with no Compute
FPGA	70.5 °C	30.1 °C	38.5 °C
Transciever Title 0	81.5 °C	35.1 °C	48.5 °C
Transciever Title 1	81 °C	37.7 °C	52 °C
Transciever Title 2	75.5 °C	43.6 °C	46 °C

Stratix 10 temperature measurements with Air-cooled heat sink and microfluidic-cooled heat sink

(b)

Figure 11.22 Measurement results and comparison with air-cooled solutions. [33]

cool the region above the FPGA die (Figure 11.21b). The pitch and diameter were scaled by the relative power density of the transceiver and FPGA dice to create the dimensions used to cool the center FPGA die. This micropin-fin design was etched to a depth of 457 μm (Figure 11.21c). This etched silicon cold-plate was attached to a 3D printed enclosure and was attached and sealed with epoxy. The assembled heat-sink (Figure 11.21d) was mounted atop the de-lidded Stratix 10 package.

The assembled heat sink and board were tested in an open-loop system with deionized water as a coolant. With the microfluidic-cooled heat sink, thermal coupling between the FPGA die and the surrounding transceiver dice was reduced compared to the air-cooled configuration, as shown in Figure 11.22a. Furthermore, the lower thermal resistance of the microfluidic heat sink enabled much higher levels of compute at lower temperatures (Figure 11.22b). In addition to the reduction in temperature and thermal coupling between the dice, a large increase in compute density was also enabled through the use of the low-profile microfluidic heat sink.

11.4.3 Monolithic Microfluidic Cooling of 3D Integration: Modelling Electrical Implications for I/Os

Thermal issues become more severe in 3D ICs due to limited heat dissipation paths in a 3D stack.The integration of microfluidic cooling with 3D ICs (Figure 11.23a) is a potential solution to these issues [34, 35]. However, prior to integration, there should be fundamental understanding and analysis of the interactions between electrical interconnects and microfluidics. In this interconnect configuration, TSV capacitive coupling occurs through the silicon dioxide, silicon, and deionized water, and thus, the electrical characteristics of the TSVs are affected by their respective material properties. While the relative permittivity of silicon is almost constant (11.9), the relative permittivity of deionized water significantly varies with frequency. The frequency response of deionized water can be captured using a Debye model to preserve causality [36]. The frequency-dependent permittivity

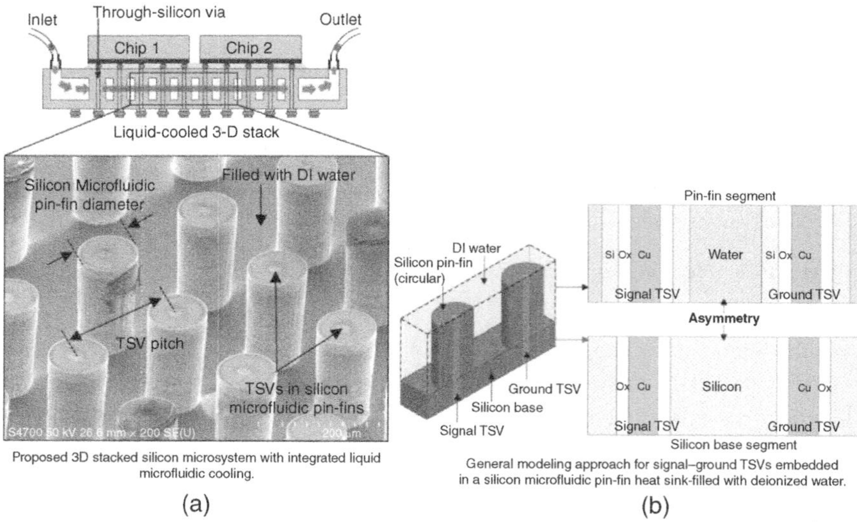

Figure 11.23 Integrated microfluidic cooling for 3D. [37]

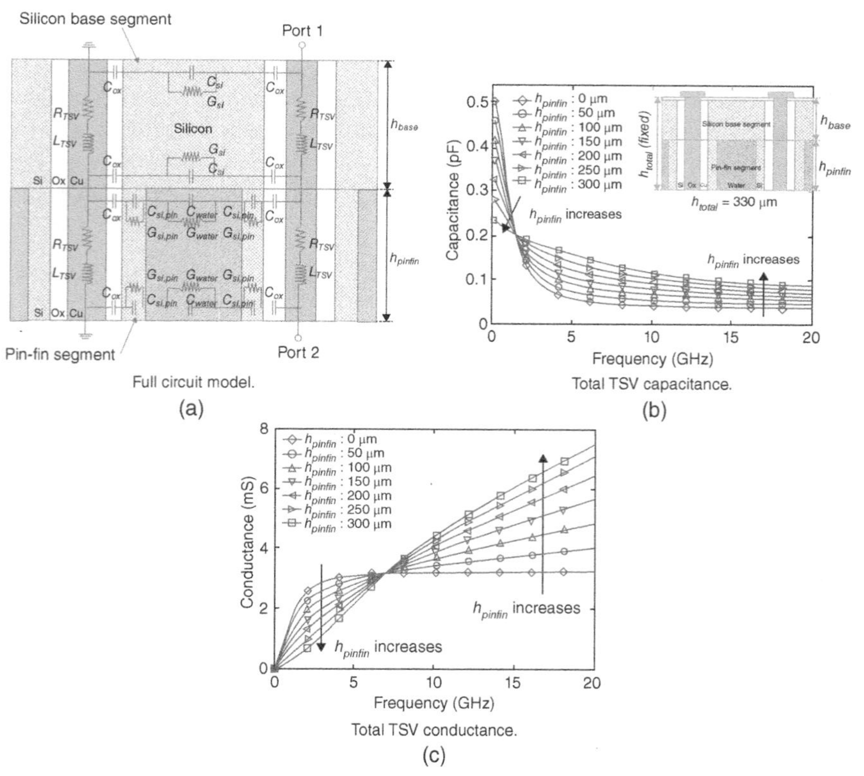

Figure 11.24 Full circuit model and results with different pin-fin heights increasing from 0 to 300 μm by an amount of 50 μm. The base thickness can be computed by subtracting the pin-fin height from a total thickness of 330 μm. [37]

of deionized water is higher than silicon and, hence, results in increased TSV parasitic capacitance and conductance.

To this end, electrical circuit models for such TSVs embedded in an asymmetric substrate (Figure 11.23b) has been explored in [37]. Results from the model (Figure 11.24a) show that the geometric design of a microfluidic heat sink can highly affect the electrical parasitic values of TSVs (Figures 11.24b and 11.24c). This points to the need for co-optimization of the geometric parameters in an inter-layer microfluidic cooling set-up to ensure optimal electrical and thermal performance.

11.5 Conclusion

In summary, this chapter presented 2.5D and 3D integration technologies for chiplets and silicon interposer technologies. Electrical and thermal design considerations were presented and highlighted the unique challenges in the implementation of such highly integrated electronics. A key challenge is cooling. To this end, this chapter presented opportunities of microfluidic cooling in both 2.5D and 3D ICs.

Acknowledgments

There are no acknowledgements for this chapter.

References

1 Su, L. (2019). "Delivering the future of high-performance computing," *DARPA ERI Summit: Opening plenary* 2019. (Presentation date: July 15, 2019).

2 Naffziger, S., Lepak, K., Paraschou, M., and Subramony, M. (2020). "2.2 AMD chiplet architecture for high-performance server and desktop products," *2020 IEEE International Solid- State Circuits Conference - (ISSCC)*, pp. 44–45.

3 Erdmann, C., Lowney, D., Lynam, A., Keady, A., McGrath, J., Cullen, E., Breathnach, D., Keane, D., Lynch, P., De La Torre, M., De La Torre, R., Lim, P., Collins, A., Farley, B., and Madden, L. (2015). "A heterogeneous 3D-IC consisting of two 28 nm fpga die and 32 reconfigurable high-performance data converters," *IEEE Journal of Solid-State Circuits* 50(1) pp. 258–269.

4 Mahajan, R. Sankman, R., Patel, N., Kim, D., Aygun, K., Qian, Z., Mekonnen, Y., Salama, I., Sharan, S., Iyengar, D., and Mallik, D. (2016).

"Embedded multi-die interconnect bridge (EMIB) – a high density, high bandwidth packaging interconnect," *2016 IEEE 66th Electronic Components and Technology Conference (ECTC)*, pp. 557–565.

5 Gomes, W., Khushu, S., Ingerly, D. B., Stover, P. N., Chowdhury, N. I., O'Mahony, F., Balankutty, A., Dolev, N., Dixon, M. G., Jiang, L., Prekke, S., Patra, B., Rott, P. V., and Kumar, R. (2020). "8.1 Lakefield and mobility compute: A 3D stacked 10nm and 22ffl hybrid processor system in 12 × 12mm2, 1mm package-on-package," *2020 IEEE International Solid- State Circuits Conference - (ISSCC)*, pp. 144–146.

6 Cong, J., Ghodrat, M. A., Gill, M., Grigorian, B., Gururaj, K., and Reinman, G. (2014). "Accelerator-rich architectures: Opportunities and progresses," *Proceedings of the 51st Annual Design Automation Conference, DAC '14*, (New York, NY, USA), pp. 1–6, Association for Computing Machinery, 2014.

7 Kim, D. H., Athikulwongse, K., Healy, M. B., Hossain, M. M., Jung, M., Khorosh, I., Kumar, G., Lee, Y., Lewis, D. L., Lin, T., Liu, C., Panth, S., Pathak, M., Ren, M., Shen, G., Song, T., Woo, D. H., Zhao, X., Kim, J., Choi, H., Loh, G. H., Lee, H. S., and Lim, S. K.(2015). "Design and analysis of 3D-MAPS (3D massively parallel processor with stacked memory)," *IEEE Transactions on Computers* 4(1), pp. 112–125.

8 Panth, S., Samadi, K., Du, Y., and Lim, S. K. (2013). "High-density integration of functional modules using monolithic 3D-IC technology," *2013 18th Asia and South Pacific Design Automation Conference (ASP-DAC)*, pp. 681–686.

9 Jo, P. K., Kochupurackal Rajan, S., Gonzalez, J. L., and Bakir, M. S. (2020). "Polylithic integration of 2.5-D and 3-D chiplets enabled by multi-height and fine-pitch CMIs," *IEEE Transactions on Components, Packaging and Manufacturing Technology* 10, pp. 1474–1481.

10 Hossen, M. O., Gonzalez, J. L., and Bakir, M. S. (2018). "Thermomechanical analysis and package-level optimization of mechanically flexible interconnects for interposer-on-motherboard assembly," *IEEE Transactions on Components, Packaging and Manufacturing Technology* 8, pp. 2081–2089.

11 Jo, P. K., Zhang, X., Gonzalez, J. L., May, G. S., and Bakir, M. S. (2018). "Heterogeneous multi-die stitching enabled by fine-pitch and multi-height compressible microinterconnects (CMIs)," *IEEE Transactions on Electron Devices* 65, pp. 2957–2963.

12 Jo, P. K. (2019). "Polylithic integration of heterogeneous multi-die enabled by compressible microinterconnects." PhD thesis, Georgia Institute of Technology, Atlanta, Georgia, US, Dec. 2019.

13 Jacquinot, H., Arnaud, L., Garnier, A., Bana, F., Barbe, J. C., and Cheramy, S. (2017). "RF characterization and modeling of 10 mm fine-pitch cu-pillar on a high density silicon interposer," *2017 IEEE 67th Electronic Components and Technology Conference (ECTC)*, (Orlando, Florida, US), May 2017, pp. 266–272.

14 Zheng, T., Jo, P. K., Rajan, S. K., and Bakir, M. S. (2020). "Polylithic integration for RF/MM-Wave chiplets using stitch-chips: Modeling, fabrication, and characterization," *2020 IEEE/MTT-S International Microwave Symposium (IMS)*, (Los Angeles, California), Aug. 2020, pp. 1035–1038.

15 Yang, H. S., Zhang, C., and Bakir, M. S. (2014). "Self-aligned silicon interposer tiles and silicon bridges using positive self-alignment structures and rematable mechanically flexible interconnects," *IEEE Transactions on Components, Packaging and Manufacturing Technology* 4, Nov. 2014, pp. 1760–1768.

16 Serafy, C., Bar-Cohen, A., Srivastava, A., and Yeung, D. (2016). "Unlocking the true potential of 3-D CPUs with microfluidic cooling," *IEEE Transactions on Very Large Scale Integration (VLSI) Systems* 24(4), pp. 1515–1523.

17 Zhang, Y., Zhang, Y., Sarvey, T., Zhang, C., Zia, M., and Bakir, M. (2016). "Thermal isolation using air gap and mechanically flexible interconnects for heterogeneous 3-D ICs," *IEEE Transactions on Components, Packaging and Manufacturing Technology* (6) 1, pp. 31–39.

18 Oprins, H., and Beyne, E. (2014)."Generic thermal modeling study of the impact of 3D-interposer material and thickness options on the thermal performance and die-to-die thermal coupling," *Fourteenth Intersociety Conference on Thermal and Thermomechanical Phenomena in Electronic Systems (ITherm)*, pp. 72–78, 2014.

19 Zhang, X., Lin, J. K., Wickramanayaka, S., Zhang, S., Weerasekera, R., Dutta, R., Chang, K. F., Chui, K.-J., Li, H. Y., Wee Ho, D. S., Ding, L., Katti, G., Bhattacharya, S., and Kwong, D.-L. (2015). "Heterogeneous 2.5D integration on through silicon interposer," *Applied Physics Reviews* 2(2), p. 02130.

20 Black, B.. Annavaram, M., Brekelbaum, N., DeVale, J., Jiang, L., Loh, G. H., McCaule, D., Morrow, P., Nelson, D. W., Pantuso, D., Reed, P., Rupley, J., Shankar, S., Shen, J., and Webb, C. (2006). "Die stacking (3D) microarchitecture," *2006 39th Annual IEEE/ACM International Symposium on Microarchitecture (MICRO'06)*, pp. 469–479.

21 Cong, J., Luo, G., and Shi, Y. (2011). "Thermal-aware cell and through-silicon-via co-placement for 3D ICs," *Proceedings of the 48th Design Automation Conference*, DAC '11, (New York, New York, US), pp. 670–675, Association for Computing Machinery, 2011.

22 Samal, S. K., Panth, S., Samadi, K., Saedi, M., Du, Y., and Lim, S. K. (2014). "Fast and accurate thermal modeling and optimization for monolithic 3D ICs," *Proceedings of the 51st Annual Design Automation Conference*, DAC '14, (New York, New York, US), pp. 1–6, Association for Computing Machinery, 2014.

23 Wei, H., Wu, T. F., Sekar, D., Cronquist, B., Pease, R. F., and Mitra, S. (2012). "Cooling three-dimensional integrated circuits using power delivery networks," *2012 International Electron Devices Meeting*, pp. 14.2.1–14.2.4.

24 Zhang, Y., Sarvey, T. E., and Bakir, M. S. (2017). "Thermal evaluation of 2.5-D integration using bridge-chip technology: Challenges and opportunities," *IEEE Trans. Compo. Packag. Manuf. Technol.* 7(7), pp. 1101–1110.

25 Zhang, Y., Zhang, Y., and Bakir, M. S. (2014). "Thermal design and constraints for heterogeneous integrated chip stacks and isolation technology using air gap and thermal bridge," *IEEE Trans. Compo. Packag. Manuf. Technol.* 4(12), pp. 1914–1924.

26 Zhang, Y., Sarvey, T. E., and Bakir, M. S. (2014). "Thermal challenges for heterogeneous 3D ICs and opportunities for air gap thermal isolation," *2014 International 3D Systems Integration Conference (3DIC)*, pp. 1–5.

27 Sohn, K., Yun, W., Oh, R., Oh, C., Seo, S., Park, M., Shin, D., Jung, W., Shin, S., Ryu, J., Yu, H., Jung, J., Lee, H., Kang, S., Sohn, Y., Choi, J., Bae, Y., Jang, S., and Jin, G. (2017). "A 1.2 v 20 nm 307 GB/s HBM DRAM with at-speed wafer-level IO test scheme and adaptive refresh considering temperature distribution," *IEEE Journal of Solid-State Circuits* 52(1), pp. 250–260.

28 Kaul, A., Rajan, S. K., Obaidul Hossen, M., May, G. S., and Bakir, M. S. (2020). "BEOL-embedded 3D polylithic integration: Thermal and interconnection considerations," *2020 IEEE 70th Electronic Components and Technology Conference (ECTC)*, pp. 1459–1467.

29 Kaul, A., Peng, X., Rajan, S. K., Yu, S., and Bakir, M. S. (2020). "Thermal modeling of 3D polylithic integration and implications on BEOL RRAM performance," *2020 IEEE International Electron Devices Meeting (IEDM)*, pp. 13.1.1–13.1.4.

30 Tuckerman, D. B., and Pease, R. F. W. (1981). "High-performance heat sinking for VLSI," *IEEE Electron Device Letters* 2(5), pp. 126–129.

31 Sarvey, T. E., Zhang, Y., Cheung, C., Gutala, R., Rahman, A., Dasu, A., and Bakir, M. S. (2017). "Monolithic integration of a micropin-fin heat sink in a 28-nm FPGA," *IEEE Transactions on Components, Packaging and Manufacturing Technology* 7(10), pp. 1617–1624.

32 Zimmermann, S., Tiwari, M. K., Meijer, I., Paredes, S., Michel, B., and Poulikakos, D. (2012). "Hot water cooled electronics: exergy analysis and waste heat reuse feasibility," *International Journal of Heat and Mass Transfer* 55(23–24), pp. 6391–6399.

33 Sarvey, T. E., Kaul, A., Rajan, S. K., Dasu, A., Gutala, R., and Bakir, M. S. (2019). "Microfluidic cooling of a 14-nm 2.5-D FPGA with 3-D printed manifolds for high-density computing: Design considerations, fabrication, and electrical characterization," *IEEE Transactions on Components, Packaging and Manufacturing Technology* 9(12), pp. 2393–2403.

34 Zhang, Y., Zheng, L., and Bakir, M. S. (2013). "3-D stacked tier-specific microfluidic cooling for heterogeneous 3-D ICs," *IEEE Transactions on Components, Packaging and Manufacturing Technology* 3(11), pp. 1811–1819.

35 Zhang, Y., Oh, H., and Bakir, M. S. (2013). "Within-tier cooling and thermal isolation technologies for heterogeneous 3D ICs," *2013 IEEE International 3D Systems Integration Conference (3DIC)*, pp. 1–6.

36 Kaatze, U. (1989). "Complex permittivity of water as a function of frequency and temperature," *Journal of Chemical and Engineering Data* 34(4), pp. 371–374.

37 Oh, H., Swaminathan, M., May, G. S., and Bakir, M. S. (2020). "Electrical circuit modeling and validation of through-silicon vias embedded in a silicon microfluidic pin-fin heat sink filled with deionized water," *IEEE Transactions on Components, Packaging and Manufacturing Technology* 10(8), pp. 1337–1347.

Index

Embedded and Fan-Out Wafer and Panel Level Packaging Technologies for Advanced Application Spaces: High Performance Compute and System-in-Package, First Edition. Edited by Beth Keser and Steffen Kröhnert.

g

h

i

j

k

l

Printed and bound by CPI Group (UK) Ltd, Croydon, CR0 4YY